small arms survey 2006

unfinished business

HEI

A Project of the Graduate
Institute of International Studies,
Geneva

OXFORD

UNIVERSITY PRESS

Great Clarendon Street, Oxford OX2 6DP

Oxford University Press is a department of the University of Oxford.
It furthers the University's objective of excellence in research, scholarship,
and education by publishing worldwide in

Oxford New York

Auckland Cape Town Dar es Salaam Hong Kong Karachi
Kuala Lumpur Madrid Melbourne Mexico City Nairobi
New Delhi Shanghai Taipei Toronto

With offices in

Argentina Austria Brazil Chile Czech Republic France Greece
Guatemala Hungary Italy Japan Poland Portugal Singapore
South Korea Switzerland Thailand Turkey Ukraine Vietnam

Oxford is a registered trade mark of Oxford University Press
in the UK and in certain other countries

Published in the United States
by Oxford University Press Inc., New York

British Library Cataloguing in Publication Data

Data available

Library of Congress Cataloguing-in-Publication Data

Data available

Typeset in Garamond Light
by Exile: Design & Editorial Services

Printed in Great Britain
on acid-free paper
by Bath Press Ltd, Bath, Avon

ISBN 0-19-929848-3   978-0-19-929848-8

1 3 5 7 9 10 8 6 4 2

FOREWORD

I became President of Africa's oldest republic in the wake of a devastating civil war that took untold numbers of lives and laid the economy to waste. Today, post-conflict Liberia is just beginning to regain some measure of social, political, and economic stability. But security, while vastly improved, remains a concern. After 14 years of armed conflict, and despite important weapons reduction efforts, Liberia remains awash with small arms.

The Small Arms Survey and others have shown that such an abundance of weapons is a recipe for trouble, especially in a region teeming with armed groups. Liberia cannot take on this challenge alone. Nor should other countries struggle with such far-reaching small arms concerns on their own.

It is in this context that I look forward to the United Nations Review Conference on Small Arms and Light Weapons in June–July 2006, during which Member States will join forces to define the next steps in the global small arms process. Undeniably, we still face huge challenges at the national, regional, and international levels.

The *Small Arms Survey 2006: Unfinished Business* explores these challenges and presents its most recent research on issues that the international community will have to grapple with if it is to make any headway in preventing the misuse of small arms. As the subtitle implies, much work remains to be done.

This volume offers initial answers in response to key questions we must continue to ask as we proceed. Why do people acquire, use, and misuse weapons? How can we measure the real costs of gun violence? What is the relationship between small arms control and reduction measures and the broader security sector picture? And regarding my own region: How can we tackle the problem of armed groups in West Africa?

Matching relevant theory with facts from the field, this book advances our understanding of the small arms issue and offers crucial insight for policy-makers and other stakeholders. I invite you to join me in using the *Small Arms Survey 2006* as a vital resource, and as an essential tool in navigating the future of the international small arms process.

Ellen Johnson-Sirleaf
President of the Republic of Liberia
April 2006

CONTENTS

ABOUT THE SMALL ARMS SURVEY

The Small Arms Survey is an independent research project located at the Graduate Institute of International Studies in Geneva, Switzerland. Established in 1999, the project is supported by the Swiss Federal Department of Foreign Affairs, and by sustained contributions from the governments of Belgium, Canada, Finland, France, the Netherlands, Norway, Sweden, and the United Kingdom. The Survey is also grateful for past and current project support received from Australia, Denmark, and New Zealand. The Small Arms Survey collaborates with research institutes and non-governmental organizations in many countries, including Brazil, Canada, Georgia, Germany, India, Israel, Jordan, Norway, the Russian Federation, South Africa, Sri Lanka, Sweden, Thailand, the United Kingdom, and the United States.

The objectives of the Small Arms Survey are: to be the principal source of public information on all aspects of small arms; to serve as a resource centre for governments, policy-makers, researchers, and activists; to monitor national and international initiatives (governmental and non-governmental) on small arms; and to act as a clearing house for the sharing of information and the dissemination of best practices. The Survey also sponsors field research and information-gathering efforts, especially in affected states and regions. The project has an international staff with expertise in security studies, political science, law, economics, development studies, and sociology. It collaborates with a worldwide network of researchers, partner institutions, non-governmental organizations, and governments.

NOTES TO READERS

Abbreviations: Lists of abbreviations can be found at the end of each chapter.

Chapter cross-referencing: Chapter cross-references appear capitalized in brackets throughout the text. For example, in Chapter 10 on Armed Groups: 'In such cases, one can point to a correlation between the availability of weapons or ammunition and the degree of control exerted over them (LORD'S RESISTANCE ARMY).'

Exchange rates: All monetary values are expressed in current US dollars (USD). When other currencies are additionally cited, unless otherwise indicated, they are converted to USD using the 365-day average exchange rate for the period 1 September 2004 to 31 August 2005.

Small Arms Survey: The plain text—Small Arms Survey—is used to indicate the overall project and its activities, while the italicized version—*Small Arms Survey*—refers to the publication. The *Survey,* appearing italicized, refers generally to past and future editions.

Web site: For more detailed information and current developments on small arms issues, readers are invited to visit the Small Arms Survey Web site at www.smallarmssurvey.org.

Small Arms Survey
Graduate Institute of International Studies
47, Avenue Blanc, 1202 Geneva, Switzerland

Tel.: +41 22 908 5777
Fax: +41 22 732 2738
Email: smallarm@hei.unige.ch
Web site: www.smallarmssurvey.org

ACKNOWLEDGEMENTS

Like the previous editions of the *Small Arms Survey*, this sixth volume is a collective product of the staff of the Small Arms Survey project, which is based at the Graduate Institute of International Studies in Geneva, Switzerland. A large number of researchers in Geneva and around the world have contributed to this volume, and it has benefited from the input and advice of numerous government officials, activists, experts, and colleagues from the small arms research community.

The principal chapter authors were assisted by a large number of in-house and external contributors who are acknowledged in the relevant chapters.

In addition, detailed reviews on the chapters were provided by: Philip Alpers, Peter Bartu, Jurgen Brauer, Michael Brzoska, Alex Butchart, Philip J. Cook, Steven R. Costner, Wendy Cukier, David DeClerq, David Diaz, Sinclair Dinnen, Louise Doswald-Beck, Sarah Garap, Jeremy Ginifer, William Godnick, David Hegarty, Iain Hall, Maximo Halty, David Hemenway, Dylan Hendrickson, Michael Kimmel, Göran Lindgren, Lora Lumpe, Tamir Magal, Ron May, David Meddings, Pierre Mills, Caroline Moser, Daniel Muños-Rojas, Sharmala Naidoo, Ryan Nichols, David Polber, Pablo Policzer, Michael Pugh, Antonio Rangel, Stella Sabiti, Robert Scharf, Yiftah S. Shapir, Christiana Solomon, Bruce Stout, Juana Téllez, Reinhilde Weidacher, Jeremy Weinstein, Siemon Wezeman, Adrian Wilkinson, and Deanna Wilkinson.

Eric G. Berman, Keith Krause, and Glenn McDonald were responsible for the overall planning and organization of this edition. Emile LeBrun provided editorial insight. Tania Inowlocki managed the editing and production of the *Survey;* she, Michael James, and Alex Potter copy-edited the book. Jillie Luff produced the maps; Rick Jones provided the layout and design; Donald Strachan proofread the *Survey;* and Lisa Kenwright of Indexing Specialists (UK) compiled the index.

Dominic Byatt and Lizzy Suffling of Oxford University Press provided support throughout the production of the *Survey.* Barbara Gimelli, Anne-Kathrin Glatz, Sahar Hasan, Jonas Horner, Kaitlyn Powles, Matthew Tarduogno, and Liliane Zossou assisted with fact-checking. Fridrich Štrba and Carole Touraine provided administrative support. Michael Cheng and Mohamed Hojeij helped with research.

The project also benefited from the support of personnel of the Graduate Institute of International Studies, in particular Philippe Burrin, Jasmine Champenois, Andrew Clapham, Wilfred Gander, Oliver Jütersonke, and Monique Nathoo.

We are extremely grateful to the Swiss Government for its generous financial and overall support of the Small Arms Survey project, in particular Heidi Grau, Thomas Greminger, Roman Hunger, Fabienne Juilland Metzger, Laurent Masmejean, Peter Maurer, Giancarlo de Picciotto, Anton Thalmann, Stefano Vescovi, and Othmar Wyss. Financial support for the project was also provided by the Governments of Canada, Finland, the Netherlands, Norway, Sweden, and the United Kingdom. In addition, the project has received financial support for various research projects from the Geneva International Academic Network, the Organisation Internationale de la Francophonie, the UN Development Programme, the UN Institute for Disarmament Research, the South Eastern Europe Clearinghouse for the Control of Small Arms and Light Weapons, and the World Health Organization. The project further benefits from the support of governmental and international agencies, including the International Committee of the Red Cross, the UN Department for Disarmament Affairs, the UN Office for the Coordination of Humanitarian Affairs, and the UN High Commissioner for Refugees.

In Geneva, the project has received support and expert advice from: David Atwood, Peter Batchelor, Christophe Carle, Martin Griffiths, Randall Harbour, Magnus Hellgren, Peter Herby, Luc Lafrenière, Patricia Lewis, Patrick McCarthy, David Meddings, Jennifer Milliken, Sola Ogunbanwo, Daniël Prins, and Fred Tanner.

Beyond Geneva, we also received support from a number of colleagues. In addition to those mentioned above, and in specific chapters, we would like to thank: David de Beer, Ilhan Berkol, Michael Cassandra, Mohamed Coulibaly, David Isenberg, Meir Itzchaki, Edward Laurance, Nicholas Marsh, Sarah Meek, Lisa Misol, Yeshua Moser-Puangsuwan, Abdel-Fatau Musah and Alex Vines.

Our sincere thanks go out to many other individuals (who remain unnamed) for their continuing support of the project. We also express our apologies to anyone we have failed to mention.

Keith Krause, Programme Director
Eric G. Berman, Managing Director

Small Arms Survey 2006

Editors
Eric G. Berman, Keith Krause, and Glenn McDonald

Publications Manager
Tania Inowlocki

Editorial Consultant
Emile LeBrun

Design and layout
Richard Jones, Exile: Design & Editorial Services

Cartography
Jillian Luff, MAP*grafix*

Copy-editors
Tania Inowlocki, Michael James, and Alex Potter

Proofreader
Donald Strachan

Principal chapter authors

Introduction	Eric G. Berman, Keith Krause, and Glenn McDonald
Chapter 1	James Bevan
Chapter 2	Aaron Karp
Chapter 3	Pablo Dreyfus and Anna Khakee with Anne-Kathrin Glatz
Chapter 4	Glenn McDonald
Chapter 5	Christina Wille
Chapter 6	Anne-Kathrin Glatz and Robert Muggah
Chapter 7	Nicole Haley and Robert Muggah
Chapter 8	Nicolas Florquin
Chapter 9	Katherine Aguirre, Robert Muggah, Jorge A. Restrepo, and Michael Spagat
Chapter 10	Louisa N. Lombard
Chapter 11	James Bevan
Chapter 12	James Bevan and Nicolas Florquin

Nigerians of Bakin Ciyawa, Plateau State, prepare to defend their town with locally produced hunting rifles and other weapons in May 2004.
© Pius Utomi Ekpei/AFP/Getty Images

Introduction

Ten years ago, the proliferation and misuse of small arms and light weapons was far from the international community's agenda. Fewer than a handful of measures had even been instituted. One innovative practical disarmament programme was launched under United Nations auspices in Mali in 1994, at the urging of then president Alpha Oumar Konaré. Shortly thereafter, UN Secretary-General Boutros Boutros-Ghali called attention to the need for 'practical disarmament in the context of the conflicts the United Nations is actually dealing with and of the weapons, most of them light weapons, that are actually killing people in the hundreds of thousands'.[1] A few experts and advocates were also drawing attention to small arms issues, but their work was only beginning.[2]

In 2006, the picture looks very different. Dozens of practical initiatives have been launched, ranging from projects to disarm, demobilize, and reintegrate ex-combatants in post-conflict zones, to measures to improve the security of government stockpiles, to large-scale destruction of surplus weapons stocks, and the negotiation of an international instrument for the marking and tracing of weapons. Innovative projects to improve local security and safety have now been established in Albania, Australia, Brazil, El Salvador, Fiji, South Africa, and countless communities around the world. International organizations as diverse as the World Health Organization, the UN Development Programme (UNDP), and the Organisation for Economic Co-operation and Development have acknowledged the importance of small arms and armed violence issues to their work, and the need to devote adequate financial and human resources to the issue. More than 700 non-governmental organizations in all corners of the world are active on the issue.

These steps have only scratched the surface of the problems posed by small arms proliferation and misuse. But compared to the situation only a decade ago, progress has been enormous. Our understanding of small arms and armed violence issues has grown significantly, and we now know much about the global distribution of victims of armed violence, the negative impact of arms availability and misuse on social and economic development, the worldwide nature of stockpiles and holdings, and the critical importance of embedding small arms work in the broader context of armed violence reduction and security sector reform initiatives.

But we have not yet done much to reduce the human cost of armed violence.

Agreement in 2001 on the United Nations *Programme of Action to Prevent, Combat and Eradicate the Illicit Trade in Small Arms and Light Weapons in All Its Aspects* was an important achievement. By itself, if fully implemented, the *Programme of Action* would represent significant progress, and the first step states must take at the UN Review Conference in 2006 is to commit to its full implementation.

Stopping there, however, would mean refusing to recognize that our understanding of key elements of the small arms agenda has grown since 2001. Over the past five years, issues such as the threat posed by the proliferation of man-portable air defence systems (MANPADS), the importance of ammunition as the 'fuel' that fires contemporary conflicts, the need to match supply-side initiatives with demand reduction measures, and the link between small arms proliferation and misuse, development, and broader security sector governance have only just begun to be tackled. Whatever their fate at the Review Conference, these issues, with which the *Programme of Action* does not deal in detail, represent key future challenges.

Chapter highlights

The publication of the first *Small Arms Survey* coincided with the 2001 UN Small Arms Conference and the adoption of the *Programme of Action*. This edition, the sixth, continues in the tradition of its predecessors by offering new information and insight on both emerging and established small arms issues. Its aim is not to sum up or take stock of past efforts, but instead to advance existing knowledge wherever possible. This year's *Survey* offers new or updated information on small arms production, stockpiles, transfers, and measures, including a review of important international developments such as the *International Tracing Instrument*.

The second part of this year's *Survey* concentrates on some of the challenges that are essential to future progress on the small arms issue. Thematic chapters and case studies examine the demand for weapons, the gender dimension of perpetrators of armed violence, small arms and security sector reform, and the economic costs of small arms use. While anchored in particular chapters, our cross-cutting themes are further explored throughout the volume. A case in point is the issue of demand, whose core aspects are presented in a dedicated chapter. In addition, the Production chapter uses a demand-side approach in estimating global military procurement of small arms and light weapons, while the Papua New Guinea, Lord's Resistance Army, and Armed Groups chapters highlight the need to take demand into account in crafting effective armed violence prevention strategies. These chapters also address the difficulties of curbing demand for the acquisition and use of guns in regions awash with weapons. Similarly, we recognize that a thorough understanding of small arms violence—and its antidotes—requires examining perpetrators of armed violence based on gender and age. The Angry Young Men chapter considers why young men account for the lion's share of global small arms violence, while the Colombia study highlights the gender-specific nature of gun violence in that country.

Update chapters

Chapter 1 (Production): Using a demand-side approach, this chapter estimates procurement of small arms and light weapons by the world's armed forces, as well as the numbers of weapons that must be produced or transferred from surplus stocks to meet these needs. The chapter uses data on troop levels and ages of small arms in 151 countries as well as detailed information on procurement patterns for 32 of those states.

The results of this projection suggest that the world's militaries procure around one million small arms and light weapons annually. Of these, an estimated 0.7 to 0.9 million are newly produced weapons and the remainder are surplus weapons, transferred from one state to another. The chapter highlights one major trend of concern: a number of wealthy states seem set to launch new procurement initiatives in the coming five to ten years. Global military production will meet this increased demand and other countries will seek parity with these modernizing militaries. Large amounts of surplus small arms and light weapons will thus become available. The potential for a cascade effect, whereby surplus stocks are displaced from richer states to poorer ones, or to non-state armed groups, is great.

Chapter 2 (Stockpiles): In reviewing the global distribution of firearms among law enforcement agencies and armed forces around the world, this chapter develops holdings estimates for these categories of ownership. It also points to the danger of catastrophic loss of control over large stockpiles and the constant problem of routine theft from state-run small arms inventories.

The chapter estimates the global inventory of law enforcement firearms at more than 26 million weapons. Militaries worldwide hold at least 200 million modern firearms, of which the large majority are automatic rifles. Armed forces also store large numbers of older weapons. Global military firearm inventories are concentrated among relatively few countries, with just ten countries controlling around two-thirds of the weapons. Governments have officially declared a total of only 16 million military-held firearms, less than ten per cent of the estimated total.

Chapter 3 (Transfers): In addition to providing new information on major exporters and importers and an updated Small Arms Trade Transparency Barometer, this chapter features a case study on illicit small arms and light weapons transfers in South America based on police and customs seizure data.

In 2003, the top exporters of small arms and light weapons—based on customs data from UN Comtrade, national arms export reports, and estimates—were the Russian Federation, the United States, Italy, Germany, Brazil, and China. The top importers—based on customs data—were the United States, Cyprus, and Germany.

As the case of South America illustrates, harmonization of national laws is crucial since criminals often exploit legal loopholes of neighbouring countries to purchase and smuggle small arms across national borders.

Chapter 4 (Measures): Since December 2005, UN Member States have been bound by the *International Tracing Instrument*. The result of intense negotiations, it reflects a series of compromises, the foremost of which is that the *Instrument* is only politically, not legally, binding. Nor does it cover small arms and light weapons ammunition. Despite these compromises, the *International Tracing Instrument* constitutes a significant step forward in multilateral efforts to address the small arms problem.

This chapter explores the contents of the *Instrument*, highlighting the contribution it makes to the international regulation of marking, record-keeping, and tracing. It is too early, however, to proclaim the UN tracing process a success; this will depend on the effective implementation of the *Instrument* by UN Member States, as well as the extent to which it spurs further normative development. In late 2005, many states had yet to determine how to bring national laws and practices into compliance.

Thematic chapters and case studies

Chapter 5 (Cambodia): Cambodia was awash with weapons during the 1990s, following several decades of conflict. In recent years, the international community, the Cambodian government, and non-governmental organizations have collaborated to address the problem through specific small arms control programmes. As a result, more than 120,000 weapons have been removed from uncontrolled circulation, and 180,000 confiscated and surplus stockpiles have been destroyed. National firearms laws have also been revised and widespread public awareness campaigns have made the Cambodian people conscious of new policies regarding gun possession and use.

Based on newly collated data, this chapter provides the first assessment of the impact of these programmes, concluding that they have contributed to a considerable improvement in human security for the Cambodian people. Yet this picture, positive as it appears, is incomplete. The chapter looks beyond small arms control to consider broader approaches to post-conflict stabilization, in particular security sector reform (SSR). It concludes that opportunities to link the two agendas in Cambodia have largely been unexploited. The continuing misuse of official weapons in Cambodia underlines the utility of incorporating SSR concepts into small arms planning.

Chapter 6 (Demand): This chapter finds that a careful treatment of the motivations and means that condition acquisition and possession of firearms is as important as supply-side interventions that regulate manufacturing, transfer, and civilian possession. Interventions are most likely to prove sustainable when they are tailored to the particular circumstances of groups and individuals and when they involve public–private partnerships.

Drawing on field research carried out in Brazil, Colombia, Papua New Guinea, the Solomon Islands, and South Africa, the chapter reveals a number of cross-cutting factors that must be taken into account if small arms reduction and control are to be successful. Factors such as the historical and social environment help shape collective and individual preferences to acquire and possess firearms, as do dramatic structural shifts in governance and the macro-economy. Moreover, the sudden and unexpected availability of weapons in a volatile environment can also induce demand by lowering the real and relative price of acquiring weapons.

Chapter 7 (Papua New Guinea): In response to chronic armed violence in its urban centres and highland provinces, the Papua New Guinea government initiated a far-reaching review of its firearms legislation and controls in 2005. Firearms, while few in number relative to the population, were found to make a major contribution to real and perceived insecurity—and pose a persistent obstacle to good governance, improved security, and sustainable development. This chapter presents the findings of an armed violence assessment administered in the country's National Capital District and Southern Highlands Province by the Small Arms Survey with support from UNDP.

Extensive household surveys reveal, for example, that victimization rates among households in the regions under review are higher than previously reported. More than half of all households surveyed reported being victimized in the preceding six months, mainly with weapons—including firearms. The chapter also observes that ammunition prices are lower than previously reported, and appear to be declining. Furthermore, the chapter finds that weapons reduction programmes do not have widespread support among the citizens, who call for improvements in law and order before they will agree to disarm.

Chapter 8 (Costing): This chapter examines the impact of small arms violence from an economic perspective. In developed countries, high costs are frequently used to justify more spending on violence prevention. The same is not yet true of developing countries, some of which cannot or do not provide extensive care for victims. Comparing costs with levels of armed violence can help identify in which societies victims are the most vulnerable.

Misconceptions about the costs of armed violence abound and methodologies need careful refinement to ensure a better understanding of the global costs of gun violence. Nevertheless, evidence clearly shows that small arms misuse accounts for an excessive proportion of the costs of violence. This finding is confirmed by two case studies on Brazil and Colombia, where the medical treatment of a firearm injury costs between 1.7 and 3 times more than that of a stabbing. These studies utilize draft estimation guidelines developed by the Small Arms Survey for the World Health Organization (WHO) and the United States Centers for Disease Control and Prevention (CDC).

Chapter 9 (Colombia): This chapter reveals that violence arising from Colombia's protracted armed conflict as well as common and organized crime has claimed the lives of more than 475,000 civilians and combatants since 1979—almost 17,600 per year. More than 80 per cent of all homicides since the late 1970s have been perpetrated with guns. This percentage has steadily increased—from about 60 per cent in the 1980s to more than 85 per cent in 2002. By 2005 more than 15 per cent of *all* deaths, from any cause, were firearm-related. Criminal violence perpetrated with firearms is also gender-specific, with men suffering more than 90 per cent of all gun deaths.

Most weapons currently in circulation in Colombia are illegal and unregistered. The number of legal and illegal weapons combined is estimated at 2.3 to 3.9 million, suggesting an ownership rate of 5 to 8 firearms per 100 inhabitants. This chapter also shows that the country witnessed a significant reduction in conflict-related civilian deaths in 2003 and 2004. It provides evidence that gun control measures have been quite successful in reducing violence in major cities such as Bogotá, Cali, and Medellín.

Chapter 10 (Armed Groups): Armed groups, equipped mainly with small arms and light weapons, constitute a persistent and growing security threat worldwide. This chapter provides an overview of the dynamics of small arms acquisition, management, and control by armed groups (including pro-state militias) in the Economic Community of West African States (ECOWAS) region, where they are a particular concern. Many West African armed groups are unpredictable, with motivations, allegiances, and scope of operations shifting over time. They tend to have easy access to the region's robust market for small arms and light weapons. Though weapons collection programmes in West Africa have had some success, the quality of weapons collected is often questionable.

The chapter finds that attempting to influence armed groups through persuasion is an important but insufficient means of engaging with West African armed groups. Since a lack of alternative employment opportunities may drive demobilized ex-fighters to return to fighting, it will often prove more effective to offer incentives such as new jobs and security sector reform.

Chapter 11 (Lord's Resistance Army): The Lord's Resistance Army (LRA) is a non-state armed group that operates in the north of Uganda and neighbouring countries. It is notorious for its abduction of children, who are forced to become fighters under unusually cruel conditions. The LRA is dependent on small arms for almost all of its violent activities. It requires few resources to sustain these, other than abducted 'recruits' and the small arms and ammunition with which to equip them. In a region already saturated with small arms, the well-armed LRA is able to capture additional arms, ammunition, and recruits in a self-sustaining cycle of violence.

The chapter argues that the LRA, despite reports to the contrary, does not face crippling shortages of arms or ammunition. In contrast with members of many other armed groups, LRA fighters are disciplined—albeit brutalized—into caring for their weapons and using ammunition sparingly. Small arms are a crucial determinant of the LRA's capacity to continue fighting; their relative abundance indicates that the LRA is far from finished as a fighting force.

Chapter 12 (Angry Young Men): Young men account for a disproportionately high percentage of the perpe-trators (and victims) of lethal small arms violence. This chapter asks what role small arms play in the lives of these 'angry young men' and what measures can be taken to disrupt this deadly relationship.

Biological and demographic explanations of the problem do not take into account the critical social and economic factors that appear to encourage or inhibit some young men's recourse to violence. The chapter argues that gender identity constructions condition many marginalized young men around the world to view armed violence as a legiti-mate means of achieving manhood and respect. Small arms are seen as useful coercive tools, as well as powerful symbols underpinning attractive masculine identities. The chapter concludes that controlling young men's access to small arms has proven effective in reducing firearm mortalities from youth violence. Countering socially constructed associations that link small arms to violence, power, and masculinity will also be pivotal to future violence prevention strategies.

Conclusion

As the *Small Arms Survey* goes to press, the outcome of the UN Review Conference remains unclear. Yet whatever its results, the small arms issue will remain with us, in all of its complexity, for years to come. The international commu-nity does not yet appear ready to adopt a comprehensive and consensual approach to the many dimensions of the small arms problem. But most practical measures to promote greater security and safety from armed violence only require greater and sustained efforts at the local, community, national, or regional level, with effective partnerships among NGOs, aid agencies, international organizations, and governments. The Small Arms Survey, for its part, will continue to support a diverse range of initiatives while pointing the way to workable solutions to reduce armed violence. ◾

Endnotes

1   United Nations General Assembly. 1995. *Supplement to the Agenda for Peace.* A/50/60-S/1995/1, paras. 60–61. 3 January.
2   Key contributions include Jeffrey Boutwell et al., *Lethal Commerce* (Cambridge, MA: American Academy of Arts and Sciences, 1995); Jasjit Singh, ed., *Light Weapons and International Security* (Delhi: Indian Pugwash Society and BASIC, 1995); Christopher Louise, *The Social Impacts of Light Weapons Availability and Proliferation,* Discussion Paper 59 (Geneva: UN Research Institute for Social Development with International Alert, 1995); the *Managing Arms in Peace Processes* series by UNIDIR's Disarmament and Conflict Resolution project; and BASIC's occasional papers.

ACKNOWLEDGEMENTS

Principal authors

Eric G. Berman, Keith Krause, and Glenn McDonald

L85A1 SA80 rifles in the armoury at the Land
Warfare Centre, Warminster, United Kingdom.
© Military Picture Library

Military Demand and Supply

PRODUCTS AND PRODUCERS

1

INTRODUCTION

Small arms are a durable good, but like any commodity they are subject to replacement. The global stock of small arms is therefore never static, but in constant flux. Demand for new stock varies according to need, political will, and the availability of finance; the global small arms production rate is consequently never fixed, but fluctuates to meet changes in demand.

This observation is of fundamental importance for any analysis of small arms production. It is not sufficient simply to note that production this year increased in contrast to last year, or the reverse. A demand-side approach explains why production rates change over long periods of time. It can establish trajectories, and hence predictions, of small arms production.

The following analysis estimates procurement and production for a 'global' sample of 151 countries' militaries.[1] It focuses on the acquisition of assault rifles and carbines—the most numerous small arms in military service. These 151 countries were the ones for which stock ages and force levels were available—key factors in the calculation of procurement rates, as will become apparent. By using ratios of each type of weapon stocked by the militaries of 32 countries in the sample for which detailed procurement data was available, it projects assault rifle and carbine procurement data to estimate pistol and light and heavy machine gun acquisition in the 151 states.[2] In so doing, the chapter covers the procurement of the majority of infantry weapons in use today.

From estimates of procurement, the chapter subsequently presents an estimate of global military production to meet this demand. It derives this estimate by gauging the likely ratios of newly produced weapons acquired by states to the transfer and acquisition of surplus weaponry. The approach adopted this year complements production-centred research in past editions of the *Small Arms Survey*. The following are among the chapter's most important conclusions:

- The world's militaries procure around 50 million small arms and light weapons over a 50-year period, or around 1 million units annually.
- Not all of this acquisition is of new stocks.
- Global production of military small arms and light weapons over a 50-year period ranges between 36 and 46 million weapons and averages 0.7–0.9 million annually.
- Production is not constant but cyclical and responds to the demands of the world's wealthier states.
- The world's poorer states often rely on surplus stocks displaced by wealthy state procurement programmes.
- This trade and transfer of surplus stocks to militaries across the globe could number up to 14 million units over a 50-year period.
- Some of the world's largest procurers will launch major procurement programmes in the next 10–15 years.

- Global military production periodically peaks and is projected to do so in the coming 20 years as wealthy states modernize their small arms.

- Unless measures are taken to remove weapons from circulation, this peak is likely to displace yet more surplus stock to the world's poorer states.

MILITARY DEMAND AND SUPPLY

Previous editions of the *Small Arms Survey* note that analysis of production trends necessitates analysis of demand. The 2004 edition, for instance, observes that, while the market for small arms appeared to be in decline during the previous four years, demand had also increased in some areas (Small Arms Survey, 2004, pp. 7–41). In response, the 2005 edition defines particular sectors in the small arms industry, so that variations in demand, and hence production, can be assessed more accurately. It notes that particular sectors respond differently according to the requirements of the consumer groups that purchase the majority of their products (Small Arms Survey, 2005, pp. 39–69).

This chapter focuses on military demand and supply, or 'Sector 5' production, as defined in the *Small Arms Survey 2005*. Sector 5 supplies primarily the armed forces of the world's states. Its products, including assault rifles, carbines, and light and heavy machine guns, are collectively the most numerous small arms and light weapons in military service. They also constitute the group of weapons that are most likely to be of 'main concern' in contemporary armed conflict (Small Arms Survey, 2005, p. 40; UN, 1997, para. 13). Of these military small arms and light weapons, state armed forces, totalling some 20 million active and 33 million reserve forces worldwide (IISS, 2004, p. 358), probably procure the lion's share.

Indian soldiers march during the country's 57th Republic Day parade in New Delhi, January 2006.
© T.C. Malhotra/Central Imaging Agency LLC/WPN

Box 1.1 Definition of small arms and light weapons

The Small Arms Survey uses the term `small arms and light weapons´ broadly to cover both military-style small arms and light weapons as well as commercial firearms (handguns and long guns). When possible, it follows the definition used in the United Nations´ Report of the Panel of Governmental Experts on Small Arms (United Nations, 1997):

Small arms: revolvers and self-loading pistols, rifles and carbines, assault rifles, sub-machine guns, and light machine guns.

Light weapons: heavy machine guns, hand-held under-barrel and mounted grenade launchers, portable anti-tank and anti-aircraft guns, recoilless rifles, portable launchers of anti-tank and anti-aircraft missile systems, and mortars of less than 100 mm calibre.

The Survey uses the terms `firearm´ and `gun´ to mean hand-held weapons that fire a projectile through a tube by explosive charge. The terms `small arms´ and `light weapons´ are used more comprehensively to refer to all hand-held, man-portable, explosively or chemically propelled or detonated devices. Unless the context dictates otherwise, no distinction is intended between commercial firearms (like hunting rifles) and small arms and light weapons designed for military use (such as assault rifles).

The UN definition was agreed through consensus by government officials. It was negotiated, in other words, to serve practical political goals that differ from the needs of research and analysis. While the UN definition is used in the *Survey* as a baseline, the analysis in this and subsequent chapters is broader, allowing consideration of weapons such as home-made (craft) firearms that might be overlooked using the UN definition. The term small arm is used in this chapter to refer both to small arms and light weapons (i.e. the small arms industry) unless otherwise stated, whereas light weapon refers specifically to light weapons.

In 2002, the Small Arms Survey estimated that military production reached around 815,000 weapons in 2000 (Small Arms Survey, 2002, p. 13). The aim of this chapter is to determine whether this figure is a defensible one when calculated from a different set of data—procurement by the world's militaries. The approach outlined below therefore triangulates with previous approaches, in the hope of reinforcing the validity of our data. By doing so, it aims to generate a better understanding of the current military demand for small arms. Moreover, by investigating past patterns of procurement, it indicates possible future trends in demand and supply. If military demand for small arms can be estimated with some degree of accuracy, then we will have achieved a significant step towards a defensible estimate of global small arms production.

The dynamics of small arms procurement

The world's militaries procure small arms frequently. Each country tries to maintain small arms in sufficient quantities to equip its forces and reserves. A country must therefore repair or replace damaged, lost, or obsolete small arms regularly. However, countries periodically modernize their weapons *en masse,* replacing large numbers of weapons when they believe that their armed forces are in need of modernization.

In terms of the global demand for small arms and light weapons, this equates to two different dynamics; a slow and constant turnover of weapons to replace lost and damaged stocks—an annual low rate—punctuated by major procurement programmes whereby countries replace a large proportion of their existing stocks over a relatively short space of time—a peak procurement rate.

The case of the United States is illustrative (see Figure 1.1). That country began to procure the M16A2 assault rifle in 1982 to replace the M16A1, which it had adopted in 1967. While no data is available for the years 1987–92, or for 1995, the graph displays a distinctive pattern. The M16A2 was adopted in 1982, which prompted initial trials, followed by large-scale procurement. The Gulf War of 1991 provided a final impetus to replace aging stocks of M16A1 rifles

Figure 1.1 **United States procurement of the M16A2 rifle and subsequent derivatives, 1982–2003**

UNITS

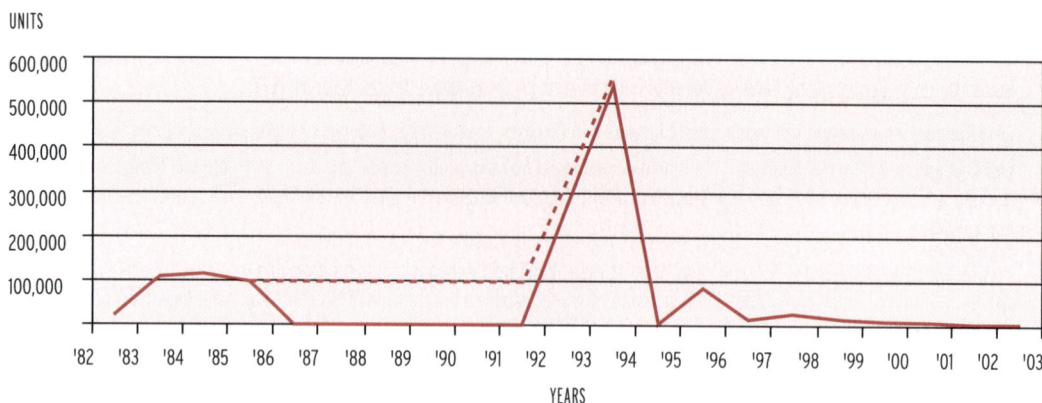

Notes: Online access to US Department of Defense Procurement Programs (P-1) data is only available for 1997 onwards. Data for 1982–86 is from Ness (1995) and Watters (2005). The red dotted line indicates an estimate.

Sources: US DoD (1997; 1998; 1999; 2000a; 2001a; 2002a; 2004; 2005a); Ness (1995); Watters (2005)

with later variants and this is clearly demonstrated by high procurement rates in 1993 and 1994, and is probably true of 1995 also. This was then followed by low-level procurement, fluctuating between 5,000 and 12,000 units per annum, which constitutes the annual low rate of replacement for damaged, destroyed, or lost weapons.

Between 1982 and 2004, the United States procured around 1.9 million M16A2 assault rifles and M4 carbines, totalling around 136 per cent of its active armed forces personnel. That is, 136 weapons were procured for every 100 serving members. The number was similar in 1967, when the M16A1 was introduced. It is also worth noting that the projected procurement of the Objective Individual Combat Weapon (OICW), Increment 1, is of 1.3 million weapons for the Army alone (US DoD, 2005c, p. 11).

Like the United States, most countries periodically review their armed forces' small arms. This may be prompted by an external event, such as a war or a deteriorating security situation, or it may simply be in response to the fact that the negativities of existing weapons begin to add up sufficiently to necessitate a change. It may also be due to countries adopting a different cartridge, for example 5.56 mm instead of 7.62 mm.[3] As Figure 1.2 demonstrates, for a selection of NATO countries, these major procurement programmes have occurred several times in the past 50 years.

The data in Figure 1.2 was derived from numerous sources, but the collection began with extensive searches of the *Jane's Online Reference* database (Jane's Information Group, 2005). Searches were made by type of assault rifle or carbine, for instance, by the Canadian C7, the British SA80, or the Singaporean SAR. These searches revealed various articles referring to the development, sales, and transfers of the weapons concerned and, importantly, government contracts. A judgement was made for each, based on the scale of the order, as to whether or not each contract comprised part of a major procurement initiative. Thus, for instance, the procurement of around 500 M14 rifles in 2002 by the United States does not constitute a major initiative, when gauged against major procurement initiatives beginning in 1967 and 1982, each of more than one million M16 rifles (DoD, 2005a; Ness, 1995; Small Arms Survey, 2004; Watters, 2005).

The research then entailed further searches for previous procurement initiatives. For instance, Canada acquired 94,135 C7 rifles and 2,365 C8 carbines in a procurement programme launched in 1983. The previous initiative had been

Figure 1.2 **Major procurement of assault rifles and carbines by selected NATO countries, 1956–2006**

PROCUREMENT
(UNITS APPROX.)

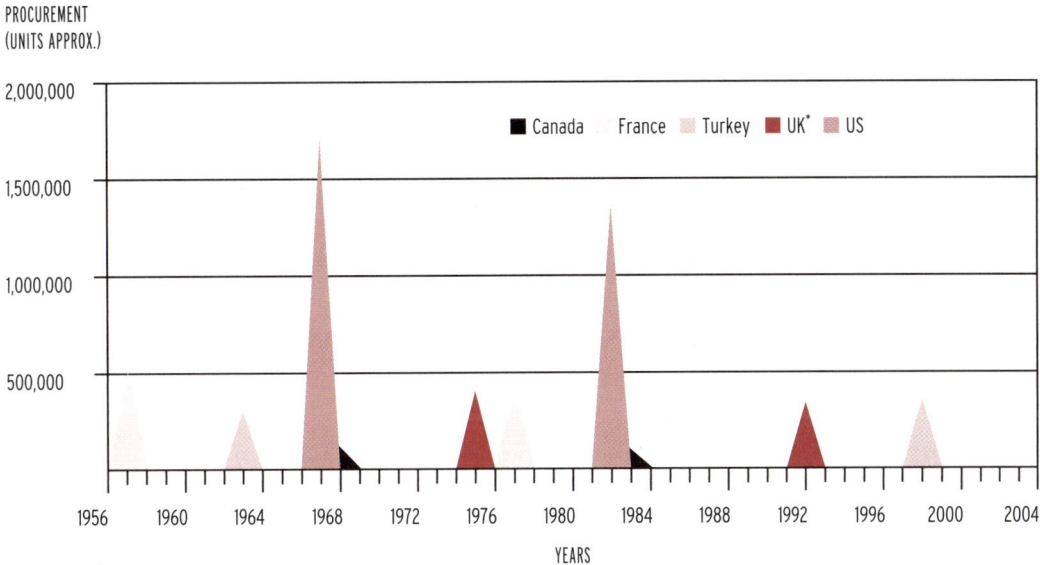

Note: Total units procured are condensed into the year in which the procurement initiative began. The graph must therefore be read as a stylized presentation of procurement trends.

* Information regarding the United Kingdom was provided in a letter from the UK Defence Logistics Organization (UKDLO) to the Small Arms Survey concerning the procurement of the SA80 weapon system, 9 November 2005.

Sources: Canada (2005); *DFASP* (1998); US DoD (1997; 1998; 1999; 2000a; 2001a; 2002a; 2004; 2005a); Forecast International (2005); Heyman (2001); *Jane's Defence Weekly* (1995; 1996); Ness (1995); Sariibrahimoglu (1998); Watters (2005)

in 1968, when Canada adopted the FN FAL (see Figure 1.2); thus there was a 15-year period between major procurement programmes (Heyman, 2001; *Jane's Defence Weekly*, 1996, p. 21). Sources used included the Small Arms Survey's hard copies of news reports on small arms production, historical works on the development of particular weapons, LexisNexis searches of defence industry publications, and government procurement programme reviews.

Derived from this data, Figure 1.2 demonstrates that countries initiate major procurement initiatives quite infrequently, which explains why small arms change very little in design over long periods of time (Small Arms Survey, 2005, p. 58). Countries also appear to replace similar numbers of weapons during every major procurement initiative because, in the last 50 years, there have been few major force reductions or increases, despite the end of the cold war (IISS, all cited).[4] Also, major procurement initiatives happen fairly regularly. Canada, France, the United Kingdom, and the United States, for example, appear to initiate such initiatives every 15–20 years.

However, there are also notable differences among countries in the scale of procurement and in the frequencies of major procurement initiatives. Turkey is a case in point. In contrast to the other four countries in Figure 1.2, the period between Turkey's last two major procurement programmes lasted approximately 30 years. These differences increase markedly when considering states other than NATO members. State armed forces differ in doctrine, in numbers of personnel, and in the resources that each country is willing or able to commit to procuring armaments, including small arms. This poses considerable problems for estimating global procurement.

FILLING IN THE BLANKS

While data on the production of military small arms is sparse, procurement information is relatively accessible for a number of states. This section presents a method of estimating world procurement. It uses data on major procurement initiatives of assault rifles and carbines from 32 countries over the last 50 years. Of the 151 countries in the global sample, these were the only states found to have sufficiently reliable procurement data to project a global rate of procurement.

The countries in the 32-country sample host around 6.5 million active troops, or around one-third of the world's active armed forces. They are arguably a geographically, economically, and politically representative cross section of the world's states.

The method employed calculates major procurement initiatives as a percentage of each country's active armed forces. This yields a rate of procurement with which countries can be ranked and compared. By using a proxy for rates of procurement, in this case the age of stock in each country's arsenal, the analysis extrapolates the procurement rates from the 32 countries to the majority of the world's remaining states.

Calculating procurement rates

Referring back to Figure 1.1, the case of the United States is arguably representative of acquisition dynamics for many states. States intermittently procure large numbers of assault rifles and carbines to replace the primary assault rifle in infantry service at the time. One example is the United Kingdom's replacement of SLR rifles with 332,092 SA80 assault rifles beginning in 1984.[5] Another is Venezuela's decision to replace its stocks of FN FAL rifles with 100,000 Russian AK-103 rifles in 2005 (*Jane's Defence Weekly*, 2005; Kirk, 2005) (TRANSFERS).

These major procurement initiatives constitute peak procurement (Pp); i.e. the largest by volume procurement a state will initiate in a period of, for example, 20 or 30 years. However, states also procure at low volume per annum between periods of peak procurement to replace lost, damaged, or worn weapons—an annual low rate of procurement (Lr).

In this analysis, data on procurement covered only the acquisition of assault rifles and carbines—the standard personal weapons of infantry troops. However, from this data, the ratios developed in Table 1.1 from detailed information supplied by five countries also allow procurement estimates to be made of numbers of pistols and heavy and light machine guns. We can say that for every 84 assault rifles and carbines procured, a country would be expected to procure 11 pistols, 4 light machine guns, and 1 heavy machine gun—in total around 19 per cent of assault rifle and carbine acquisition. This percentage is then added to the figure for assault rifles and carbines to calculate total procurement of small arms and light weapons.

Collectively, the weapons referred to in the previous paragraph are the most numerous infantry small arms and light weapons, whereas other infantry weapons, such as anti-tank guided weapons (ATGW), man-portable air defence systems (MANPADS), and recoilless rifles, tend to be deployed in far fewer numbers. The weapons listed in Table 1.1 therefore provide the basic complement of weapons for any infantry force.

In the analysis, procurement data was obtained for the 32-country sample, using the same methods used for the five countries in Figure 1.2. Each major procurement initiative was plotted chronologically and by units procured. For all 32 countries, at least two procurement initiatives were recorded, giving an interval between major initiatives (peak procurement) for each country.

Table 1.1 Ratios of small arms and light weapons in state armed forces in five countries, 2001–03

Country	Assault rifles & carbines		Pistols		Light machine guns		Heavy machine guns		Total small arms & light weapons
Canada[1]	195,803	84%	25,125	11%	11,667	5%	1,354	1%	233,949
Finland[2]	514,600	96%	8,000	1%	7,500	1%	3,523[6]	1%	533,623
Togo[3]	11,020	87%	1,184	9%	360	3%	70	1%	12,634
Switzerland[4]	450,000	83%	70,000	13%	21,469[6]	4%	3,076[6]	1%	544,545
US[5]	1,146,920	75%	232,565	15%	118,582	8%	27,525	2%	1,525,592
Median		**84%**		**11%**		**4%**		**1%**	

Notes: Figures for Canada are from 2000; Switzerland, Togo, and the United States from 2001; and Finland from 2003. US numbers are for the US Army only and exclude other services. The five countries were chosen simply because these are the only ones that made available detailed information on stocks to the Small Arms Survey. Percentages are rounded.

1 Source: Small Arms Survey (2001, p. 74).
2 Source: Letter from Pauli Järvenpää, director general, Finish Ministry of Defence, 21 August 2003.
3 Source: Togo (2001).
4 Source: Small Arms Survey correspondence with the General Staff of the Swiss Armed Forces, 5 and 6 December 2001; 1 February 2002.
5 Source: Small Arms Survey (2002, p. 84).
6 Indicates dummies.

In order to make procurement comparable across countries, the analysis used each country's most recent volume of peak procurement (Pp) and the number of persons serving in its armed forces (S) to develop an up-to-date rate of acquisition for that country. Thus, for example, the acquisition of 150 assault rifles for a force of 100 active troops would equate to a 150 per cent rate of peak procurement (see Table 1.2).[6] This rate of acquisition enables comparisons between how much per capita procurement each country engages in (in the example 1.5:1). Personnel data was derived from the International Institute for Strategic Studies (IISS) *Military Balance* series.

As Table 1.2 illustrates for a selection of the 32 countries, most appear to launch major procurement initiatives every 15–30 years. Certainly, no states for which data was available did so more frequently than every 15 years (see Appendix 4). A large number did so every 25–30 years. It is likely, however, that many of the world's poorer states

Table 1.2 Rates of per capita procurement (%) for 6 of the 32-country sample

Country	Current active forces	Last major procurement (units)	Frequency of major procurement (years)	Major procurement per capita (Pp/S)
Canada	52,300	96,500	15	185%
France	259,050	435,000	30	168%
Spain	150,700	120,000	30	80%
Taiwan	290,000	130,000	25	45%
Venezuela	82,300	100,000	30	122%

simply do not engage in major procurement initiatives. This may be due to lack of funds or trade restrictions. In contrast to wealthier states, they replace stocks on an ad hoc basis, according to need and ability—a potentially low rate, but high frequency.

Age of stock as a proxy variable for procurement

While reasonably good data was available for the 32-country sample, this was not the case for the remainder of the world's countries. The analysis therefore necessitated using a proxy variable to project the procurement rates of the 32 countries to the states of the rest of the world.

However, despite some ostensibly plausible proxies, such as military expenditure and gross domestic product (GDP), there were too many intervening variables that prevented defence and economic indicators from being used to determine small arms procurement (see Appendix 1). Such variables include expenditure on larger items of military hardware such as artillery or ballistic missiles; variations in armed force type, composition, and doctrine; or the impact of corruption on reducing direct expenditure on equipment. Or, like Iceland, a state may not even field an army, but channel its GDP into non-military expenses (IISS, 2004, p. 277).

These observations suggest the need for a proxy variable that represents a state's direct commitment to the procurement of small arms and light weapons, rather than other military costs.

One such proxy—age of stock—was hypothesized to be a good gauge of a state's commitment to procuring small arms and light weapons, potentially revealing important information about the scale and frequency of its procurement.

Countries that procure small arms frequently and in large numbers should, on average, have relatively young stocks of weapons. In contrast, those that procure infrequently or in fewer numbers should stock older weapons. In mathematical terms, the rate of procurement (Pp/S) should be a function of the inverse age of the stock (A):[7]

$$Pp/S = f(A^{-1})$$

Using this proxy necessitated logging the age of a selection of a particular state's small arms and light weapons. However, for reasons of data availability, weapon ages were not recorded by date of manufacture, but by the date at which the design first became available on the world market for military weapons. This is not strictly, therefore, the age of a weapon, but the age of a design. Some weapons are newer than recorded in this analysis, having been produced after the date at which the type first entered service. However, age of stock arguably remains a good indicator of the modernity of a state's weapons.

One example illustrates this. A single rocket launcher developed in 1950 and still in service with a state's armed forces may tell us little about procurement. However, when all of a state's weapons are of that vintage, we can say with some surety that the state expends little on procurement. By contrast, states that stock numbers of the latest guided missile launchers and similar weapons clearly procure more frequently and at greater expense. Thus, while this method does not yield the 'real' age of a country's arsenal, when the same method is applied to all countries, it produces a comparable relative age.

In this analysis, infantry support weapons (including anti-tank, anti-bunker, and multi-purpose rocket launchers) and MANPADS—both of which fall under the UN definition of small arms and light weapons (UN, 1997)—were used to produce a combined age that was representative of a country's small arms and light weapons stocks. In contrast to assault rifles and carbines, there are fewer similarities among types of infantry support weapons and MANPADS, which makes them far easier to differentiate by age.[8] They are a valid indicator of age of stock because, in countries

Figure 1.3 **Average age of weapons stocked and procurement rates in major procurement initiatives for 32 countries**

Pp/S: PERCENTAGE OF WEAPONS PROCURED PER ACTIVE MEMBER OF ARMED FORCES

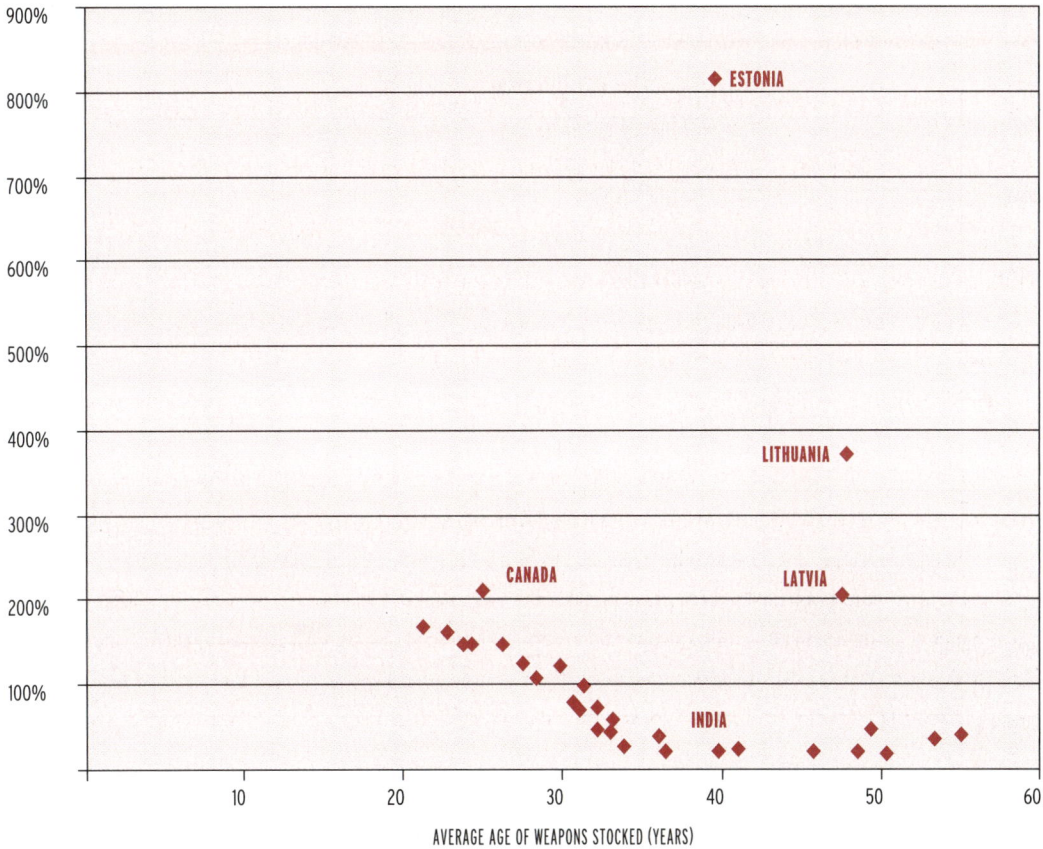

AVERAGE AGE OF WEAPONS STOCKED (YEARS)

ranging from the United States, to Turkey, to Uganda, their ages mirror those of other infantry weapons, including assault rifles and carbines.

Data from *Jane's Infantry Weapons,* the IISS *Military Balance* series, and the *Small Arms Survey 2004* was used to determine the age of these weapons, and for the 151 countries, to produce a single average age for each state's arsenal (IISS, 2004, pp. 14–249; Jones and Cutshaw, 2004, pp. 409–71; Small Arms Survey, 2004, p. 82). A portion of this data was then plotted against the known procurement rates for the 32-country sample, to determine whether age of stock was a reliable proxy for procurement (see Figure 1.3).

Figure 1.4 reproduces Figure 1.3, but without the outliers of the three Baltic states, to make the pattern clearer.

What age of stock tells us about procurement

The world's wealthier states appear to procure newly produced weapons at regular intervals. By contrast, the world's poorer states procure on an ad hoc basis. Discounting transfers of surplus stocks to states in conflict or those recovering from conflict, these states probably procure more frequently, but they acquire fewer weapons in any one initiative. Poorer state acquisitions are very often of older surplus weapons and this is one further reason why these states have far higher stock ages than their richer counterparts.

Figure 1.4 **Average age of weapons stocked and procurement rates in major initiatives for 29 countries**

Pp/S: PERCENTAGE OF WEAPONS PROCURED
PER CAPITA ACTIVE MEMBER OF ARMED FORCES

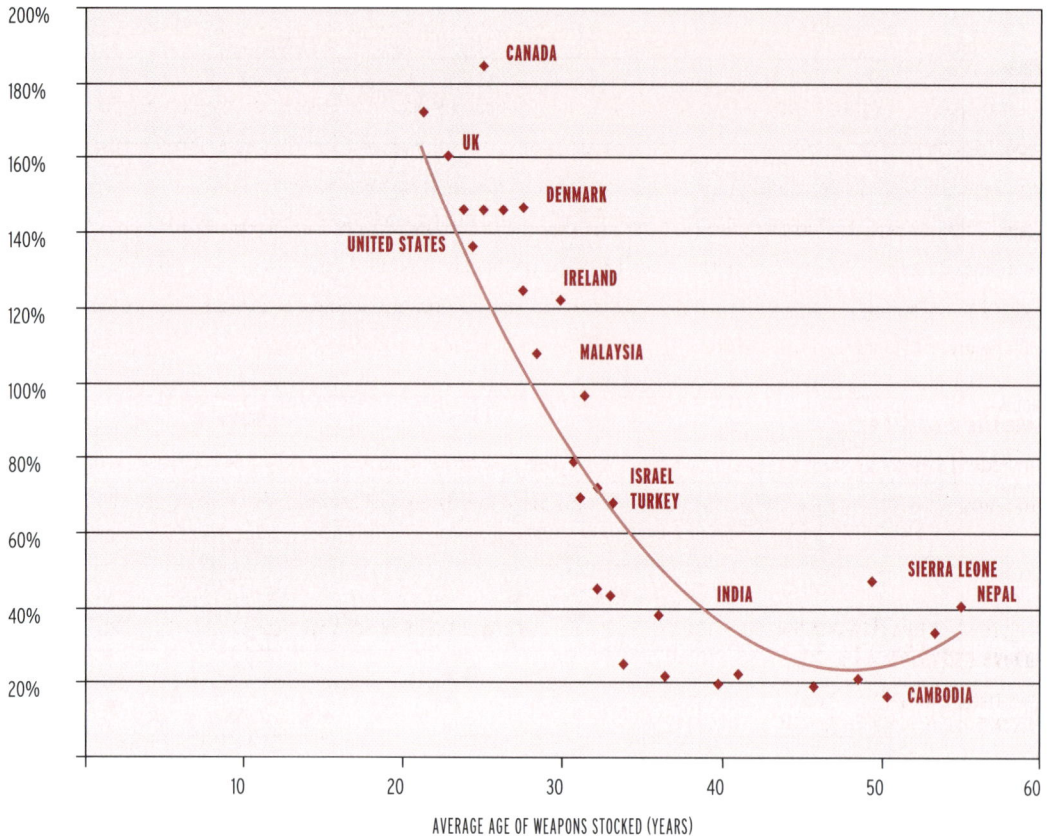

AVERAGE AGE OF WEAPONS STOCKED (YEARS)

For the largely wealthy countries with average stock age of less than 40 years (to the left of Figures 1.3 and 1.4), the relationship to procurement rates appears to be a close one. These countries procure increasingly in proportion to declining average ages of stocked weapons. Their arsenals are newer because they replace large numbers of weapons with newly produced stock. Thus, for instance, when Canada wanted to replace its stocks of FN FAL rifles, it replaced nearly all of them with new Diemaco C7s and C8s (Heyman, 2001; *Jane's Defence Weekly,* 1996; Canada, 2005).

States with an average age of stock in excess of 40 years (to the right of Figures 1.3 and 1.4) are unable to replace stocks on such a scale. They retain older varieties of infantry weapons and their procurement of assault rifles and carbines reflects this. A number of these countries procure entirely from surplus stocks. In India's case, acquisition of the Indian-developed INSAS assault rifle has been delayed by financial and technological setbacks, which has meant that only a fraction of the number of INSAS rifles desired have entered service. The Indian Army continues to rely on older designs of weapons, and relatively small numbers of weapons procured from overseas in the interim (Bedi, 2001; *Jane's Defence Weekly,* 1997).[9]

PROJECTING A GLOBAL ESTIMATE OF PROCUREMENT

The section above has demonstrated that, for the 32 sample countries, average age of stock is a valid indicator of procurement rates. Subtracting the three outlying Baltic states from the sample leaves 29 states. The procurement rates of these states can be projected, with the caveats expressed below, to yield procurement rates for the remaining 122 states in the 151-state study.

The projection is not without problems. As Figure 1.5 demonstrates, introducing a dummy variable for the 74 countries with stocks in excess of 40 years of age (to the right of the graph) illustrates how projections made from the 7 countries that fall into the 40+ year grouping (an admittedly small number, marked in black) out of the 32-country sample are invalid.[10] Excluding outliers such as the three Baltic states (see Figure 1.3), most countries in this part of the sample procure at rates below 50 per cent. In fact, in the admittedly very small sample of seven countries, the average rate of procurement was around 28 per cent. However, the projections for countries with stocks younger than 40 years of age are, by contrast, relatively strong (see Appendix 2 for regression statistics).

Method and results

In order to estimate the number of units procured by each country, state armed force numbers were multiplied by the projected rate of procurement (Pp/S) (see Table 1.3). In the absence of a strong correlation (see Figure 1.5), for

Figure 1.5 **Projected and known procurement for 151 countries (sample split into countries with stock aged above and below 40 years)**

Pp/S: PERCENTAGE OF WEAPONS PROCURED
PER CAPITA ACTIVE MEMBER OF ARMED FORCES

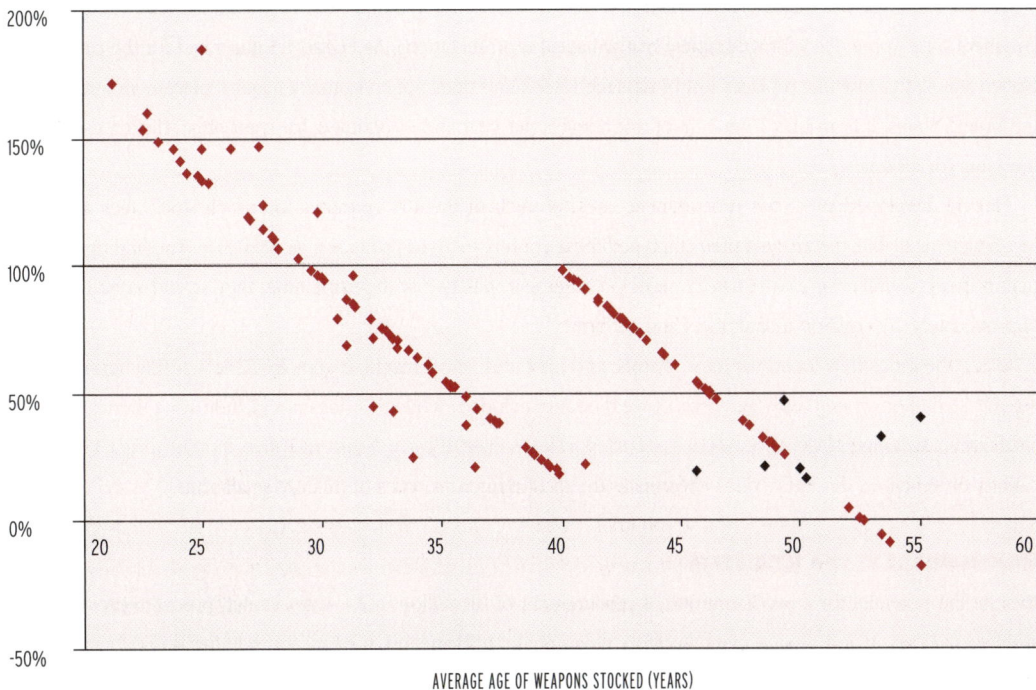

AVERAGE AGE OF WEAPONS STOCKED (YEARS)

Table 1.3 **Projecting assault rifle and carbine procurement rates: a selection of the 151 states studied**					
Country	**Average age of stocks**	**Major procurement rate (Pp/S)**	**Number of active troops (S)**	**Projected assault rifles and carbines procured (major initiative)**	**Annual low-rate procurement (Lr/S = 1%)**
China	30	98%	2,225,000	2,211,655	22,500
Gabon	46	28%	4,750	1,330	48
Philippines	49	28%	108,500	30,380	1,085
Qatar	28	110%	7,500	8,286	75
Sweden	23	153%	27,600	42,268	276
US	24	136%	1,433,600	1,948,421	14,336

Note: Projected assault rifles and carbines procured are subject to standard deviation of the Pp/S of 0.287 for countries with stock below 40 years of age (see Appendix 2). In this selection of states, Gabon and the Philippines have stock aged over 40 years and are therefore allocated a fixed Pp/S of 28 per cent.

the 74 countries whose average age of stocks was over 40 years, the average rate of procurement for the seven countries discussed above of 28 per cent was allocated.

The potential impact of using this fixed 28 per cent, in terms of distorting world procurement data, is attenuated by the fact that these 74 countries, whilst hosting around 50 per cent of the world's active armed forces, almost certainly comprise the lowest procuring states. By contrast, the bulk of the world's procurement of assault rifles and carbines is by countries whose weapons average less than 40 years of age. The procurement rates for these latter countries were calculated using the regression equation (see Appendix 2 for regression statistics).

For all countries, the major procurement rate (Pp/S) was then combined with an annual low rate (Lr/S) of procurement (see Appendix 3 for a complete mathematical representation). As Figure 1.1 illustrated for the case of the United States, this low rate replaces lost or damaged stock and does not constitute a major replacement initiative. In the United States, it is around 1 per cent of troop levels per year and is assumed, for want of conflicting data, to be the same for all states.

Having developed projected procurement rates for each of the 151 countries for which stock ages and force levels were available, the analysis then combined these country totals to produce a global figure. The findings suggest that, if every country were to initiate a major procurement initiative at the same time, they would in total acquire approximately 17.5 million assault rifles and carbines.[11]

Using the ratios developed for rifles, pistols, and light and heavy machine guns in Table 1.1, this figure rises to around 21 million when these categories of weapon are included. While this does not include any other category of small arms, including MANPADS, ATGWs, recoilless rifles, sniper rifles, anti-material rifles, mortars under 100 mm in calibre, or shotguns, the figure does encompass the most numerous types of infantry small arms.

Procurement in the long term

Despite the potential for a maximum rate of procurement of 21 million units, states launch procurement initiatives at different times. World procurement therefore never peaks at 21 million. Instead, this number is distributed over a number of years.

In the 50-year period for which data was available, none of the 32-country sample were found to have launched major procurement initiatives more frequently than at 15-year intervals. Many procured every 20 or 30 years. Many of the world's poorest states—which are under-represented in this sample—procure little in volume, but do so frequently.

With these observations in mind, estimates of procurement in a 50-year period were based on the shortest procurement cycles of around 15 years for the 74 countries with stock aged over 40 years. For the 77 countries whose stock was younger than 40 years, the average procurement cycle length of 24 years was used (see Appendix 4). The results of this method suggest that the countries of the world procure an estimated 50 million assault rifles, carbines, pistols, and light and heavy machine guns over a 50-year period, or around 1 million units per annum (see Appendix 3).[12]

FROM PROCUREMENT TO ESTIMATING PRODUCTION

This section projects world military demand of newly produced and surplus weapons. It does so by estimating the proportion of surplus to new acquisition and applying these ratios to the procurement estimates generated in the previous section.

As we saw, when countries procure stocks of small arms and light weapons, the wealthiest states procure new equipment, whether from abroad or from domestic suppliers. Less wealthy states often have to procure a mixture of new and used stocks because of financial constraints. The poorest countries, and those in the most urgent need of small arms, often re-equip with sales or gifts of surplus stock.

Developing an estimate of global small arms and light weapons production from procurement data is therefore hampered by the fact that the transfer of old stocks distorts projections. Unfortunately, there is no precise way of alleviating this problem. The only sure way to do this would be to assess every state's procurement on a case-by-case basis and, from there, to develop some country-specific ratio of new-to-old weapons procured. The fact that only 32 countries were found to have sufficiently reliable procurement data to develop the projections used in this chapter is evidence that this method is, at present, unfeasible.

A rough means of assessing production is therefore to develop a series of thresholds, from the 32 cases, at which countries procure varying ratios of new-to-old stock. This approach stands the risk of considerable arbitrariness, but this is alleviated by the fact that the world's highest procurers are, for the most part, the wealthiest. Put simply, the largest part of world production appears to supply the wealthiest countries and, as Figures 1.3, 1.4, and 1.5 demonstrate, these comprise the countries for which the most reliable data is available.

The countries of the world procure about 1 million military small arms per year.

Such observations suggest that age of stocks should again be used, in this case to determine the likelihood that a country will procure old or new stocks, or a mixture of the two. As was clearly demonstrated in Figure 1.3 in the case of the Baltic states, high average ages of stocks appear to correlate strongly with high volumes of surplus stock procurement.

Countries with the oldest stocks: Of the nine countries in the 32-country sample (including the Baltic States) that have stocks of an average age of more than 40 years, none was found to have procured new assault rifles and carbines in a recent major procurement initiative. This does not mean, however, that they did not recently acquire new weapons, but rather that they did not do so in large quantities. For these countries, and in fact for most, very little data is available on low-level procurement.

Countries with the youngest stocks: For countries with stock aged 30 years and below, the 11 countries of the 32-country sample that fell into this category all procured 100 per cent newly produced weapons. This figure was distributed evenly throughout the sample; from countries with the newest stocks, to those with stocks approaching 30 years of age. On this basis, all countries with stocks younger than 30 years of age were deemed to procure 100 per cent new weapons in major procurement initiatives (see Table 1.4).

Countries with mid-age stocks: For countries with stocks of an average age of between 30 and 40 years, there were considerable differences. The ratio of new-to-old procurement did not closely relate to variations in age of weapons stocked by this group. For instance, Israel, with an average age of weapons of 32 years, procured around 60 per cent new equipment (Ben-David, 2003; NISAT, 2005; Pineo and Lumpe, 1996). In contrast, Thailand, whose stocks averaged 36 years, procured 100 per cent new stocks in its last major procurement initiative (*Bangkok Post,* 2005; Haug et al., 2002; Thai Press Reports, 2005). The rate of new procurement in the 12 countries of the 32-country sample that fell into this category averaged 81 per cent. However, the sample was arguably skewed towards the acquisition of new weapons.

Eight out of the 12 states in this category, for instance, each acquired 100 per cent new equipment in their last major initiative, but it is highly plausible that this number over-represents acquirers of new equipment. This is simply because of the newsworthiness of contracts with major small arms producing firms, in contrast to the transfer of excess defence articles or used equipment.

With this in mind, developing a range—rather than a single figure—of new–old procurement appears more appropriate. The lowest rate of procurement for this set of 12 countries was found to be 30 per cent acquisition of new stock. The maximum was 100 per cent, but due to the fact that the sample was probably skewed, the sample average—80 per cent—was used as a conservative estimate of maximum new stock acquisition.

Table 1.4 extrapolates the figures for procurement given above regarding the 32-country sample onto the remainder of the 151 countries for which information about force size and stock age is available and reflects figures for the entire 151-state sample.

Applying these figures to the projected procurement rates of each country gave an estimated range of assault rifles and carbines produced to meet major procurement initiatives each year. Conversely, it also generated a rate of procurement for surplus or used weapons. An amount of 19 per cent was again added to the figures to encompass production and surplus transfer of pistols and light and heavy machine guns.

The results plotted in Figures 1.6 and 1.7 suggest that within a given 50 years, world production of assault rifles, carbines, pistols, and light and heavy machine guns ranges between 36 million and 46 million units. Average annual production ranges from 0.7 million (see Figure 1.6) to 0.9 million (see Figure 1.7) weapons. It is nonetheless

Table 1.4 **Age of stock and the proportion of new to surplus stocks procured by 151 states**				
Group of states	**Number of states in group**	**Average age of stock**	**% new weapons procured**	**% surplus weapons procured**
Youngest stocks	29	21–30	100	0
Mid-age stocks	57	31–40	30–80	20–70
Oldest stocks	65	41+	0–30	70–100
Total	**151**			

Figure 1.6 **Numbers of weapons produced (low estimate) or transferred from surplus (high estimate) to meet the annual procurement needs of 151 states**

NUMBERS OF WEAPONS PRODUCED
OR TRANSFERRED FROM SURPLUS

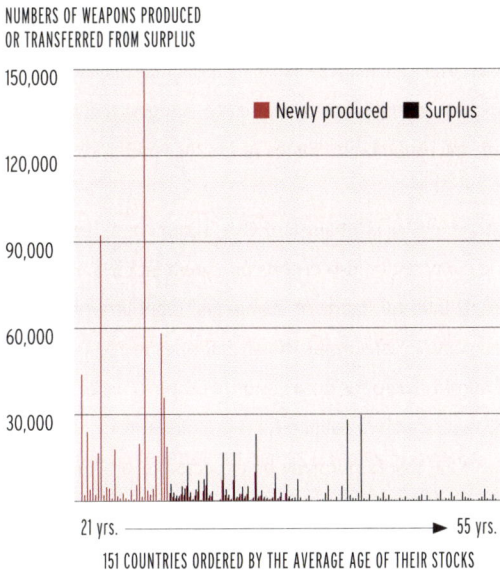

151 COUNTRIES ORDERED BY THE AVERAGE AGE OF THEIR STOCKS

Note: The method used to produce the data assumes countries with stock aged 31–40 years procure 30 per cent new small arms and light weapons and 70 per cent surplus, and countries with stocked aged over 40 years procure only surplus weapons. See accompanying text for details.

Figure 1.7 **Numbers of weapons produced (high estimate) or transferred from surplus (low estimate) to meet the annual procurement needs of 151 states**

NUMBERS OF WEAPONS PRODUCED
OR TRANSFERRED FROM SURPLUS

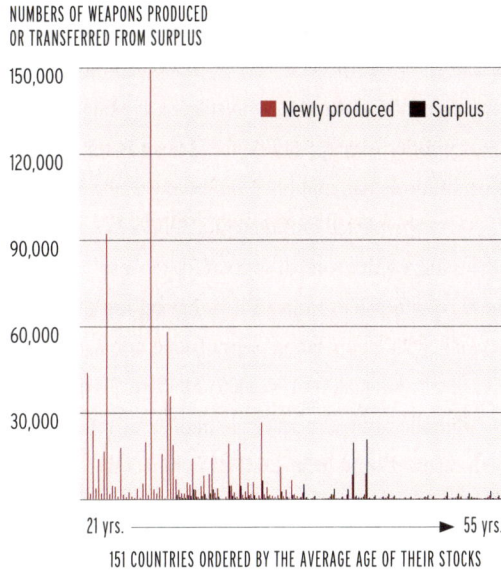

151 COUNTRIES ORDERED BY THE AVERAGE AGE OF THEIR STOCKS

Note: The method used to produce the data assumes countries with stock aged 31–40 years procure 80 per cent new small arms and light weapons and 20 per cent surplus, and countries with stocked aged over 40 years procure 30 per cent new weapons and 70 per cent surplus. See accompanying text for details.

important to stress that, because this estimate is derived from the procurement projection, for each parameter, the same standard deviation of 0.287 applies (see Appendix 2).

WHERE THE MODEL MEETS REALITY

The projected demand of 36–46 million units produced for the world's militaries in a 50-year period is likely to be a reasonable approximation of the true level. However, a yearly production rate of 0.7–0.9 million units is unlikely to hold for any given year. Projections over long periods (in this case 50 years) mask short-term fluctuations in periods such as five or ten years. Countries' procurement cycles may be unsynchronized, or they may be clustered into short periods of time, creating peaks in world procurement.

In reality, global procurement does follow some broad trends and there are clear reasons why certain groups of states procure at approximately the same time. In short, some countries' procurement cycles *are* synchronized and this has important consequences for the evolution of future world procurement and production levels.

1) Political alliances and changes of calibre: Change of calibre is an important push factor in procurement and can be a collective one. Being a member of a military alliance frequently necessitates interoperability of weapons and ammunition among member state armed forces. Thus, for the case of new NATO members, this requirement has, and will, push them to procure weapons that utilize the NATO 5.56 x 45 mm and 7.62 x 51 mm cartridges, rather than the 7.62 x 39 mm Warsaw Pact calibre (NATO, 2001; Atlantic Council, 2001, p. 24).

2) Wars and insurgencies: States experiencing war and armed opposition procure at rates that differ from 'normal' procurement. For example, Nepal procured around 30,000 assault rifles from India, Israel, the United Kingdom, and the United States in a ten-year period. This was not part of a concerted effort to change Nepal's primary infantry weapon. The weapons procured were of varying calibres, from 5.56 mm and 7.62 mm NATO to 5.56 mm Indian-produced weapons (Bedi, 2005a; Davis, 2005a; Heyman, 2001; Hill, 2004; Press Trust of India, 2002). Patterns of procurement such as this, with states accepting grants of weapons when finances are scarce, suggest that quantity takes priority over quality or the demands for within-military interoperability. States in conflict therefore tend to procure more than and often differently to those at peace.

3) Independence and domestic political change: The experiences of Afghanistan, East Timor, Iraq, and Sierra Leone suggest that foreign support for reconstructing a state's security sector also distorts procurement rates. In effect, these countries start from scratch, having lost their state arsenals into the hands of insurgents, rebel groups, or the general civilian population. Sierra Leone, for instance, received some 2,520 surplus British SLR rifles in 2000 (London Press Association, 2000; UK, 2000). Similarly, Iraq has received surplus weapons from Eastern Europe to replace stocks lost with the disbandment of the Iraqi Army following the 2003 invasion by coalition forces (Holdanowicz, 2005). In both cases, procurement exceeds numbers that a state in peacetime would normally be expected to procure.

4) Competitive and cooperative modernization: Criticism has been levelled at decisions to replace stocks of assault rifles with new designs, given the degree to which rifles can be improved by relatively minor alterations (Steadman, 1994). However, the fact remains that the world's wealthier nations frequently procure newly developed rifles. Analysis demonstrates that this does not necessarily occur on a random basis, but in response to developments in other countries, particularly those with a similar level of investment in infantry weapons. As is the case with selecting calibres, some of these programmes take place within a framework designed to ensure interoperability among alliance members. In NATO, for instance, small arms programmes are under way in Belgium, Canada, Denmark, France, Germany, Italy, the Netherlands, Norway, the United Kingdom, and the United States and are regulated by the NATO Army Armaments Group. Similar initiatives are under way in Australia, Singapore, South Africa, and Sweden (Gourley, Janssen, and Pengelley, 2002; *Jane's International Defence Review,* 2005).

Annual world production of military small arms is around 0.7–0.9 million weapons.

These are just four brief examples of why events can produce some patterning to global procurement rates. At any one time, a number of similar—connected or unconnected—events, such as wars, participation in alliances, or regime changes, can make groups of states increase procurement above 'normal' or average rates. Projecting future trends in procurement and production therefore necessitates knowing where we are in respect of these potentially global procurement cycles.

Current trends and projected demand in the near future

What can the findings in this chapter tell us about the future? The most important observation is that procurement is never static and fixed at one rate. Consequently, neither is production, nor the growth of and trade in surplus stock.

While it is not within the capacity of this short chapter to explore these trends in great detail, a number of recent events suggest that, for some states, major procurement initiatives will occur in the coming few years. This will certainly increase global production and will also have the potential to increase the world's transfers of surplus stocks.

Major increases in global procurement follow the acquisition policies of the world's wealthiest states—the highest volume procurers of newly produced small arms and light weapons. At least 13 states are currently developing or evaluating new infantry rifles, the majority of them NATO members (Gourley, Janssen, and Pengelley, 2002). As

Figure 1.8 **Duration of development of personal weapons in the United Kingdom, France, and the United States: past, present, and projected**

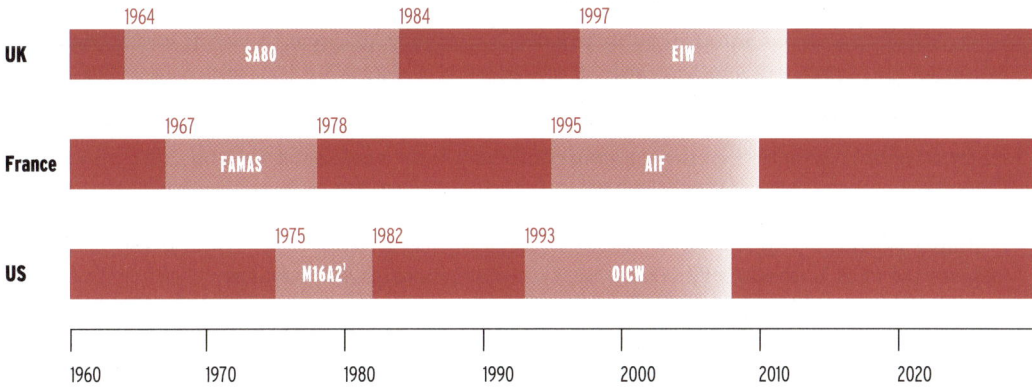

Note: Date markers indicate known start and end of development phases.

1 The M16A2, in contrast to the SA80 and FAMAS, was a revised version of the M16A1. Its development phase was consequently shorter.

Sources: Cutshaw and Pengelley (2000); *Jane's International Defence Review* (1995); Jones and Cutshaw (2004)

Figure 1.8 illustrates, based on the duration of past development and procurement initiatives, weapons currently under development in NATO countries are likely to enter service in the next 5–10 years. As in past initiatives, procurement is unlikely to be condensed into a single year, but phased over a long period. This trend is best represented by US procurement of M16A2 rifles, charted previously in Figure 1.1.

Procurement by these 13 states is also likely to spur procurement by states seeking to maintain infantry parity with them. Since 1945, Russian and Chinese adoption of new assault rifles and carbines has mirrored US procurement cycles, usually following by around 5–10 years and 10–15 years, respectively. Assuming a similar pattern emerges in the next round of procurement, we should expect to see a collective period of procurement for the world's wealthiest states beginning in around 2010–15, and ending in 2030–35. Based on the projections developed in this study, this

Figure 1.9 **Projected demand for new production from countries currently undertaking modernization programmes, other wealthy states, and Russia and China**

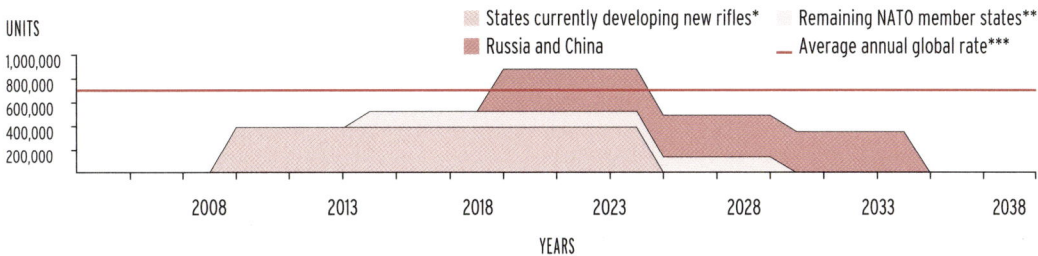

Note: The x-axis shows the estimated period in which the states concerned begin to procure weapons (but does not include the development period). The y-axis indicates expected average annual procurement.

* Includes Australia, Belgium, Denmark, France, Germany, Italy, the Netherlands, Norway, Singapore, South Africa, Sweden, the United Kingdom, and the United States.

** Includes Canada, Ireland, Greece, Portugal, and Spain.

*** Lower estimate (700,000).

Box 1.2 Lengthy developments: the Objective Individual Combat Weapon

The development of the OICW is illustrative of the lengthy development stages of small arms. The costs and decision-making involved indicate why states launch major procurement initiatives only infrequently.

The OICW requirements were first approved in December 1993, in response to a US Army requirement for an infantry weapon 'to engage primary targets, such as personnel protected with body armor or in improvised fortifications and tactical vehicles, and secondary targets, such as light armor and slow moving aircraft' (US DoD, 2005c, pp. 1–4). In February 2002, the 'Objective Individual Combat Weapon Operational Requirements Document' (OICW ORD or ORD) outlined the specifications of 'a dual-weapon system that [would] combine high explosive air bursting munitions and kinetic energy munitions' (US DoD, 2005c, p. 37). The OICW was envisaged as a weapon that could fire munitions akin to spin-stabilized grenades fused to explode above a target in addition to conventional cartridge-based munitions common to contemporary assault rifles.

However, by 2005, the programme had not been able to meet the requirements of ORD. In fact, the most developed stage of the weapon, known as 'Increment I' or the XM8, does not feature an air bursting capability, but is simply a new type of carbine, designed to fire the same NATO 5.56 mm ammunition as the current M16 variants used by the United States. The XM8 programme has become separated from the original OICW programme. The latter, slower-moving programme now comprises the XM25 stand-alone air bursting weapon and the original specification integrated air bursting and kinetic energy weapon specified in ORD (Kucera and Gourley, 2005). The XM8 programme, by contrast, consists of a family of four weapons–carbine, special compact, designated marksman, and light machine gun–which have been designed for the US Army alone (Kucera, 2005a). By 2004, the programme had cost around USD 122 million (see Table 1.5), with the XM8 (Increment I) expected to replace an estimated 1.3 million weapons in the US Army (US DoD, 2005c, p. 11).

Table 1.5 Cost (USD '000) of the OICW development programme, 2000–04

Year	Cost
2000	10,319
2001	25,273
2002	29,860
2003	26,898
2004	29,906
Total	**122,256**

Sources: US DoD (2000b; 2001b; 2002b; 2003b; 2004; 2005b)

However, the development phase has been mired in controversy. The army has been accused by competitors of favouring the XM8's developers, Heckler & Koch (USA) (Kucera and Gourley, 2005). In addition, the considerable differences between the OICW, as originally envisaged in ORD, and the XM8 have incurred much criticism. As the army deputy general counsel (acquisition) noted in September 2004, the XM8 differs so greatly from the OICW that it constitutes 'a materially different [new] requirement . . . More specifically, the OICW ORD does not appear to support the XM8 development to date' (US DoD, 2005c, p. 37). In addition, rather than being a successor to the widely deployed M16 series, the XM8 was being developed for the US Army alone, rather than for the other branches of the armed services as well, and without the input of the latter.

In response to these criticisms, the US Army announced in July 2005 that it was suspending the competition for the next generation of infantry weapons in favour of a programme for all US services, including the Marine Corps, navy, and air force (Gourley, 2005). Furthermore, in November 2005, the army announced that the programme had been suspended to allow the army and other services time to integrate lessons learned from Iraq and Afghanistan into the specifications for the new family of weapons (Kucera, 2005b).

These events, however, are not unusual in the development stages of small arms and light weapons. The programme's suspension is unlikely to be made permanent. The OICW/XM8 programme has not been scrapped, but its development is likely to be longer than first expected. The programme will now be open to other competitors for the design and production of the weapon and is also notably likely to include FN Herstal, which is reportedly planning to bid for the OICW Increment 1 (Kucrea, 2005b).

would herald procurement of around 10 million new weapons over approximately 20 years (see Figure 1.9). This would push procurement for these states alone above the average global rate of procurement in a 50-year period, creating a peak in world procurement.

If this prediction is accurate, global production of small arms will increase significantly as the world's largest procurers re-equip with new stock. However, because procurement of new weapons is likely to displace current stocks from active service, this could have important implications for the trade in surplus stock.

Procurement and the cascade of surplus weapons

Large procurement initiatives in wealthy states can have a major impact on the trade in surplus weapons. Unless major procuring states adopt a systematic policy of destroying surplus stocks, or only agree to supply surplus weapons with the proviso that the recipients' existing stocks are destroyed on a one-to-one basis, the kind of overlapping procurement initiatives displayed in Figure 1.9 can generate a cascade of surplus weapons.

This dynamic occurs because major procurement initiatives by wealthy states replace existing stocks with newly produced weapons. If not destroyed, these weapons are held in reserve. Subsequently they may be redistributed, whether through sales at second-hand rates to poorer countries, or as grants to aid countries with few resources with

---

Box 1.3 The temptation to transfer surplus stocks

The impact of new procurement initiatives on generating surplus stock is clearly evidenced by past events. The decades following the end of the cold war demonstrate some notable examples of major transfers from states that had either downsized their armed forces or begun to re-equip with more modern small arms and light weapons. Because of the duration of small arms development and procurement cycles, some of this readjustment is still in progress.

In 2001, for instance, Bulgaria's largest small arms producer, Arsenal:

proposed that the Ministry of Defence provide the company with its outdated weapons, which Arsenal subsequently intended to sell on international markets. The revenues of the transactions would have been used to produce new, NATO-standard weaponry for the national armed forces (Kiss, 2004, p. 35).

Although this offer was reportedly never taken up by the Bulgarian government, it is a good example of how countries seeking to modernize may be easily tempted to offset the cost of modernization by exporting surplus stocks. Tacitly recognizing this tendency, the *OSCE Document on Small Arms and Light Weapons* notes that 'any small arms identified as surplus to national requirement should, by preference, be destroyed' (OSCE, 2000, IV Introduction; IV C1).

Today, the temptation for states to engage in large-scale surplus transfers is still very real. Prospective NATO member states, for instance, working under the PfP, are switching to NATO standard weapons because of the demands for weapon interoperability within the alliance (NATO, 2001). Unfortunately, and despite NATO aid to help manage and destroy stockpiles, the economic attractiveness of selling weapons to offset the cost of re-equipping has nonetheless been strong. Fears over such transfers led the advocacy organization Human Rights Watch to address its concerns directly to NATO, stating:

Yuri Orlov (Nicolas Cage), illicit arms dealer, beside his most recent acquisition: a vast arsenal of surplus Kalashnikov assault rifles, depicted in the film *Lord of War*.
© Lions Gate Films/Zuma Press

While new PfP programs regarding disposal of surplus small arms are optional, we feel that PfP countries should be actively encouraged to take advantage of all programs that will help address the proliferation of these weapons. In addition, where appropriate, NATO member states should arrange exchanges, by which the transfer of newer military equipment to PfP or new NATO states would be contingent on the recipient country's responsible disposal of quantities of surplus weapons. Such arrangements could do much to prevent weapons from ending up in the hands of abusive military forces (HRW, 2000).

Unfortunately, there are now many producers of NATO standard weapons other than current NATO member states, and unless there is some mechanism to ensure one-to-one replacement and destruction, states–whether prospective NATO entrants or not–are likely to be attracted by the prospect of part-financing new procurement by selling old stocks.

which to re-equip. According to the estimates generated in this chapter, such surplus trade and transfer could total some 14 million units in a 50-year period, or around 280,000 units per annum (see Figures 1.6 and 1.7).

Importantly, the countries that procure surplus stocks are generally the poorest and consequently more likely to be states that suffer armed conflict. The links between procurement of new stock and the displacement of old stock illustrate a critical dynamic—unless destroyed, the arsenals of today's wealthier states are likely to become those of poorer states tomorrow. If the projection in Figure 1.9 proves to be an accurate one, then acquisition of new stock by the world's richer states could potentially displace a further ten million units of surplus stock into the global market place in the coming decades.

The impact of large numbers of surplus weapons was clear in the transfers from Eastern Europe to the world's conflict zones in the 1990s (Faltas and Chrobok, 2004; HRW, 2002; Lumpe, 1999). The poorest states of the world clearly demand weapons above their purchasing power, to say nothing of non-state actors. Rising supply of surplus stocks in the coming years could meet this demand at rates that ensure even larger military small arms transfers. In real terms, this will signify yet more stocks leaving the arsenals of the world's richer and politically robust states, to reside in arsenals where stocks are far less secure.

CONCLUSION

The figures presented and patterns described in this chapter point to one major trend of concern: wealthy states (NATO members, Australia, Japan) will initiate new procurement initiatives in the coming 5–10 years. This will have a number of consequences.

- Global military production will rise to meet the demands of modernizing militaries.
- These initiatives will lead to other countries seeking to maintain parity in weapon quality and numbers (China, Russia).
- Together, these factors could boost world consumption by around 10 million units in the next 25 years.
- This will in turn make large amounts of surplus small arms and light weapons available in the arsenals of these modernizing nations.

Global surplus trade and transfers average around 280,000 military small arms per year.

Past experience suggests that many of these weapons will not be destroyed, but will be transferred to the poorer regions of the world, where stockpiles are less secure, whose militaries are more prone to disintegration, and in which intra-state armed conflict is more common. Without efforts to destroy surplus weapons, the global stock of small arms will continue to grow year by year.

The models presented in this chapter are far from complete. A sample of 32 countries is admittedly a small one and there are critical problems with projecting data from these countries to those for which data is scarce. Nevertheless, for the world's wealthier states, procurement data is relatively good. For these states, the relationship between average age of stock and procurement appears to be a powerful tool for understanding both demand and supply.

With these findings, we are further advanced in understanding global demand and supply for small arms and light weapons. In 2002, the Small Arms Survey estimated that production of weapons for the civilian market in the year 2000 was around seven million units. Military production was estimated to total around 815,000 for the same year (Small Arms Survey, 2002, p. 13). The estimates of military production presented in this year's edition of the *Small Arms Survey,* which are importantly derived by different means, support the 2002 estimate of military production.

The fact remains, however, that military production probably remains a small component of overall production in any given year. Understanding non-state actor demand for weapons is the next big challenge. ◢

APPENDIX 1. PROBLEMS WITH PROXY VARIABLES

State procurement appears to be conditioned by financial constraints. It is therefore plausible that GDP and military expenditure could provide some indication of (a) a country's financial ability to procure frequently and (b) its investment in defence, and hence in the procurement of small arms. In broad terms, this appears to be true (see Figure 1.10). In a ranked sample of the 32-country sample, the highest (by volume per capita) procurers were the states of Western Europe and North America. Middle-ranking countries included the larger South American countries and wealthier South-East Asian nations, such as Thailand and India. The lowest-ranking countries were states such as Cambodia, Sierra Leone, and Sri Lanka.

Unfortunately, however, this trend was not found to be sufficiently regular across the 32 countries to be used as a proxy. In the case of GDP, for instance, countries such as Iceland have a high GDP but no armed forces, while the similar GDPs of Canada and the United Kingdom cannot account for the disparity in the numbers of weapons each procures (see Appendix 4). In the latter cases, Canada procured 96,500 assault rifles from 1998 for a force of 52,300 active troops, giving a per capita procurement rate of 185 per cent (Canada, 2005). In contrast, the United Kingdom acquired only 332,092 assault rifles and carbines for a force of 207,630, which gives a lower per capita procurement rate of 160 per cent (UK, 2005). The same disparities are true of military expenditure, as the cases of Nepal and Uruguay demonstrate (see Table 1.6). The approximate relationship between GDP and military expenditure, on the one hand, and procurement, on the other, suggests that, for the majority of states, modernizing weapons is conditioned, *but not determined,* by financial considerations.

As Figure 1.10 illustrates, one problem with using GDP per capita as a proxy is that the relationship is weak for the wealthier states. Because these states are the largest procurers, using GDP as a proxy stands a greater risk of distorting a world procurement projection than using age of stock.

Country	Major procurement per capita (active troops)	GDP per capita (USD)	Current active armed forces	Military expenditure (USD millions)	Military expenditure per capita (USD)
France	168%	29,410	259,050	45,695	765
Turkey	68%	3,399	514,850	11,649	165
Iceland	–	36,377	–	–	–
Nepal	41%	237	69,000	110	21
Uruguay	19%	3,308	24,000	103	30

Table 1.6 **Cross section of selected countries from the 32-country sample for which procurement data was available ranked by rates of procurement**

Note: Iceland does not have a standing army. Its 'security budget' is listed as USD 2.6 bn in IISS (2004, p. 277) and described as 'mainly for Coast Guard'.

Sources: IISS (2004, pp. 353–7); UNDP (2005, pp. 266–9)

It is also plausible that states with large armed forces procure more small arms. This could be argued from the perspective that a large armed force indicates a commitment to maintaining an effective military capability, which should reflect in the scale of procurement. However, numbers of troops appear to be no gauge of the 'quality' of a state's armed forces, or of its willingness or capacity to modernize or improve them through small arms procurement. For the 32-country sample, there appeared to be no correlation between size of armed forces and volume of procurement. The disparity between France and Turkey (Table 1.6) is just one example.

Figure 1.10 **Procurement in relation to GDP per capita**

PROCUREMENT RATE

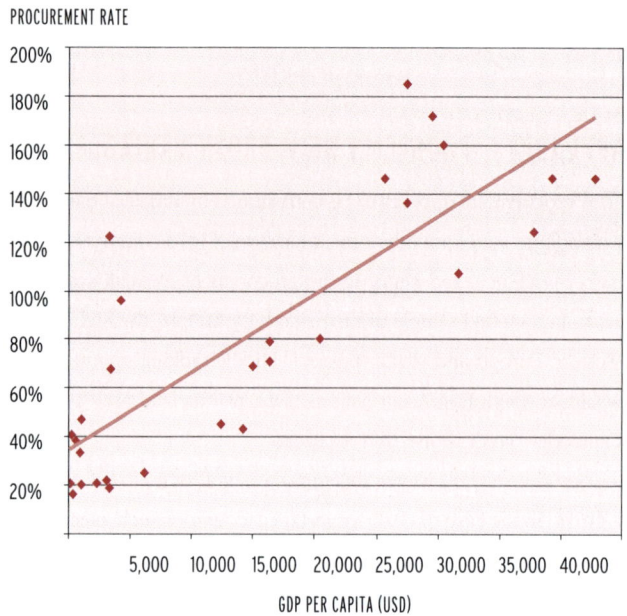

GDP PER CAPITA (USD)

APPENDIX 2. REGRESSION STATISTICS FOR 29 COUNTRIES FROM THE 32-COUNTRY SAMPLE (FIGURE 1.5)

Multiple correlation coefficient	0.86506527
Coefficient of determination R^2	0.74833792
R^2 adjusted	0.72897929
Standard deviation	0.28737293
Observations	29
F-statistic	38.65656981
P-value of the t-statistic for the variable average age of stocks (A)	3.49745E-07

APPENDIX 3. MATHEMATICAL PRESENTATION OF GLOBAL PROCUREMENT CALCULATION, AND SUBSEQUENTLY PRODUCTION AND SURPLUS TRANSFER

1) For each country (i), the ratio of peak procurement (Pp) divided by the number of active personnel (S) is a function of the inverse of the average age of the weapons it stocks (A):

$$Pp/S = f(A_i^{-1})$$

2) This relationship can be estimated for the 32 countries for which data is available and is positive and linear above the threshold where stock is an average age of 40 years. For stock above 40 years, Pp/S is conservatively estimated at a fixed 28 per cent.

3) These numbers allow an estimate of each country's annual procurement over a long—in this case 50-year—period (t). The number of times that peak procurement occurs (m) for each country (i) is equal to the long period (t), divided by the length of the procurement cycle (a):

$$m_i = t/a_i$$

4) Annual procurement is equal to peak procurement (Pp), multiplied by the number of times (m) it occurs (in this case in cycles of 15 and 24 years) over the long period (t), plus the annual low rate of procurement.

$$Pa_i = (Pp_i \times m_i/t) + (Lr_i)$$

5) This can be calculated for each country for which Pp has been estimated and the length of procurement cycle (a) is known/estimated.

6) Global procurement (Pg) for all countries (i, with i running from 1 to n) is therefore:

$$Pg = \Sigma Pa_i$$

7) Each country potentially procures both newly produced weapons (N) and used, surplus stock (U):

$$Pa_i = (Na_i + Ua_i)$$

8) The numbers of newly produced small arms and light weapons procured (N) is a function of the inverse of the number of used, or surplus, small arms and light weapons procured (U):

$$Na_i = (Ua_i^{-1}) \text{ or } Ua_i = (Na_i^{-1})$$

10) Global production (Ng) for all countries (i, with i running from 1 to n) is therefore:

$$Ng = (\Sigma Pa_i - \Sigma Ua_i)$$

APPENDIX 4. DATA FOR THE 32 COUNTRIES USED TO GENERATE GLOBAL PROJECTIONS OF PROCUREMENT AND PRODUCTION

Country	Average age of stock (years) (A)	Current active personnel (S)	Peak procurement (Pp)	Procurement ratio (Pp/S)	Procurement cycle (years)
France	21	259,050	435,000	168%	30
UK	23	207,630	332,092	160%	16
Switzerland	24	205,400	300,000	146%	29
US	24	1,433,600	1,950,000	136%	20

Australia	25	51,800	75,674	146%	30
Canada	25	52,300	96,500	185%	15
Spain	26	150,700	120,000	80%	30
Denmark	27	21,180	31,000	146%	20
Ireland	28	10,460	13,000	124%	28
Netherlands	28	53,130	57,000	107%	33
Venezuela	30	82,300	100,000	122%	30
Greece	31	170,800	134,870	79%	30
Portugal	31	44,900	31,000	69%	30
Malaysia	31	110,000	106,000	96%	20
Israel	32	168,000	119,909	71%	20
Taiwan	32	290,000	130,000	45%	25
Brunei	33	7,000	3,000	43%	20
Turkey	33	514,850	350,000	68%	35
Mexico	34	192,770	48,178	25%	30
India	36	1,325,000	499,500	38%	30
Thailand	36	306,600	65,000	21%	15
Indonesia	40	302,000	60,000	20%	20
Estonia	40	4,980	40,500	813%	30
Belize	41	1,050	230	22%	30
Uruguay	46	24,000	4,562	19%	30
Latvia	48	4,880	10,000	205%	30
Lithuania	48	13,510	50,000	370%	30
Sierra Leone	49	12,000	2,520	21%	15
Philippines	49	106,000	50,000	47%	15
Cambodia	50	124,300	20,000	16%	15
Sri Lanka	53	151,000	50,000	33%	15
Nepal	55	69,000	28,500	41%	15
Average **Total**	**35** –	– **6,470,190**	– **5,324,035**	**82%*** –	**24** –

* Sample Pp/S calculated using the sample totals of S and Pp: thus, 5,324,035/6,470,190. By contrast, an average of country Pp/S would yield 115%.

LIST OF ABBREVIATIONS

ATGW	anti-tank guided weapon	OICW	Objective Individual Combat Weapon
GDP	gross domestic product	ORD	Operational Requirements Document
IISS	International Institute for Strategic Studies		(for OICW)
MANPADS	man-portable air defence system	PfP	Partnership for Peace (NATO)
NATO	North Atlantic Treaty Organisation	UKDLO	UK Defence Logistics Organization

ENDNOTES

1. A number of countries did not appear in the sample either because of insufficient data on small arms stocks, problematic personnel data, or problems with procurement figures. These include Antigua & Barbuda, Bahamas, Barbados, Bhutan, Costa Rica, Equatorial Guinea, Fiji, the Gambia, Iceland, Iraq, Jamaica, Panama, Papua New Guinea, and the Solomon Islands.

2. See Appendix 4 for a list of these 32 countries.

3. Other factors may also intervene in decisions to procure weapons, such as business or politico–military–industrial interests. Like any industry, arms production can create and sustain employment. Sometimes, therefore, procurement programmes are perceived to have a social welfare function. Obvious examples of this phenomenon were the arms production policies of Eastern European states in the 1990s (Kiss, 2004).

4. There are some exceptions, of which the most notable is probably China, whose military has decreased in numbers by around 800,000 personnel since 1991 (IISS, all cited).

5. Letter from the UKDLO to the Small Arms Survey, 9 November 2005.

6. Only active personnel levels were used for two important reasons. Firstly, there are variations in how reserve forces are calculated. For example, in some cases paramilitary forces are included in reserve figures, while in others they are not. Secondly, it is unclear whether the majority of some reserve forces are in fact armed, particularly in states that are known to experience small arms shortages. Whether or not this is the case for all states, the rationale is that a state's front line of defence—its active troops—are more likely to be armed than its reserves.

7. For a complete mathematical model of the approach used to estimate procurement in this chapter, please refer to Appendix 3.

8. Using assault rifles and carbines to calculate age of stocks is difficult because of the ubiquity of certain varieties, such as the Kalashnikov and M16 series. This ubiquity makes it harder to differentiate among the ages of various countries' arsenals because of the degree of uniformity in stocks. Gauging age of stock is made even more problematic because assault rifles and carbines are frequently referred to in generic terms. For example, types of Kalashnikov vary considerably in age of design, ranging from the AK-47 of 1947 vintage, to the AK-74 of 1974, to the latest AK-100 series. In many reports, however, all of these weapons are simply referred to as AK-47s.

9. In the case of the three Baltic states (the outliers in Figure 1.3), between 1995 and 1998, they received surplus M14 rifles from US excess defence articles stocks (Lumpe, 1997). Because the three states effectively started from zero small arms stocks upon independence, the volume of weapons transferred—100,500 rifles for the three—was clearly beyond their expected purchasing power at the time. Moreover, the sheer number of old weapons received, in this case, suggests that age of stock is a good predictor of a state's dependence on surplus transfers. The case of Sierra Leone—another outlier—is similar to that of the Baltic states. In 2000, the British government supplied Sierra Leone with surplus 2,520 SLR rifles as part of a training programme for the country's armed forces, which had disintegrated during the war (*DFASP*, 1999; London Press Association, 2000; UK, 2000). Similarly, Nepal received assistance from the United States and India between 2003 and 2004 for its operations against Maoist insurgents (Bedi, 2005a; Heyman, 2001; Hill, 2004). As was the case with the Baltic states, the transfers to Sierra Leone and Nepal have been of surplus weapons from foreign supporting nations.

10. These countries are Belize, Uruguay, Sierra Leone, Philippines, Cambodia, Sri Lanka, and Nepal.

11. With standard deviation of 0.287, the estimate is more accurately a range—between 13 million and 22 million. It is important to note that this standard deviation only applies to the 77 countries whose stocks were younger than 40 years and whose procurement rates (Pp/S) were calculated from the regression equation. For the remaining 74 countries, a fixed rate of procurement of 28 per cent was used to project numbers of weapons procured. The range of 13–22 million assault rifles and carbines thus reflects the range 11.5–20.5 million for the 77 countries, combined with a fixed value of 1.5 million for the remaining 74 states. The same method is used to calculate all further figures presented in the chapter.

12. Figure includes 1 per cent annual low rate of acquisition. Note that with standard deviation, the range over a 50-year period could be some to 37–61 million weapons, with annual procurement rates accordingly of around 0.8–1.2 million weapons.

BIBLIOGRAPHY

Aljazeera. 2005. 'Thailand Buys US Arms to Fight Rebels.' Aljazeera.Net. 13 July. Accessed 23 January 2006.
<http://www.english.aljazeera.net/NR/exeres/AA81EB14-ADA1-495E-91C6-2259B3C5DF10.htm>

Atlantic Council. 2001. *The Bulgarian Defense Industry Strategic Options for Transformation, Reorientation & NATO Integration.* Policy Paper. Washington, DC: Atlantic Council of the United States.

Bangkok Post. 2005. 'Military Seeks Nod for B640m Arms Purchase.' 25 October, p. 10.

Barreira, Victor. 2005. 'Portuguese MoD Announces Light Weapons Buy.' *Jane's Defence Weekly.* 12 January.

BBC (British Broadcasting Corporation) Worldwide Monitoring. 2004. 'Brazil: Arms Firms Question Award of Contract for New Security Forces Rifles.' Reading: BBC Worldwide Monitoring, BBC Monitoring Latin America. 24 August.

Bedi, Rahul. 2001. 'Problems Plague Indian Rifle.' *Jane's Defence Weekly.* 19 December.

——. 2005a. 'India and Nepal in Rifle Dispute.' *Jane's Defence Weekly.* 31 August.

——. 2005b. 'Indian Army to Receive Tavor Rifles.' *Jane's Defence Weekly.* 23 February.

Ben-David, Alon. 2003. 'Rifle Output may Rise in Israel with $4.5m in Funding.' *Jane's Defence Weekly.* 28 May.

——. 2004. 'IMI Sells Light Weapons Division.' *Jane's Defence Weekly.* 25 February.

Canada. Ministry of Supply and Services. 2005. *National Defence. 1997–98 Estimates: Part III Expenditure Plan.* Ottawa.
<http://www.tbs-sct.gc.ca/est-pre/19971998/CDND97E.PDF>

Capie, David. 2002. *Small Arms Production and Transfers in Southeast Asia.* Canberra Papers on Strategy and Defence No. 146. Canberra: Strategic and Defence Studies Centre, Australian National University.

Chrobok, Vera. 2004. 'Germany.' In Sami Faltas and Vera Chrobok, pp. 41–55.

Connolly, Niamh. 2001. 'Second Investor may Save FLS.' *Sunday Business Post* (Cork). 25 November.

Connors, Shaun. 2003. 'Irish Army Continues Re-equipment Drive.' *Jane's Defence Weekly.* 17 September.

Cutshaw, Charles and Rupert Pengelley. 2000. 'Infantry Weapons Aim at Integration Age: The Next Round of Requirements for Individual Weapons Take Shape.' *Jane's International Defence Review.* 1 October.

Davis, Anthony. 2005a. 'Nepal Buys Ammunition from China, Pakistan.' *Jane's Defence Weekly.* 5 October.

——. 2005b. 'Bulgaria Sends Arms to Afghan National Army.' *Jane's Defence Weekly.* 4 September.

Davis, Ian. 2004. 'United States.' In Sami Falta and Vera Chrobok, pp. 17–29.

DFASP (Defense & Foreign Affairs Strategic Policy). 1987. 'Significant Recent International Transfers of Defense Goods and Services.' April, p. 50.

——. 1988a. 'Significant Recent International Transfers of Defense Goods and Services.' March, p. 42.

——. 1988b. 'Significant Recent International Transfers of Defense Goods and Services.' June, p. 42.

——. 1989. 'Significant Recent International Transfers of Defense Goods and Services.' September, p. 50.

——. 1996. 'Significant Recent International Transfers of Defense Goods and Services.' August, p. 23.

——. 1998. 'Arms Transfer Tables: Significant Recent International Transfers of Defense Goods and Services in the Past Six Months.' February, p. 20.

——. 1999. 'Arms Transfer Tables: Significant Recent International Transfers of Defense Goods and Services in the Past 12 Months.' November, p. 14.

——. 2001. 'Arms Transfer Tables: Significant Recent International Transfers of Defense Goods and Services in the Past 12 Months.' April/May, p. 18.

——. 2004. 'Arms Transfer Tables: Significant Recent International Transfers of Defense Goods and Services in the Past 24+ Months.' October, p. 14.

Erbe, Jurgen. 1995. 'Germany Enters the RDF Arena.' *Jane's International Defence Review.* 1 September.

Faltas, Sami and Vera Chrobok, eds. 2004. *Disposal of Surplus Small Arms: A Survey of Policies and Practices in OSCE Countries.* Bonn: Bonn International Center for Conversion, British American Security Information Council, Saferworld, and Small Arms Survey. January, p. 19.

Federation of American Scientists Fund. 1999. *Arms Sales Monitor,* No. 41. December. <http://www.fas.org/asmp/library/asm/asm41.pdf>

Forecast International. 2005. *Ordnance and Munitions Forecast.* January.

Gourley, Scott, Joris Janssen, and Rupert Pengelley. 2002. 'Soldier Modernization Programs Put More Fight into the Warrior.' *Jane's International Defence Review.* 1 December.

Haseman, John. 2003. 'Indonesia Expands Arms Purchasing.' *Jane's Intelligence Review.* 1 December.

Haug, Maria, et al. 2002. *Shining a Light on Small Arms Exports: The Record of State Transparency.* Geneva: Small Arms Survey. January, p. 35.

Heyman, Charles, ed. 2001. *Jane's World Armies.* Coulsdon: Jane's Information Group.

Higuera, José. 2005. 'Colombia, US Worried by Venezuelan Build-up.' *Jane's Defence Weekly.* 23 March.

Hill, John. 2004. 'Royal Nepalese Army Adapts to Counterinsurgency Role.' *Jane's Intelligence Review.* 1 July.

Holdanowicz, Grzegorz. 2005. 'Iraqi MoD Purchases More Equipment before Elections.' *Jane's Defence Weekly.* 2 February.

HRW (Human Rights Watch). 2000. 'NATO Should Promote Responsible Arms Sales: Open Letter to North Atlantic Treaty Organization (NATO) Foreign Ministers.' New York. 19 May. Accessed 12 January 2006. <http://www.hrw.org/press/2000/05/nato0519.htm>

——. 2002. 'The NATO Summit and Arms Trade Controls in Central and Eastern Europe.' Human Rights Watch Briefing Paper. 15 November. Accessed 12 January 2006. <http://www.hrw.org/backgrounder/arms/nato1115-bck.pdf>

IISS (International Institute for Strategic Studies). 1980. *The Military Balance 1980–1981.* London: Brassey's.

——. 1981. *The Military Balance 1981–1982.* London: Brassey's.

——. 1985. *The Military Balance 1985–1986.* London: Brassey's.

——. 1987. *The Military Balance 1987–1988.* London: Brassey's.

——. 2000. *The Military Balance 2000–2001.* Oxford: Oxford University Press.

——. 2002. *The Military Balance 2002–2003.* Oxford: Oxford University Press.

——. 2003. *The Military Balance 2003–2004.* Oxford: Oxford University Press.

——. 2004. *The Military Balance 2004–2005.* Oxford: Oxford University Press.

Ing, David. 1999. 'Spain Will Adopt H&K G36 5.56mm Assault Rifle.' *Jane's Defence Weekly.* 24 February.

Jane's Defence Weekly. 1991. 'First Batch of 1000 STEYR 5.56mm AUG Assault Rifles for Malaysian Armed Forces Completed by SME Tools in Malaysia.' 5 October.

——. 1994. 'C7 Rifles Chosen for Netherlands.' 12 February.

——. 1995a. 'Luxembourg Army Selects Steyr AUG 5.56mm Assault Rifle.' 22 July.

——. 1995b. 'Turkey Shortlists Contenders for Rifle Contest.' 16 December.

——. 1996. 'Armed Forces Update: Danes Order Extra C7A1s.' 24 January.

——. 1997. 'Country Briefing: India, Army Must Rethink Financial Priorities.' 12 February.

——. 1998. 'Greek Defence Industry: Army Programmes.' 30 September.

——. 2002. 'Malaysia Orders More Steyr Rifles.' 24 April.

——. 2005. 'Venezuela Seeks New Assault Rifles.' 20 June

Jane's Information Group. 2005. *Jane's Online Reference.* Database. Coulsdon: Jane's Information Group. Accessed September–December 2005. <http://www.janes.com/>

Jane's International Defence Review. 1995. 'France Launches AIF Weapon Program.' 1 December.

——. 2005. 'Germany Advances Plans for Soldier Modernisation Systems.' 1 January.

Jones, Richard and Charles Cutshaw, eds. 2004. *Jane's Infantry Weapons 2004–2005.* Coulsdon: Jane's Information Group.

Karniol, Robert. 2005. 'Back to Basics: The Philippines: Internal Security Concerns.' *Jane's Defence Weekly.* 19 January.

Kemp, Ian. 1994. 'RDM Set to Uprate Canadian M101S.' *Jane's Defence Weekly.* 2 April.

——. 1995. '$22M MINIMI Order Ends 10-year SARP.' *Jane's Defence Weekly.* 22 July.

——. 2003. 'Spain: Coming of Age.' *Jane's Defence Weekly.* 10 December.

Kirk, Mark. 2005. '100,000 AK-103 Russian Assault Rifles for Venezuela: 36 Members of Congress Call for Urgent OAS Meeting.' Web site of Mark Steven Kirk, Congressman. 25 July. Accessed 26 October 2005. <http://www.house.gov/apps/list/press/il10_kirk/venezuelarifles.html>

Kiss, Yudit. 2004. *Small Arms and Light Weapons Production in Eastern, Central and Southeast Europe.* Occasional Paper No. 13. Geneva: Small Arms Survey. October.

Kucera, Joshua. 2005a. 'US Army Suspends OICW Programme.' *Jane's Defence Weekly.* 27 July.

——. 2005b. 'US Rifle Programme Hits Further Delay.' *Jane's Defence Weekly.* 16 November.

—— and Scott Gourley. 2005. 'US Army Decides to Compete OICW.' *Jane's Defence Weekly.* 25 May.

Lenaerts, Jacques. 1984. 'The SIG SG550: A New Assault Rifle for the Swiss Army.' *Jane's International Defence Review.* December.

Lintner, Bertil. 2002. *Report on the Involvement of Organised Crime in the Illicit Trade in Small Arms in Southeast and East Asia.* Small Arms Survey Background Paper. Geneva: Small Arms Survey. 18 February, p. 9.

London Press Association. 2000. 'UK Warns Sierra Leone not to Give Arms to Child Soldiers.' 24 May.

Lumpe, Lora. 1997. 'US Policy on Small/Light Arms Exports.' Paper prepared for the American Academy of Arts and Sciences Conference on Controlling Small Arms. Washington, DC, 11–12 December.

——. 1999. 'The Legal Side of a Dirty Business.' *Washington Post.* 24 January.

Martov, Andrei. 1994. 'Baltic States Emerge from Cold War Shadow.' *Jane's International Defence Review.* 1 May.

NATO (North Atlantic Treaty Organisation). 2001. 'Programmes and Activities: Key Logistics Functions: Logistics Interoperability and Standardization.' In *NATO Handbook,* chapter 8. Brussels: NATO. 22 August.

Ness, Leyland. 1995. 'US Army's Industrial Base on Road to Nowhere.' *Jane's International Defence Review*. 3 May.

NISAT (Norwegian Initiative on Small Arms Transfers). 2005. *Database of Small Arms Production and Transfers: US–Israel*. Accessed 10 November 2005.
   <http://www.nisat.org>

OSCE (Organization for Security and Co-operation in Europe). 2000. *OSCE Document on Small Arms and Light Weapons*. FSC.DOC/1/00. Vienna: OSCE.
   24 November.

Pineo, Paul and Lora Lumpe. 1996. 'Recycled Weapons: American Exports of Surplus Arms, 1990–1995.' Washington, DC: Federation of American Scientists.
   May. Accessed 27 October 2005. <http://www.fas.org/asmp/library/publications/recycle.htm>

Press Trust of India. 2002. 'UK's Arms Sale to Pak Doubled Last Year.' 20 July.

Pyadushkin, Maxim and Ruslan Pukhov. 2004. 'Russia.' In Sami Faltas and Vera Chrobok, pp. 107–18.

Sariibrahimoglu, Lale. 1998. 'Turkey Selects Israel, UK for F-5, Rifle Contracts.' *Jane's Defence Weekly*. 14 January.

Schulze, Carl and Torsten Verhulsdonk. 1996. 'Armed Forces Update: German Army's New Rifle Ready for Action.' *Jane's Defence Weekly*. 4 September.

Small Arms Survey. 2001. *Small Arms Survey 2001: Profiling the Problem*. Oxford: Oxford University Press.

——. 2002. *Small Arms Survey 2002: Counting the Human Cost*. Oxford: Oxford University Press.

——. 2004. *Small Arms Survey 2004: Rights at Risk*. Oxford: Oxford University Press.

——. 2005. *Small Arms Survey: Weapons at War*. Oxford: Oxford University Press.

Sofia News Agency. 2004. 'India's Government Refutes Bulgarian Rifle Affair.' 6 April. Accessed 23 January 2006.
   <http://www.novinite.com/view_news.php?id=33106>

Steadman, Nick. 1994. 'Is that New Rifle Really Necessary?' *Jane's International Defence Review*. 1 March.

TaiwanMilitary.Org. 2003. 'ROCA to Buy over 100k T91 Rifles Starting 2004.' *TaiwanMilitary.Org*. Web site. Accessed 11 November 2005.
   <http://www.taiwanmilitary.org/phpBB2/viewtopic.php?t=1483&>

Thai Press Reports. 2005. 'Thailand: Defence Ministry Seeks Cabinet Approval on Arms Procurement.' 27 October.

Togo. 2001. *Republic of Togo Report to the UN Arms Register*. Submitted 3 July.

UK (United Kingdom). Parliament. 2000. 'Written Answers: Sierra Leone Forces: Arms Supply. Answer given by Baroness Ramsay of Cartvale to a
   question from Lord Roberts of Conwy.' Hansard, Vol. 614. London: UK Parliament. Accessed 10 November 2005.
   <http://www.publications.parliament.uk/pa/ld199900/ldhansrd/vo000629/text/00629w01.htm>

UN (United Nations). 1997. *Report of the Panel of Experts on Small Arms*. A/52/298 of 27 August.

UNDP (United Nations Development Programme). 2005. *Human Development Report 2005*. New York: UNDP.

US DoD (United States Department of Defense). 1997. *Procurement Programs (P-1): Department of Defense Budget, Fiscal Years 1998/1999*. Washington,
   DC: Department of Defense, Office of the Under Secretary of Defense. February. Accessed 14 November 2005.
   <http://www.defenselink.mil/comptroller/defbudget/fy2005/#>

——. 1998. *Procurement Programs (P-1): Department of Defense Budget, Fiscal Year 1999*. Washington, DC: Department of Defense, Office of the Under
   Secretary of Defense. February. Accessed 14 November 2005. <http://www.defenselink.mil/comptroller/defbudget/fy2005/#>

——. 1999. *Procurement Programs (P-1): Department of Defense Budget, Fiscal Years 2000/2001*. Washington, DC: Department of Defense, Office of the
   the Under Secretary of Defense. February. Accessed 14 November 2005. <http://www.defenselink.mil/comptroller/defbudget/fy2005/#>

——. 2000a. *Procurement Programs (P-1): Department of Defense Budget, Fiscal Year 2001*. Washington, DC: Department of Defense, Office of the
   Under Secretary of Defense. February. Accessed 14 November 2005. <http://www.defenselink.mil/comptroller/defbudget/fy2005/#>

——. 2000b. *Army RDT&E budget item justification (R2 Exhibit). 4 – Demonstration and Validation: 0603–02A - Weapons and Munitions - Adv Dev*. Fort
   Belvoir: Department of Defense, Defense Technical Information Center. February. Accessed 13 January 2006.
   <http://www.dtic.mil/descriptivesum/Y2001/Army/0603802A.pdf>

——. 2001a. *Procurement Programs (P-1): Department of Defense Budget, Fiscal Year 2002*. Washington, DC: Department of Defense, Office of the
   Under Secretary of Defense. June. Accessed 14 November 2005. <http://www.defenselink.mil/comptroller/defbudget/fy2005/#>

——. 2001b. *Army RDT&E budget item justification (R2 Exhibit). 4 – Dem Val: 0603–02A - Weapons and Munitions - Adv Dev*. Fort Belvoir: Department
   of Defense, Defense Technical Information Center. June. Accessed 13 January 2006.
   <http://www.dtic.mil/descriptivesum/Y2002/Army/0603802A.pdf>

——. 2002a. *Procurement Programs (P-1): Department of Defense Budget, Fiscal Year 2003*. Washington, DC: Department of Defense, Office of the
   Under Secretary of Defense. February. Accessed 14 November 2005. <http://www.defenselink.mil/comptroller/defbudget/fy2005/#>

——. 2002b. *Army RDT&E budget item justification (R2 Exhibit). 4 - Advanced Component Development and Prototypes: 0603–02A - Weapons and
   Munitions - Adv Dev*. Fort Belvoir: Department of Defense, Defense Technical Information Center. February. Accessed 13 January 2006.
   <http://www.dtic.mil/descriptivesum/Y2003/Army/0603802A.pdf>

——. 2003. *Army RDT&E budget item justification (R2 Exhibit). 4 - Advanced Component Development and Prototypes: 0603–02A - Weapons and Munitions - Adv Dev.* Fort Belvoir: Department of Defense, Defense Technical Information Center. February. Accessed 13 January 2006. <http://www.dtic.mil/descriptivesum/Y2004/Army/0603802A.pdf>

——. 2004. *Procurement Programs (P-1): Department of Defense Budget, Fiscal Years 2004/2005.* Washington, DC: Department of Defense, Office of the Under Secretary of Defense. February. Accessed 14 November 2005. <http://www.defenselink.mil/comptroller/defbudget/fy2005/#>

——. 2005a. *Procurement Programs (P-1): Department of Defense Budget, Fiscal Year 2005.* Washington, DC: Department of Defence, Office of the Under Secretary of Defense. February. Accessed 14 November 2005. <http://www.defenselink.mil/comptroller/defbudget/fy2005/#>

——. 2005b. *Army RDT&E budget item justification (R2 Exhibit). 4 - Advanced Component Development and Prototypes: 0603–02A - Weapons and Munitions - Adv Dev.* Fort Belvoir: Department of Defense, Defense Technical Information Center. February. Accessed 13 January 2006. <http://www.dtic.mil/descriptivesum/Y2006/Army/0603802A.pdf>

——. 2005c. *Acquisition: Acquisition of the Objective Individual Combat Weapon. D-2006-004.* Arlington: Department of Defense, Office of the Inspector General. Accessed 13 January 2006. <http://www.dodig.mil/audit/reports/FY06/06-004.pdf>

US Department of State. 2003. *End-Use Monitoring Report.* Washington, DC: Bureau for International Narcotics and Law Enforcement Affairs. December. <http://www.state.gov/p/inl/rls/rpt/eum/2002/27612.htm>

US DSCA (United States Defense Security Cooperation Agency). 1999. 'DSCA Foreign Military Sales: Detailed Deliveries for Fiscal Year 98.' *DSCA Factbook.* Washington, DC: DSCA, p. 39.

Valmas, Theodore. 2005a. 'Greece: Future on Hold.' *Jane's Defence Weekly.* 16 February.

——. 2005b. 'Greek Industry Progresses Privatisation.' *Jane's International Defence Review.* 1 August.

Watters, Daniel. 2005. 'The 5.56 x 45mm: 1995–1999: A Chronology of Development.' The Gun Zone Web site. Accessed 10 November 2005. <http://www.thegunzone.com/556dw-5.html>

Whelan, Lawrence. 1996. 'Scandal: Latin Arms Shipped to Croatia.' *Jane's Intelligence Review.* 1 August.

ACKNOWLEDGEMENTS

Principal author

James Bevan

Contributors

Michael Brzoska, David Capie, Michael Cheng, Nicolas Florquin, Barbara Gimelli Sulashvili, Aaron Karp, Keith Krause, Emile LeBrun, Lora Lumpe, Glenn McDonald, and Ruxandra Stoicescu

China's People's Armed Police is a heavily armed domestic security force of roughly 1.5 million men and women. Here members drill in a railway station in Chongqing in January 2006. © Liu Chan/Xinhua/AP Photo

Trickle and Torrent

STATE STOCKPILES

2

INTRODUCTION

It was a small event. In October 2005, Yemeni officials revealed that two AK-47 rifles used in a major terrorist attack in Jeddah the year before had been traced to the Yemeni Ministry of Defence (*Yemen Observer,* 2005). The underlying problem is not exceptional, for no country is immune to loss of its official weapons. In most countries the process is a trickle, in others a steady stream, and in some it can degenerate into an outright torrent.

The sight of public arsenals undermining public security is a reminder of the fundamental importance of stockpile management and security. The risks are widely acknowledged, as the 2001 *UN Programme of Action* testifies. But policy-making has been retarded, in the first instance by the lack of basic information. How many military and law enforcement small arms are there in the world? Which counties have the largest arsenals? Which are at greatest risk of loss?

This chapter reviews the global distribution of firearms among state-sponsored armed forces. Its findings include the following:

- Official reports confirm the existence of 910,615 million law enforcement firearms. Approximately 25.4 million others are estimated to exist, for a combined global total of at least 26.3 million law enforcement weapons.

- Extrapolation techniques conservatively show there are approximately 200 million modern, official military firearms worldwide.

- Approximately 141 million military firearms are automatic rifles, some 27 million are pistols and revolvers, some 12 million are machine guns, and roughly 20 million are other types. These figures do not include older weapons still in storage.

- Global military firearms inventories are concentrated among relatively few countries, with approximately two-thirds controlled by just ten countries.

- A key measure of international transparency, 16,328,000 military firearms have been publicly declared by their governments, roughly eight per cent of the suspected total.

- Many governments are willing to provide data on official small arms, but there is no international mechanism to facilitate reporting.

- Better security over official stockpiles requires greater *transparency* and *anticipatory action* to counter known and potential dangers.

These estimates of global law enforcement firearms inventories and military stockpiles are the most comprehensive ever. They reveal a substantially higher figure for law enforcement inventories than earlier assumed. Previously estimated at 18 million (Small Arms Survey, 2001, p. 88; 2002, p. 104), more complete research shows there are at least 26.3 million firearms. The actual total is almost certainly higher, since the size of major domestic security agencies (and their weapons stocks) remains unknown in many countries.

Previous military stockpiles estimates advanced by the *Small Arms Survey* totalled 226 million–241 million weapons (Small Arms Survey, 2001, p. 88; 2002, p. 104). This yearbook introduces a system of tailored estimating ratios, based on readily observable military doctrine. The result is greater certainty and accuracy. Global military inventories are estimated conservatively to hold at least 200 million firearms. This approach also permits the first breakdown of military stockpiles by weapons type.

The government-owned small arms identified in this chapter constitute the largest coherent stockpiles in the world. Correspondingly, they also create large-scale risks. The guns in civilian possession, by contrast, are typically dispersed among millions of owners in small individual holdings. Government-owned small arms, concentrated among a few owners, can move in vast quantities. They will always be tempting targets for theft, illegal diversion, and questionable sales. Their very presence also creates a more serious danger of sudden, catastrophic loss of control.

Government stockpiles are targets for theft, illegal diversion, and questionable sales.

STATE ARSENALS: BIG AND RISKY

Government-owned firearms are significantly outnumbered by the inventories owned by civilians, but the former represent the largest category stored in coherent stockpiles. Unlike civilian-held weapons, which are distributed among millions of owners, official weapons, especially military weapons, tend to be stockpiled. This makes them tempting targets for theft. Loss of individual small arms appears to be relatively common. Massive haemorrhaging of weapons, including the loss of entire state stockpiles, is not rare.

In the worst scenarios, hundreds of thousands or millions of firearms have been looted from government facilities in single incidents. The collapse of South Vietnam in 1975 allowed more than 2.1 million American-supplied small arms to fall into North Vietnamese control, creating a pool for uncontrolled exports that showed up around the world for more than a decade (E. Ezell, 1988, p. 444). On a smaller scale, it was the looting of Ugandan military arsenals in 1979 that precipitated the collapse of Idi Amin's hold on power and enabled dissident groups to establish their autonomy (Goldsmith, 2005). Better known was the loss in 1991 of several hundred thousand small arms, pillaged as the government of Somalia ceased to exist. Even more infamous was the loss of 640,000 small arms taken from Albanian government arsenals in 1997, and the estimated 4.2 million looted in Iraq in 2003 (UN, 2000; Small Arms Survey, 2004, pp. 42–48).

Less dramatic losses continue unabated. Unauthorized losses, small and great, from state-sponsored armed forces, announced in the previous year include the following:

- The UK government reported in 2005 that the British Army lost ten weapons, including one SA-80 automatic rifle and four pistols (Kirkup, 2005).
- In Uganda, 86 out of 872 newly recruited Local Defence militiamen ran off with their weapons while still in training (Butagira and Yumbe, 2005)
- China reported in April 2005 that since 1996 it has recovered 30,000 stolen military weapons, although it did not reveal the full scale of its losses (Xu, 2005).
- The Russian Federation reported that its Ministry of Defence lost 66,679 'rifled arms' during 2004 alone (*Novosti*, 2004).
- Weapons supplied to the re-created Iraqi army reportedly disappear continuously (Galbraith, 2005).

None of these reports is unprecedented or exceptional. Confirming the cliché that the worst thing most governments can do in the postmodern era is to collapse, the largest catastrophes were the direct result of sudden loss of authority. The smaller incidents illustrate the dangers inherent in stockpiles, even in stable societies. The worst small arms dangers are always a possibility, an inherent consequence of the simple existence of concentrated official stockpiles. This risk reinforces the imperatives of rigorous stockpile management and of the rapid destruction of surplus weapons, a task that can be accomplished only if excess inventories can be identified.

LAW ENFORCEMENT: LOST IN PLAIN SIGHT

Law enforcement small arms prominently illustrate the difficulty of evaluating weapons inventories even though they are routinely in public view. In some countries, to be sure, law enforcement agencies maintain large concealed arsenals. More commonly, though, law enforcement weapons are elusive exactly because they are in plain sight. Like Edgar Allen Poe's purloined letter, most law enforcement weapons are readily visible. Police agencies often do not maintain large stockpiles. Most of their weapons equip sworn officers and patrol cars; what you see is what they have (Small Arms Survey, 2001, pp. 68–70). There are incentives not to hide law enforcement weapons; their deterrent impact is facilitated by open display. But the operational display of weapons is not the same thing as institutional transparency. In most countries, total law enforcement small arms inventories are concealed by official reticence.

Even when governments want to contribute to international small arms transparency, law enforcement inventories can confound good intentions. Among the most serious barriers to estimating law enforcement small arms totals:

- Federalism and devolution often create *layered jurisdictions* and legally independent law enforcement agencies, separately armed, without centralized record-keeping.
- Multifaceted law enforcement requirements and *institutional specialization* lead to the establishment of numerous agencies, each with independent small arms requirements, procurement, and inventory control.
- In many countries, jurisdictions, and specific agencies, law enforcement officers must *privately purchase* all or part of their armament.

Because many law enforcement agencies permanently issue most of their weapons instead of maintaining formal stockpiles, their inventories tend to be less vulnerable than the armed forces to catastrophic losses. But, because individual officers routinely carry their weapons, often taking them home, or have access to local storage facilities, small-scale pilferage is more likely. Anecdotal evidence, usually individual crime reports, suggests that there is a steady trickle of law enforcement weapons into civilian hands. Reports about lost or illegally transferred police weapons often include:

- break-ins against police offices for the purpose of stealing weapons;
- assaults against law enforcement officers in which a weapon is lost;
- officers selling their government-issued weapons (most serious in poorer countries);
- officers pilfering weapons from crime scenes and evidentiary storage (anecdotal reports suggest this is a problem everywhere); and
- officers abusing gun-buying privileges by making straw purchases for unauthorized individuals.

Table 2.1 Known law enforcement firearms inventories

Country/Territory	Organization	Population	Sworn officers	Population /officer	Proportion armed	Total small arms	Ratio firearms per officer	Year	Sources
Belgium	police	10,364,000	17,767	583		23,953	1.3	2000	SAS (2001, p. 69)
Bulgaria	police	7,450,000	27,000	276		59,400	2.2	2004	Rynn, Gounev, and Jackson (2005, p. 13)
	fire fighters		5,000			5,500			Rynn, Gounev, and Jackson (2005, p. 13)
	gendarmerie		2,000			4,400			Rynn, Gounev, and Jackson (2005, p. 13)
	prison administration		5,850			6,435			Rynn, Gounev, and Jackson (2005, p. 13)
Central African Republic	police	3,900,000	1,685	2,315	0.46	1,072	0.6	2003	SAS (2005, pp. 306, 307, 312)
England and Wales	police	54,218,000	180,107	301	0.10	36,000	0.2	2004	Faltas and Chrobok (2004, p. 32)*
Germany–Saarland	Saarland police	1,061,000	2,750	386	1.00	5,344	1.9	2004	Faltas and Chrobok (2004, p. 43)
Jamaica	police	2,735,000	8,500	322	1.00	9,000	1.1	2004	Correspondence with the Small Arms Survey, 4 November 2004
Japan	police	127,400,000	240,000	530		250,000	1.0	2004	Correspondence with the Small Arms Survey, May 2004
Macedonia	police	700,000	7,500	93		23,000	3.1	2003	Grillot (2004, p. 13)
Montenegro	police	2,049,000	4,227	485		5,157	1.2		Florquin and Stoneman (2004, p. 7)
New Zealand	police	4,035,000	7,038	573		2,000	0.3	2005	Alpers and Twyford (2003, p. 15)

Norway	police	4,593,000	7,500	612		9,000	1.2	2000	SAS (2001, p. 70)
Papua New Guinea	police	5,545,000	5,000	1,109		4,800	1.0	2004	Alpers (2005, p. 48)
	prison guards		1,279			3,000		2004	Alpers (2005, pp. 48–49)
Philippines	police	87,857,000	117,376	749	0.70	82,000	0.7	2004	Clapano (2004)
Scotland	police	5,062,000	14,810	342	0.10	3,000	0.1		Faltas and Chrobok (2004, pp. 32–33); interview with Mo Poole, June 2004*
Serbia	police	8,104,000	35,400	229	0.45	53,100	1.5	2004	Taylor, Phillips, and Bogosavljevic (2005, pp. 17–18)
Sierra Leone	police	5,300,000	7,000	757		3,200	0.5	2004	Interview with Mo Poole, June 2004
South Africa	police	45,300,000	101,022	448		262,062	2.6	2004	Gould and Lamb (2004, pp. 150–51, 231; total is after 75,000 police guns destroyed 1999–2003; population from IISS, 2005, p. 397
United States	FBI	293,000,000	11,523		1.00	49,600		2002	US GAO (2003, pp. 3, 31)
	DEA		4,161			14,921		2002	US GAO (2003, pp. 3, 31)
	Fish & Wildlife		888			5,234		2002	US GAO (2003, pp. 3, 31)
Yugoslavia	police	23,500,000	40,000	588	1.00	290,000	7.3	1989	Firearms from Gorjanc (2000), population from Curtis (1990)
Average							**1.5**		
Average without outliers							**1.3**		

Note: population figures from then-year volumes of IISS, *The Military Balance*, unless otherwise stated.

* Firearms for England and Wales, and Scotland, based on the assumption that two firearms are maintained for each armed officer.

Data on all or part of the law enforcement weapons inventories is publicly known for 19 countries. This information relates mostly to police. Collectively, their disclosures average 1.5 police firearms per sworn officer, or 1.3 per officer if statistical outliers are excluded (see Table 2.1.). For countries where such data is unavailable, national police estimates can be developed either by multiplying the average of 1.3 by the total number of sworn officers, or, where the number of sworn officers is unknown, by estimating both the number of officers and their gun inventories. The number of police officers is known for 71 countries, and varies regionally from an average of one officer for every 901 people in sub-Saharan Africa to one officer for every 356 people in the Americas and the Middle East (see Table 2.2).

When these averages of known police armament and personnel figures are extrapolated across regions to countries for which data is lacking, a global police estimate emerges. In all, authorities have disclosed or confirmed the existence of 910,615 current law enforcement (mostly police) firearms. The total excludes data for the former Yugoslavia, which no longer exists. Extrapolating from average police arming rates and force levels suggests that approximately 25.4 million additional law enforcement weapons can be estimated to exist, for a combined total of at least 26.3 million law enforcement small arms (see Appendix I[1]). This includes the weapons of sworn police officers, as well as all known gendarmeries and paramilitary (domestic security) forces. It does not include other law enforcement agencies, such as secret police, specialized police forces, wildlife management officers, or prison guards.

Although exact data is lacking, police appear to constitute about half of all law enforcement personnel around the world. The next largest category appears to be paramilitary forces, especially gendarmeries. Designed to maintain domestic order, these forces often have major weapons systems such as armoured vehicles unavailable to ordinary police. They can be armed at much higher ratios per person as well. In China, for example, the People's Armed Police (PAP), a gendarmerie of approximately 1.5 million personnel, appears to be armed with roughly the same array of small arms as the People's Liberation Army.[2] Although evidence is very limited, the few examples available suggest that these other law enforcement agencies can be three or four times more heavily armed than ordinary police (see Table 2.3). If true, this would imply that the law enforcement firearms total suggested here of 26.3 million is a significant underestimate of the actual global total.

Another weakness of this estimate is that it fails to measure the differences between specific types of law enforcement weapons. Casual observation reveals that in some countries police carry side arms, while in others they routinely patrol with automatic rifles. There has been a marked tendency for law enforcement agencies to invest in more powerful weapons, especially since the 1980s, when criminals began to use automatic weapons with greater fre-

Table 2.2 Average number of sworn police officers for known countries, by region (people per officer)

Region	Known countries	Average	Average without outliers
Americas	18	368	356
Asia-Pacific	18	512	496
Europe	13	401	384
Middle East	9	410	356
Sub-Saharan Africa	13	840	901

Source: Appendix I

Table 2.3 **Police firearms vs. other law enforcement agencies**						
Country	**Agency**	**Personnel**	**Firearms**	**Guns/person**	**Year**	**Source**
Papua New Guinea	police	5,000	4,800	0.96	2004	Alpers (2005, pp. 48–49)
	prison guards	1,279	3,000	2.3	2004	Alpers (2005, pp. 57, 26)
United States	police	641,000	831,000	1.3	2000	SAS (2001, pp. 70–71)
	FBI	11,523	49,600	4.3	2002	US GAO (2003, pp. 3, 31)
	DEA	4,161	14,921	3.6	2002	US GAO (2003, pp. 3, 31)
	Fish & Wildlife	888	5,234	5.9	2002	US GAO (2003, pp. 3, 31)

Source: Appendix I

quency. Matching the firepower in civilian hands has led more law enforcement agencies to invest in automatic rifles and grenade launchers. In 2004, for example, Norwegian police initiated acquisition of automatic rifles (Solholm, 2004). In 2005 police in New Zealand announced their intention to replace older shotguns with AR-15 rifles (Pierard, 2005). Iranian police have been purchasing large numbers of Austrian-made sniper rifles (Jahn, 2005). Even the United Kingdom is debating routine arming of its sworn officers.

TOWARDS A CUMULATIVE UNDERSTANDING OF MILITARY STOCKPILES

Five years after the first edition of the *Small Arms Survey* was published, accurate data on military small arms possession has accumulated for 23 countries. The most important source of data has been dedicated country research, conducted in cooperation with subject governments. Table 2.4 represents the collective efforts of the community of small arms researchers and supportive governments.

Two entries refer to countries that have since disappeared; they are included exclusively for analytical purposes. Government declarations and similar highly reliable statements account for 16.3 million confirmed military firearms (see Table 2.4). This represents 23 out of the 196 sovereign states in the world today. Combined with two historical cases—East Germany and Yugoslavia—for a total of 25 countries, this is the best available foundation for global extrapolation. Only cases based on *officially provided* data are presented. Several are incomplete. Norway and Switzerland only included handguns and rifles. The figures for Australia and Venezuela cover only one type—albeit the predominant type—of automatic rifle. Most countries provide data only on military firearms, not all small arms or light weapons. For most countries, moreover, there is no information on inventories of older, obsolescent military firearms such as bolt-action rifles. Because of incomplete entries, the model has a tendency to *underestimate* actual global totals.

As official data, all these figures have been accepted at face value. None has been arbitrarily excluded, because of the risk of exposing the results to researcher bias. Personnel figures come from country case studies and the International Institute for Strategic Studies when necessary. Only *formal armed forces* are included. Gendarmerie, paramilitary, interior ministry, intelligence, customs, and border control personnel are treated as law enforcement personnel, discussed above.

Table 2.4 Military firearms inventories of 25 countries/provinces

Country	Total military personnel	Total firearms	Year	Ratio	Sources
Albania	21,500	148,742	2005	6.7	Holtom et al. (2005, pp. 36–37)
Australia	96,612	280,000	1987	2.9	E. Ezell (1988, p. 45). Requirement for AUG rifles only
Bosnia-Herzegovina	84,600	450,000	2001	5.3	Zivalj (2001)[a]
Bulgaria	100,000	504,096	2004	5.0	Rynn (2005, pp. 11–13)
Cambodia	204,000	390,000	1991	1.9	Wille (2005)[b]
Canada	102,400	233,949	2000	2.3	Canadian DFAT 2001, in SAS (2001, p. 73)
Central African Republic	4,442	5,552	2003	1.3	SAS (2005, p. 312)[c]
Czech Republic	49,450	500,000	2003	10.0	Macha (2003)
Estonia	15,300	83,550	2005	5.5	Combined statements; rifles only
Finland	462,000	531,000	2003	1.1	Letter from the Finnish MoD to the Small Arms Survey, 21 August 2003
German D.R.	460,700	1,205,725	1990	2.6	Nassauer (1995, p. 57); Faltas and Chrobok (2004, p. 41)
Jamaica	3,783	7,000	2004	1.9	Correspondence from the Permanent Mission of Jamaica to the Office of the United Nations at Geneva with the Small Arms Survey, 4 November 2004
Macedonia	33,000	85,446	2003	2.6	Grillot et al. (2004, p. 16); reserves from IISS (2004)
Malaysia	156,600	255,000	1987	1.6	E. Ezell (1988, p. 257); rifles only
Norway	248,700	295,070	2000	1.2	Norwegian MoD in SAS 2001 p. 73; rifles and pistols only
Papua New Guinea	3,100	7,200	2004	2.3	Alpers (2005, pp. 51, 52)[d]
Serbia	345,300	789,016	2004	2.3	Taylor, Phillips and Bogosavljevic (2005, pp. 18–20); reserves from IISS (2004)
South Africa	115,750	350,636	2004	3.0	Gould and Lamb (2005, p. 155)
Sweden	289,600	920,000	1999	3.2	Justiedepartementet (1999, pp. 27, 36, 74, 112–15)
Switzerland	175,000	324,484	2004	1.9	Personnel from IISS (2004); Swissinfo (2004)
Togo	6,950	12,850	2000	1.8	Togo to the UN Arms Register, 2001
Ukraine	1,187,600	7,000,000	2005	5.9	Polyakov (2005, pp. 9, 49); total may be higher
United States	2,515,300	3,054,553	2002–05	1.2	Stout (private communication, August 2005); other services, SAS (2002, p. 85)
Venezuela[e]	67,300	100,000	2005	1.5	Stratfor.com (2001); rifles only
Yugoslavia	705,000	3,115,000	1989	4.4	Gorjanc (2000)
Average for 25 known cases				**3.2**	

a Wilkinson (2005) reports 761,000 firearms, for a ratio of 4.0.

b Figures presented by Wille as ranges have been averaged.

c Does not include the Presidential Guard, with 1000 troops and 3000 small arms and light weapons.

d Large numbers of weapons lost according Alpers (2005); troop numbers from IISS (2004, p. 185).

e Personnel figures do not include the 23,000-member National Guard, a gendarmerie.

Note: The combined total is 16,328,144 for 23 existing countries, 20,648,869 including the former German Democratic Republic and Yugoslavia.

An important strength of this approach is the diversity of its foundations. It includes some the world's largest states and some of the smallest, wealthiest, and poorest, and all regions except central and north-east Asia. Of greatest importance for military stockpile analysis, it includes a diverse range of military strategies and tactical doctrines.

It is one the characteristic ironies of global small arms policy that much more is known about small arms disarmament and destruction than about actual inventories. We know more about what we get rid of, in other words, than what we have. Prominent examples include Germany and the Russian Federation. Many states that do not respond to requests for small arms data on their own armed forces readily release figures about military destruction programmes they support in other countries. Examples include France, the Netherlands, Norway, and Sweden.[3] Without corresponding data on the stockpiles of state-sponsored armed forces, it is impossible, however, to evaluate the significance of these destruction efforts. Even greater numbers of countries publicize destruction of weapons confiscated from civilian or illegal militias, including China and many African governments, but reveal nothing about their inventories of official weapons.

FOUR KEYS TO MILITARY SMALL ARMS REQUIREMENTS

In a simpler world, global military small arms possession could be extrapolated by averaging the totals from a few known countries, and applying the resulting ratio to the rest of the world. This average is readily computed for known examples at 3.1 firearms per soldier, sailor, and airman. But this is too crude for meaningful insight into the military inventories of countries as diverse as China and Paraguay, the Netherlands and Seychelles. Similarly, there is no evidence of a persuasive correlation between wealth and state demand, although this remains a valuable starting hypothesis. In practice, some of the poorest countries are among the best armed (North Korea is only the most poignant of many examples), while some of the richest intentionally shrink their armed forces (such as Norway and Switzerland).

Better accuracy can be achieved through a model based on the nuances of official preferences. This element is captured here through an emphasis on military strategy and doctrine. The defining characteristics of state armed forces, these explain why the armed forces exist and how they expect to fight. In practice, state-sponsored armed forces fall into four distinct groups, each with a distinct approach to small arms procurement and inventory management. As the examples collected in Table 2.4 will be used to demonstrate, above all it is military strategy and doctrine that determine military requirements. Small arms are primarily infantry weapons, but not every country plans to mobilize or fight with its infantry in the same way. The differences directly affect the quantity of small arms a country's armed forces require.

Much more is known about disarmament and destruction than about inventories.

To permit accurate extrapolation for 166 countries (those with a population over 400,000) it is necessary to divide the 23 known examples into four major doctrinal schools. These categories constitute four keys to estimating military firearms inventories elsewhere. All four are well known in strategic planning: Trinitarian Warfare, People's War, Constabulary, and Reserve-based forces. Although the categories originated for other purposes, they are developed here *empirically,* explaining differing national approaches to military small arms procurement.

While all state-sponsored armed forces fit into one of these categories, not all countries fit identically. In every category, at least one example has been included which differs in terms of firearms per soldier. In each case, these countries have higher or lower weapons totals than expected, typically because of rapidly changing strategic priorities or unique historical considerations. Their presence reinforces an essential point about national small arms inventories: they can be highly flexible. With a single procurement decision, a surplus destruction decision, or a change in personnel strength, basic small arms ratios can change as well. Consequently no estimating system can be highly accurate across the board. Its role, rather, is to outline *typicality*.

Trinitarian militaries: small arms as weapons of last resort

The traditional state-sponsored armed force is a military establishment designed for Trinitarian warfare. Although strategic theorists never tire of pointing to its historical uniqueness, created in response to distinctive Westphalian, European experiences, such armed forces have acquired the patina of normality (Metz, 1994–95; Villacres and Bassford, 1995). As explained almost 200 years ago by von Clausewitz, this is an approach to warfare based on a unique system of distinctions between the state, military, and citizenry. Trinitarian warfare emphasizes the armed forces as a defining element of international security, separate from the state they serve and the people they protect (Clausewitz, 1831/1991, pp. 212–13, 962, 964–65).[4] As an instrument of state policy, Trinitarian militaries are designed primarily to serve the interests of the state against its foreign foes (Paret, 1976).

Configured principally for operations against the similarly conceived armed forces of other states, Trinitarian forces tend to be meticulously trained for highly specialized operations. In the contemporary world, their operational tactics are dominated by major weapons systems, deployed in pursuit of a decisive outcome through formal battle, prosecuted with the goal of eliminating an adversary's ability to fight (van Crevald, 1991, pp. 35–42). Trinitarian militaries are distinguished by large active-duty components and primary reliance on major conventional weapon systems. Soldiers with small arms perform an important, albeit residual, role in Trinitarian warfare. This division of labour explains why many major military establishments spend much more money on major hardware than they do on infantry equipment.

Sheer numbers of personnel are useful in Trinitarian armies. And they possess concomitant small arms inventories (see Table 2.5). But firearms are not the principal weapons of Trinitarian conquest and resistance. More so today than

Table 2.5 **Examples of Trinitarian military firearms inventories**						
Country	**Active personnel**	**Reserve personnel**	**Total uniformed personnel**	**Total firearms**	**Year**	**Ratio**
Australia	70,456	26,112	96,568	280,000	1987	2.9
Canada	62,000	36,900	103,900	233,949	2000	2.3
German Democratic Republic	137,700	323,000	460,700	1,205,725	1990	2.6
Macedonia	12,000	21,000	33,000	85,446	2003	2.6
Serbia	65,300	280,000	345,000	789,000	2004	2.3
South Africa	55,750	60,000	115,750	350,300	2004	4.2
United States	1,473,960	1,290,988	2,746,948	3,054,553	2002-05	1.1
Average						**2.6**

Sources: see Table 2.4

ever before, military establishments conceived along such lines tend not to maintain massive small arms surpluses; they no longer plan for massed infantry operations. The United States is an extreme example, with armed forces tailored for network-centric operations and no expectation of large-scale infantry recruitment. Committed never to fight with massed infantry, as Secretary of Defense Donald Rumsfeld repeatedly stresses, the United States has no official military small arms stockpile and an exceptionally low ratio of firearms per soldier.[5]

Although their formal inventories tend to be relatively low, even countries with Trinitarian militaries may have hidden reserves. In 1990 this was suggested by the revelation of the covert Echelon plan for armed guerrilla resistance in western Europe after a Soviet invasion. As part of Echelon, small arms reportedly were cached throughout western Europe (Fitchett, 1990). The number and fate of these weapons have never been revealed. They are not included here.

People's War and legacy arsenals: small arms above all

A very different approach to war leads to an emphasis on small arms combat as the very foundation of national security. Starting with the guerrilla traditions of revolutionary warfare, and adapting them to defence of the state, People's War stresses mobilization of massive numbers of lightly armed infantry. Inspired—directly or indirectly—by Maoist notions of resistance and conquest, this strategy relies on infantry and partisans armed primarily with small arms. In the classic formulation, 'Without question, the fountainhead of guerrilla warfare is in the masses of the people, who organize guerrilla units directly from themselves' (Mao, 1965, ch. 5). The most relevant implication for present purposes is the greater size of their military small arms inventories. Arsenals expand over time to arm not only uniformed military personnel but also potentially all politically loyal or patriotic sectors of society.[6] The result is a national military stockpile far larger than the number of active and even reserve military personnel (see Table 2.6). In a country where anyone is a potential combatant, there must be arms for virtually everyone.

One distinctive trait of many People's War countries is huge military reserve organizations (see Box 2.1). Most—but not all—reached their peak in the mid-1980s and declined as accelerating technological change deprived

Table 2.6 **Examples of firearms inventories of People's War and legacy militaries**						
Country	Active personnel	Reserve personnel	Total uniformed personnel	Total firearms	Year	Ratio
Albania	21,500	0	21,500	145,000	2005	6.7
Bosnia-Herzegovina	24,672	60,000	84,600	450,000	2001	5.3
Bulgaria			100,000	504,000	2004	5.0
Czech Republic			49,450	500,000	2003	10.0
Estonia	6,600	8,700	15,300	83,550	2005	5.5
Yugoslavia			705,000	3,115,000	1989	4.4
Ukraine	187,600	1,000,000	1,187,000	7,000,000	2005	5.8
Average						**6.1**

Sources: see Table 2.4

People's War of much its credibility. Since the weapons associated with these reserve structures are assumed to remain, even after reorganization, older personnel totals have been used to compute likely firearms inventories.[7]

The exceptional scale of this mobilization planning has profound implications for the estimation of global military small arms; these are the countries with many of the largest national stockpiles. With the large numbers involved, special precautions have been taken to reduce the risk of error. The *legacy arsenals* of eastern Europe are especially problematic as a basis for extrapolation. Instead of the empirical average of 6.1 firearms per person typical of known legacy militaries, this analysis relies instead on a lower ratio of 4.8 firearms for every active-duty personnel. This is about 20 per cent lower, insuring that estimates remain consistently conservative. For countries where reserve organizations surpass the scale of active-duty personnel by a factor of three or more, total military firearms have been calculated using the formula:

$$T = A(4.8) + R (1.2)$$

where A refers to the highest number of active-duty personnel since 1970 and R designates the highest number of reservists for the same year. This approach has been used with core People's War militaries like China, Cuba, North Korea, and Vietnam, as well as countries that long maintained comparable mobilization plans such as Brazil, Iran, and Spain. This compensates for reports that some reserve organizations have purely logistic roles and may not be routinely armed. The procedure also tends to err on the side of caution, arbitrarily assuming a relatively modest limit for mobilization plans. Use of a smaller ratio for reserve elements lowers the estimated scale of People's War

Box 2.1 Reserve forces and small arms inventories

The world's greatest reserve forces directly contribute to the largest state-controlled small arms arsenals, a major element in global stockpiles. China and North Korea illustrate how doctrinal emphasis on massed infantry and infantry reserves creates exceptional demand for small arms. Because of their great scale, these stockpiles are a major source of international concern, raising grave risks of future resale or catastrophic loss.

North Korea: Mass mobilization remains the foundation of North Korean security strategy (Kwo'n), More so than for most major militaries, the Korean People's Army (KPA) is heavily concentrated into ground troops, with over one million active-duty ground personnel, organized into 153 divisions and brigades. For comparison, the United States Army has ten divisions. The heavy concentration on army operations is a direct manifestation of *Songun Chongch'i* [Army First] political doctrine, an approach adopted by North Korea's leader Kim Jong Il, in 1995. This aims to solve '. . . all problems arising in the revolution and construction on the principle of giving priority to the military affair and advances the overall cause of socialism relying on the army as the pillar of the revolution'.[8]

Songun Chongch'i reinforced North Korea's older *Chu'che* ideology of extreme self-sufficiency, which already led to the stockpiling of ammunition, food, and petroleum in underground facilities to sustain several months of fighting without foreign help. According to Seoul, by 1989 Pyongyang had stockpiled some 990,000 tons of ammunition (roughly 1.8 times what was found in Iraq, a country of similar population, in 2003).[9] The expansion of the KPA to some 1.2 million full-time personnel in the 1990s is another result of this outlook (*North Korea Country Handbook*, 1997, p. 33).

The creation of a massive reserve organization is a further result of ideological predispositions. According to the most authoritative source, '. . . About 30% of all North Koreans between the ages fifteen to sixty are mobilized for reserve units'. They are divided into:

'. . . Reserve Military Training Unit consist(s) of approximately 1.7 million persons (men 17–45 and unmarried women 17–30). . . mobilized. . . for a total of forty days' training out of the year.

'The Worker–Peasant Militia is a combination of older men aged 45–60, along with men ages 17–45 and unmarried women ages 17–30 who. . . train for a total of thirty days out of the year. Their current numbers stand at 4.1 million.

'The Young Red Guards consist of 1.2 million male and female Higher Middle (High) School students aged 14–16. . . subject to a mandatory four-hour drill session every Saturday. . .' (Republic of Korea, 1999).[10]

This suggests a total reserve of seven million. While most reserve organizations have pre-assigned arsenals to facilitate mass mobilization, the Youth Red Guards appears to be primarily a training organization, capable of mobilizing only three-quarters of its members (Bermudez, 2001, p. 169).

Unlike standing forces, which peaked in the 1990s and declined somewhat thereafter, reserves continued to grow. Another major source maintains that reserve elements reached 7.45 million in 1999 (Bermudez, 2001, ch. 6). According to the 2004 South Korean *Defense White Paper*, North Korean reserve organizations now number 7.7 million (Fifield, 2005). To enhance certainty, a figure of six million is used here. With 1.2 million active-duty personnel and at least six million armed reservists, North Korea would have approximately 13 million military firearms [1.2 mn (4.8) + 6 mn (1.2)], or a probability range of 10 million–16 million military firearms.

China: The armament for the People's Liberation Army (PLA) is another source of ambiguity. Maoist orthodoxy encouraged enormous small arms procurement, although actual inventories may never have matched requirements. In the early 1960s, for example, 'the Chinese peasantry was organized through the commune system into a vast people's militia. In all, about a quarter of the population was involved. The militia was given simple training, *often with wooden rifles.*'[11]

As the inheritors of Maoist tradition, Chinese commanders honour the concept of People's War, even today as they try to become more Trinitarian (Ji, 1999, ch. 1). Active forces peaked in the late-1970s at 6.1 million and gradually declined, with the most dramatic reductions beginning in 1992 (Ji, 1999, p. 33). One result of this process has been formalization of the military reserve. Previously dominated by the Chinese Communist Party, this increasingly is controlled by the armed forces themselves. While the transformation retains the outlines of previous commitments to People's War, it increasingly resembles traditional Trinitarian operations. As expressed in the 2000 Defense White Paper, 'Combining the armed forces with the people and practicing self-defense by the whole people. China adheres to the concept of people's war under modern conditions, and exercises the combination of a streamlined standing army with a powerful reserve force for national defense' (People's Republic of China, 2000, ch. II). As a result, reserves were formalized and reduced.

Today's PLA includes three reserve elements. Each armed service maintains its own reserve force, currently estimated by the International Institute for Strategic Studies at a total of 800,000 personnel (IISS, 2005, p. 270). In addition there are two national militias. The *primary militia* is a traditional reserve force; it 'comprises rapid reaction detachments, infantry detachments, specialized technical detachments and detachments with corresponding specialties. There are now 10 million primary militia members throughout the country' (People's Republic of China, 2004, ch. VI). This is supported by the six-million-strong ordinary militia. According to one prominent source, the latter does not routinely train with weapons.[12]

Based on an active-duty PLA of the mid-1980s, with six million personnel, and reserve units totalling 16 million, of which at least ten million can be armed, China is believed here to have approximately 41 million military firearms [6 mn (4.8) + 10 mn (1.2)], or a distributed probability of 31 million–52 million military firearms.

militaries by over 40 per cent. Use of a lower estimating ratio (4.8 instead of 6.1) reduces estimates of People's War militaries by another 20 per cent. The new result may be misleadingly low. In lieu of better evidence, however, such caution seems warranted.

Often, these vast stockpiles are nothing more than a *strategic remnant* of discarded war plans. Many—perhaps most—of today's largest military small arms stockpiles are left from past governments, sometimes from past countries. Inherited from the ashes of the Soviet Union, Yugoslavia, or communist rule elsewhere, these enormous inventories reveal nothing about the intentions of the governments that own them today. Long after countries abandon People's War, most of these legacy arsenals sit in storage. Even as a metaphor for the perversities of providence, though, the weapons are real enough.

In the case of Ukraine, incredible munitions dumps—estimated by NATO and the Ukraine armed forces to include 2.5 million tons of ammunition and equipment—remain from the military preparations of a country that no longer exists (Chivers, 2005). Ukraine did not deliberately choose after independence in 1991 to be responsible for at least

> **Vast stockpiles often are nothing more than a remnant of discarded war plans.**

seven million military small arms. With a military establishment authorized at 272,000 personnel (and a nominal reserve component of one million, according to the IISS), it has no use for so much equipment. Its current military doctrine stresses border security, peacekeeping, and collective security by professional soldiers, not desperate arming of the countryside to repel non-existent invaders. Its vast small arms collection was an unsought burden, much like the Soviet ballistic missile industry and the rest of its largely uncounted military stockpiles.

Even more extreme is the Czech Republic, with a military inventory of approximately 500,000 small arms and light weapons for a military that numbers fewer than 50,000 (Macha, 2003). The Czech Republic joined NATO in 1999, as did Bulgaria and Romania in 2004. All abandoned mass mobilization strategy years before. Most of the old guns still sit there, ready for export, pilferage, or perhaps catastrophic loss.

Constabulary militaries: on patrol at home

Although Hollywood inspires entertaining visions of countries littered with vast warehouses of military equipment, in reality this is the exception rather than the rule. For the armed forces of many countries, large-scale capabilities are a distant fantasy. Many countries, especially in poor parts of the planet or in regions isolated from threat of foreign attack, maintain relatively small military establishments. In other cases the armed forces are intentionally deprived of men and material as a form of 'coup proofing', insuring that they stay inferior to politically favoured internal security organizations (Quinlivan, 1999). The net result is many armed forces that are small and minimally armed compared to Trinitarian or People's War counterparts.

These are *Constabulary militaries,* organized not to fight the armed forces of other states but to preserve domestic order (Field, 2002–03). Here the term is used specifically to mean any state-sponsored armed service configured not to defeat foreign enemies but to maintain domestic stability and prevent rebellion. In practice such forces resemble gendarmeries or heavily armed police. They are distinguished not only by low ratios of weapons per solider but also by small inventories of major weapons systems. The latter are essentially irrelevant to operational responsibilities, in which battle plays little or no part. A typical constabulary military is poor in ships, tanks, planes, missiles, and the other accoutrements of Trinitarian warfare. They have little need for weapons designed to defeat the armed forces of other states. Nor do they have the generous small arms inventories for sudden expansion, as needed by practitioners of People's War. Instead, as the examples in Table 2.6 show, they tend to be equipped at levels more suited to their distinctive missions of maintaining domestic order and suppressing rebellion.

Table 2.7 Examples of firearms inventories in Constabulary militaries

Country	Active personnel	Reserve personnel	Total uniformed personnel	Total firearms	Year	Ratio
Cambodia	166,000	38,000	204,000	390,000	1991	1.9
Central African Republic	4,442	0	4,442	5,552	2003	1.3
Jamaica	2,830	953	3,753	7,000	2004	1.9
Malaysia	110,000	46,600	156,100	290,000	1991	1.6
Papua New Guinea	3,100	0	3,100	7,200	2004	2.3
Togo	6,950	0	6,950	12,850	2000	1.8
Venezuela	59,000	8,000	67,300	100,000	2004	1.5
Average						**1.9**

Sources: see Table 2.4

Reserve militaries: defence through monthly drill

The reserve-based armed forces of Europe are a special case. While most countries maintain a reserve component, mostly to save money, these countries base their military security almost entirely on reservists (Roberts, 1976). Their force planning is essentially Trinitarian, designed to defend national territory against foreign invasion, usually through network-centric operations. But for historical and constitutional reasons they approach the tailored efficiency of one soldier–one rifle. The most extreme case—Switzerland—has no standing military except for administration. Most other reserve-based forces maintain an active-duty element, mostly for training and international peacekeeping operations. But the backbone of national defence is the much larger force of reservists, who train routinely to maintain skills and receive current orders.

They can be technically sophisticated, heavily invested in major weaponry, and meticulously systematized for integration with NATO reinforcements. But, due to their emphasis on rapid mobilization of reservists, exclusively for territorial defence, these countries typically have little or no additional mobilization capability. Nor do most of them

Table 2.8 Examples of firearms inventories in Reserve militaries

Country	Active personnel	Reserve personnel	Total uniformed personnel	Total firearms	Year	Ratio
Finland	27,000	435,000	462,000	531,000	2003	1.1
Norway	26,700	222,000	248,700	295,070	2000	1.2
Sweden	27,600	262,000	289,600	920,000	1999	3.2
Switzerland	4,400	170,600	175,000	324,484	2004	1.9
Average						**1.8**

Sources: see Table 2.4

have surplus weapons for reinforcement (see Table 2.8). Unlike Constabulary militaries, Reserve militaries are configured primarily to defeat foreign invasion, with well-developed doctrine stressing network-centric operations and major weapon systems. Unlike armies configured for People's War, they do not plan for mass mobilization except for formally designated reservists; the limits on their mobilization potential are clear.

GLOBAL MILITARY FIREARMS TOTALS

Applying the doctrinal distinctions outlined above to 166 countries (all countries with a population over 400,000 that have a military) permits global estimation of the number of firearms controlled by the state-sponsored armed forces of the world (see Table 2.9). Combined with formally declared military inventories, this analysis concludes that the world harbours at least 200 million official military firearms. When distributed to compensate for a statistical margin of error (plus or minus 25 per cent), the range of global military firearms appears to be between 150 million and 250 million. The estimating procedures used here for People's War militaries are conservative, which suggests that actual global totals are more likely to be closer to the upper parameter.

Detailed estimates for all 166 countries are included in Appendix I. Although they have been overlooked here, due to practical research difficulties, state-owned small arms can be extremely important in the very smallest countries as well (Muggah, 2005). Smaller countries have been left out in recognition of their highly idiosyncratic nature; with declining size, rules of thumb like those explicated here cease to be reliable. As illustrated by a methodical examination of the microstates of the Pacific Ocean, the scale of their armed forces and armament is extremely hard to predict. Similarly, the relationship between military and law enforcement institutions tends to become increasingly obscure (Alpers and Twyford, 2003).

The regional distribution of military firearms corresponds closely to the global distribution of population; military small arms tend to be most numerous where most of the world's people live. The dominance of Asia, home to roughly 47 per cent of the world's suspected military firearms, is a direct reflection of the presence of the world's two most populous and many other large countries. It also is a consequence of the historic importance of People's War doctrine in the armed forces of several east Asian powers, most notably China, North Korea, and Vietnam. Above all, however, security planning in east Asia is dominated by the overwhelming importance of the sovereign state and

Table 2.9 **Estimated regional military firearms totals**				
	Declared	**Estimated**	**Combined**	**% of global total**
Americas	3,295,502	10,286,328	13,581,830	7
Asia-Pacific	967,200	93,777,734	94,174,534	47
Europe	11,627,550	62,302,738	73,930,288	37
Middle East	0	13,887,304	13,887,304	7
Sub-Saharan Africa	369,038	3,506,118	3,870,000	2
Global total	**16,360,000**	**184,000,000**	**200,000,000**	**100**

the remaining possibility of state-to-state conflict. This strategic priority leads directly to a doctrinal emphasis on numerically large standing forces, on a scale unmatched elsewhere in the world.

Europe is home to the world's second largest regional arsenal, roughly 37 per cent of the global total for all state armed forces. The high standing of European small arms arsenals is more surprising. It is a reality that weakens claims that European security emphasizes peaceful conflict resolution and non-military means associated with human security (Kagan, 2002). While European governments have been at the forefront of international campaigns to destroy surplus small arms elsewhere, they appear to have been slower to eliminate their own surpluses.

Other regions play much smaller roles in the distribution of global military stockpiles, with approximately 15 per cent of the world total among them. The Middle East stands out for the realist approach of its armed forces, designed to serve Trinitarian or People's War ends. The Middle Eastern total would be approximately four million higher were it not for the collapse of Iraq in 2003 and the diffusion of former President Saddam Hussein's military arsenal among civilians (Small Arms Survey, 2004, pp. 44–48). The Americas are home to smaller military organizations, either because they are on the cutting edge of Trinitarian operations (in the case of the United States) or structured for Constabulary roles (in much of the Caribbean and Central and South America). The small scale of African state military inventories is especially striking in view of the continent's severe security difficulties. Both phenomena testify to the chronic weakness of African states.

Table 2.10 **Proportions of military firearms types (in selected countries)**						
Country	**Year**	**Total firearms**	**% rifles**	**% pistols**	**% machine guns**	**% other**
Canada	2000	233,949	72	11	6	11
East Germany	1990	1,182,000	74	23	4	
Macedonia	2003	85,446	88		12	
Finland	2003	531,000	97	2	1	
Norway	2000	295,070	90	10		
Switzerland	2001	695,000	90	10		
Togo	2001	12,850	74	9	3	13
US Army	2005	1,357,616	68	14	11	7
Averages						
All countries			82	11	6	
Complete countries			72	14	6	10
Global equivalent			110–18 m	20–33 m	8–13 m	14–25 m

Notes: US Army data refers only to 'Army In Use' weapons category only. Macedonian data does not include pistols. Norwegian and Swiss data do not include machine guns or other firearms. Swiss figure is from 2001, different from the 2004 data used in Table 2.1 and used here because of its greater detail. 'Other' firearms column refers mostly to military sniper rifles, shotguns, sub-machine guns, and grenade launchers.

Sources: see Table 2.1

THE TYPES OF FIREARMS IN MILITARY INVENTORIES

One of the shortcomings of macro-analyses of total firearms is that it creates the impression that all the world's 200 million or so military firearms are the same. Reality is very different. Global, regional, and country totals conceal major differences of type and lethality. Most seriously, there is no way to systematically separate the proportion of heavy machine guns from light machine guns, fully automatic rifles from single-shot rifles, pistols from revolvers and sub-machine guns. Reports from the destruction of some eastern European military surpluses reveal substantial quantities of obsolescent armaments (Faltas and Chrobok, 2004, p. 89). The latter are not useless; they can be resold for use elsewhere and are readily usable in crime, terror, or intra-state warfare. But they do not represent the same dangers of illicit resale or use as a Kalashnikov rifle or a Markov pistol.

Authorities agree that total AK-47 production amounts to 70 to 100 million.

More can be said about the breakdowns between basic types of military firearms. The approach used to estimate the size of official military stockpiles can be used to divide the proportion of rifles, side arms, and machine guns. Data from seven states provides a basis for extrapolation (see Table 2.10). The inventory breakdowns are complete for only five cases (Canada, East Germany, Finland, Togo, and the US Army) but these also are fortunately diverse. They offer a useful, albeit imprecise, foundation for generalization. Their example suggests that rifles make up roughly three-quarters of military firearms inventories. Assuming this rule of thumb holds globally, we can begin to understand the rough division of weapon types among the global military arsenal. Of the total of 150 million–250 million military firearms in existence today, approximately 112 million–183 million can be expected to be automatic rifles and older, bolt-action types. Some 22 million–34 million are pistols and revolvers, 9 million–14 million are machine guns, and 16 million–25 million are other varieties of firearms such as shotguns, grenade launchers, and sub-machine guns.

ESTIMATING FIREARMS INVENTORIES THROUGH PRODUCTION ESTIMATES

Alternative confirmation of the stockpile estimation method outlined above comes from comparison with automatic rifle production. Independently developed production numbers provide a second window on global stockpiles. Production data often relies on analysis of serial numbers, a technique developed for estimating production of all kinds of military equipment (Ruggles and Brodie, 1947). Even so, they are complete and reliable only for military automatic rifles (see Table 2.11). Production data for pistols is available but harder to interpret since military-style pistols also have major civilian and law enforcement markets.

Production and inventory data is not identical, reflecting unresolved ambiguities. Cumulatively, the production data shows that approximately 122 million–156 million automatic rifles have been manufactured, starting with the AK-47 in 1947, with a mean total of 141 million. This compares with a the broader range of 112 million–180 million rifles whose existence can be shown through inventory analyses (above), an estimated 72 per cent of all military firearms. The largest portion of the global military rifle inventory is the AK-47 and its versions. Although there is no accurate database, authorities appear to agree that total AK-47 production by the Soviet Union, China, and their allies and clients amounts to 70 million–100 million since 1947 (see Table 2.11). This production perspective is highly consistent with the conclusion that there are at least 200 million military small arms, and probably considerably more.

If the proportions for modern firearms types in military inventories, as developed above, are applied to production, total modern military firearms production can be estimated (see Table 2.12). Adding proportionate quantities of other

Table 2.11 Production of major modern military automatic rifles

Type and variants	Country of origin	Other producers	Quantity produced	Basis of estimate	Sources
AK-47	USSR	Many	70,000,000–100,000,000	Serial numbers	McNab (2001, pp. 7, 49), Novosti (2005), Small Arms Survey (2001, pp. 62–63), Tretiakov (2005)
SKS	USSR	China	15,000,000	Serial numbers	Genovese (2002)
M16	United States	S. Korea	12,000,000	Manufacturer	Colt (2004), Tretiakov (2005)
G3	Germany	Iran, Pak.,Turkey	7,000,000	Serial numbers	UN (1997)
Type 63	China		6,000,000	Serial numbers	E. Ezell (1988, p. 104).
FAL	Belgium	Aust., Mex., UK, etc.	5,000,000–7,000,000	Serial numbers	UN (1997)
M14	United States		1,380,346	Procurement	Jane's Information Group (2004, p. 73)
Stgw 90	Switzerland		600,000	Gov. reports	Correspondence of the General Staff of the Swiss Armed Forces with the Small Arms Survey, 5 December 2001, 21 December 2001, and 1 February 2002
Stgw 57	Switzerland		600,000	Serial numbers	SAS 2002 (pp. 78, 79), E. Ezell (1988, p. 347)
INSAS	India		528,000	Requirement	Jane's Information Group (2004, p. 32)
F1	France		400,000	Procurement	Jane's Information Group (2004, p. 26)
L85	United Kingdom		400,000	Procurement	
AK5	Sweden		250,000	Requirement	Jane's Information Group (2004, p. 63)
Others*			5,000,000	Estimate	Small Arms Survey (2001, pp. 62–63)
Total			**124,000,000–156,000,000**		

* Examples include automatic rifles such as AUG, CETME, Galil, R4/5, and SIG 540.

kinds of military side arms and machine guns supports overall production of 200 million military firearms. This is identical to the inventory-based estimate, but likewise incomplete. To this total also must be added the large numbers of single-shot (bolt-action) rifles, revolvers, and older sub-machine guns preserved in many national arsenals. In this way, the production-based assessment supports the conclusion that the inventory-based approach developed above tends to underestimate actual stockpiles.

Several provisos are in order. An unknown proportion of these weapons no longer exists. Many have been destroyed or irreparably damaged over the years through wear and breakage, poor storage, and battlefield wastage.

Table 2.12 Production-based estimate of global military firearms inventories (all figures in millions)				
Type	Production proportion (%)	Estimated total production	Low parameter	High parameter
Automatic rifles	72	141	105	175
Pistols	13	27	21	34
Machine guns	6	12	9	16
Other firearms	9	20	15	25
Total	**100**	**200 m**	**150 m**	**250 m**

Sources: based on production proportions from Table 2.10

Several million have been destroyed through official surplus destruction programmes. A significant proportion of automatic rifles—at least several hundred thousand—belong not to armed forces, but to domestic law enforcement agencies, including police and gendarmeries. In some countries, especially in the Middle East and the United States, civilian ownership of automatic rifles is legal; American civilian ownership of military-style automatic rifles is estimated at some three million.[13] In addition, a major category of military firearms still cannot be accurately estimated. These are the large number of obsolescent, non-automatic firearms still preserved in many military stockpiles, as revealed by recent destruction programmes in Bulgaria and Romania. A complete estimate would subtract all the automatic weapons that do not (or no longer) belong to armed forces for all the reasons listed above, and add obsolescent non-automatics.

The number of obsolescent firearms in many stockpiles cannot be estimated.

THE LARGEST MILITARY INVENTORIES

The list of the largest military firearms inventories reveals the concentration of global stockpiles among a handful of countries (see Table 2.13). With a total of 128 million estimated military firearms between them, the top ten countries control approximately two-thirds (63 per cent) of the global total. The top 20 countries are home to roughly three-quarters (approximately 155 million) of all state-owned military firearms.

The list of the largest military small arms arsenals does not correspond closely to any orthodox ranking of national power; the United States does not rank among the top ten, France qualifies only for the top 20, while the United Kingdom does not figure at all. Instead, it is fighting doctrine that appears to dominate global hierarchies; countries relying on large ground forces or mass mobilization strategies crowd the ranks of military small arms powers. These strategies favour countries with large populations. Although there is no exact correspondence between population and military firearms stockpiles, it is no accident that three of world's five most populous countries are present in the top ten military stockpiles (China, India, and the Russian Federation). The lower position of the United States is another illustration of its distinctive military doctrine, emphasizing network-centric warfare through advanced munitions instead of massed infantry operations.

Top 20 rankings notwithstanding, inventory estimates are not reliable enough for direct comparison of countries with roughly comparable stockpiles. Rather, they permit only a sense of relative scale. Whether it is Iran or India, for example, that actually has the larger stockpile of the two cannot be determined using this methodology. But the

Table 2.13 The 20 largest estimated military firearms arsenals*

Country	Rank	Estimated or confirmed firearms	Low parameter	High parameter
China	1	41,000,000	31,000,000	52,000,000
Russian Federation	2	30,000,000	22,000,000	37,000,000
Korea, North	3	14,000,000	9,000,000	16,000,000
Vietnam	4	9,800,000	7,400,000	12,000,000
Korea, South	5	7,100,000	5,300,000	8,900,000
Ukraine	6	7,000,000	7,000,000	7,000,000
India	7	6,300,000	4,700,000	7,800,000
Taiwan	8	5,000,000	3,800,000	6,300,000
Turkey	9	4,400,000	3,300,000	5,500,000
Iran	10	3,700,000	2,800,000	4,600,000
Top ten combined total		**128,000,000**	**96,000,000**	**157,000,000**
Germany, F.R.	11	3,100,000	2,400,000	3,900,000
United States	12	3,054,553	3,054,553	3,054,553
Italy	13	3,000,000	2,200,000	3,700,000
Pakistan	14	2,900,000	2,200,000	3,600,000
Egypt	15	2,700,000	2,000,000	3,400,000
Cuba	16	2,600,000	2,000,000	3,300,000
France	17	2,400,000	1,800,000	3,000,000
Poland	18	2,300,000	1,700,000	2,800,000
Indonesia	19	2,200,000	1,600,000	2,700,000
Brazil	20	2,100,000	1,600,000	2,600,000
Top 20 combined total		**155,000,000**	**117,000,000**	**189,000,000**

* In descending order, rounded to two significant digits.

Source: Appendix II

approach is sufficiently accurate to conclude that the size of their small arms arsenals is similar in scale. Although it cannot be said which has the bigger stockpile, both clearly belong among the world's top ten largest military firearms stockpiles.

Quantity, of course, is not quality. It is not easy to generalize about the kinds of weapons predominating in each country's armed forces. Reserve inventories, in particular, can be heavy in older or obsolescent equipment, much of it cast-offs discarded by active-duty units after modernization. The largest inventories appear to include substantial proportions of older sub-machine guns and manually operated (bolt-action) rifles. Although the evidence is inconclusive, it appears that the stockpiles of countries with large reserve structures, associated with People's War strategies, often hold substantial quantities of older weapons. The best evidence comes from foreign-sponsored disarmament programmes that typically receive mostly aging weapons for destruction.

TRANSPARENCY: OVERCOMING BUREAUCRATIC INERTIA

The different kinds of law enforcement and military inventory data used in this chapter—declared and estimated—help illustrate the status of global firearms transparency. One of the most important data points to emerge from this review is the relatively thin proportion of military small arms whose existence has been formally declared by their governments (see Table 2.14). These declarations are a unique tool for evaluating transparency. Out of a conservative estimate of 200 million military firearms, only 16,360,000 have been formally acknowledged, roughly eight per cent of all military firearms. This is another way of expressing the most troubling barrier to insight and policy-making, namely, the remarkable lack of transparency on official inventories.

Before 2000, virtually no country made data on its total military small arms inventories publicly available. Today more than 20 countries have revealed this kind of information. The greatest hurdle to official transparency undoubtedly remains national security classification. Many countries turn research inquiries down or fail to respond. But many show no such preoccupation. A group as diverse as Bulgaria, Canada, Central African Republic, Jamaica, Macedonia, Papua New Guinea, Serbia, and the United States has volunteered official data (see Tables 2.1 and 2.4). They have typically done so in response to a research query. Their willingness shows that small arms data is often not highly classified. The greatest barrier to making it available, rather, appears to be lack of communication with custodial authorities, those with legal responsibility for military firearms.

For these countries the biggest problem appears to be bureaucratic barriers to reporting (see Box 2.2). Since no one is specifically responsible for making such data available, typically nothing happens. The officials with the most relevant role, the National Points of Contact, designated under the 2001 *UN Programme of Action* (UNGA, 2001), tend

Table 2.14 Regional military firearms transparency*

Region	Estimated total	Low parameter	High parameter	Confirmed	Combined total	Approximate percentage confirmed
Americas	10,286,328	7,714,746	12,857,910	3,295,502	13,600,000	24
Europe	62,302,738	46,727,054	77,878,423	11,627,550	73,900,000	16
Sub-Saharan Africa	3,506,118	2,629,589	4,382,648	369,038	3,870,000	10
Asia-Pacific	93,777,734	70,333,300	117,222,167	967,200	95,000,000	1
Middle East	13,887,304	10,415,478	17,359,130	0	14,000,000	0
Global total	**184,000,000**	**138,000,000**	**23,000,000**	**16,360,000**	**200,000,000**	**8**

* Ranked by proportion of military firearms publicly declared.

Box 2.2 Bureaucratic barriers to transparency: National Points of Contact

'Where you sit', as every public policy student quickly learns, 'is where you stand'. Agency perspectives and priorities affect the implementation of all policy issues. Small arms reporting is no different.

Under the 2001 *UN Programme of Action,* governments have undertaken to establish a National Point of Contact (NPC), to 'act as liaison between States on matters relating to [its] implementation' (UNGA, 2001, para. II.5). There has been a tendency to designate NPCs from foreign ministries, where 61 are based, including virtually all the most active ones (Kytömäki, 2004, p. 27). This makes efficient use of expertise developed through UN processes and facilitates routine national reporting with the UN. But the tendency influences the reporting process.

Instead of addressing all small arms issues comprehensively, the reliance on foreign ministries promotes emphasis on the foreign policy aspects of these issues. National reporting, in other words, often stresses only activities going on outside a country. Domestic matters—outside the competence of a foreign ministry—receive much less attention. Only 23 out of 122 countries with an NPC give this responsibility to a ministry with *custodial responsibility* or actual control over major small arms inventories (Kytömäki, 2004, p. 27). Since the officials that constitute the NPC typically lack responsibility for domestic small arms policy and do not automatically receive stockpile information, most countries report only the outlines of their domestic firearms situations, and sometimes not even that.

The foreign affairs orientation in reporting inadvertently draws attention away from many of the most important aspects of small arms. Foreign ministries typically lack ready access to detailed information on possession of small arms by the military or law enforcement agencies. Research queries about such matters addressed to NPCs often receive an honest 'we don't know'. To report on them, NPCs must request cooperation from other government agencies, especially ministries of defense, armed services, ministries of the interior, and law enforcement agencies. Anticipating this difficulty, the *Programme of Action* encourages inter-agency cooperation (para. II.4).

Bureaucratic segmentation explains the tendency for national reporting to emphasize international disarmament activities.[14] The latter is one area where foreign ministries have operational responsibilities. Domestic small arms data tends to be reported much less systematically, reflecting the mandate of the NPC and the practical limits of inter-agency coordination.

Box 2.3 Unknown unknowns: the international reporting deficit

Despite the importance of stockpile and inventory data for effective domestic management, there is no standardized international reporting mechanism for small arms and light weapons statistics. The lack of a systematic reporting system inhibits transparency, conceals surplus stockpiles, creates barriers to the prioritization of international stockpiles management, and may even discourage reporting.

Stockpile management is a major theme of the 2001 *UN Programme of Action.* Although it does not explicitly address systematic record-keeping, this appears to be an *implicit requirement* to satisfy its mandate for stockpile management, including the need 'to ensure . . . that the armed forces, police or any other body authorized to hold small arms and light weapons establish adequate and detailed standards and procedures relating to the management and security of their stocks of these weapons. These standards and procedures should, inter alia, relate to: . . . inventory management and accounting control' (UNGA, 2001, para. II.17).

Somewhat more detailed guidance can be found in some regional agreements. One of the most systematic stockpile reporting requirements can be found in the 2000 *OSCE Document on Small Arms and Light Weapons.* It calls upon participating states to share information on their *surplus small arms,* emphasizing 'the category, sub-category and quantity of small arms that have been identified as surplus and/or seized and destroyed on their territory during the previous calendar year' (OSCE, 2000, para. IV.E.1). These reports are, however, circulated only among OSCE participating states.

The lack of an international structure to encourage transparency is a significant weakness of the small arms regime. Neither the *Programme of Action* nor any regional instrument requires systematic reporting of inventory data. Nor is there a mechanism to facilitate voluntary data sharing. Even when governments want to make such data available, it is not obvious where they can submit it. In the past governments have occasionally presented small arms data under the miscellaneous category of the UN Arms Trade Register or shared it with non-governmental research organizations for publication. Destruction figures appear intermittently in national reports on *Programme of Action* implementation, but not other stockpile details. While individual governments can do more by themselves, systematic reporting probably is not possible without clearer international commitment.

to be in foreign ministries. This explains the tendency for national reporting to stress *external* aspects of small arms policy, the area of their writ. Lacking authority over *domestic* matters, such as national military stockpiles, National Points of Contact give them less attention.

No current international agreement establishes a mandate for comprehensive stockpile reporting (see Box 2.3). The key to better transparency in military small arms, then, is establishment of an international mechanism for *systematic reporting*. Without demand, motives will be weak. A reporting mechanism, even if entirely voluntary, would catalyse transparency, encouraging cooperation by creating a venue for it. Reporting is most likely to occur when structured through *custodial authorities,* the government agencies with physical control over the weapons. While this small arms problem, like many others, may be too complicated for panaceas, a systematic reporting mechanism involving custodial authorities offers the best promise of a transparency breakthrough.

CONCLUSION: SEEING AND ANTICIPATING

Whether the danger is the trickle of small-scale theft, pilferage, and individual loss, or the torrent of catastrophic loss, control over state-owned small arms inventories remains a fundamental challenge. The guns in civilian possession, dispersed among millions of owners, are usually lost in small numbers at any one time. Government-owned small arms, concentrated among a few owners, can move in vast quantities. They always will be tempting targets for theft, illegal diversion, and questionable sales. Worse, control of state arsenals can disintegrate completely, flooding society with hundreds of thousands or millions of weapons.

These dangers heighten the need for greater *transparency* in order to identify surpluses and for *anticipatory action*—especially surplus destruction and security enhancement—to eliminate the most obvious risks.

Codified in the United Nations Charter, the principle of self-defence justifies state acquisition of the tools of self-defence. This chapter has illustrated the scale of the responsibilities that come with that right. In light of the extent of their inventories, with at least 200 million military firearms and at least 26 million law enforcement guns in official hands, management and control are essential for domestic and international security.

The key to transparency is an international mechanism for stockpile reporting. Better stockpile management is a major theme of the 2001 *Programme of Action* and a foreign-policy goal for many countries. Catastrophic losses of vast inventories almost always occur unexpectedly. As the examples reviewed in the introduction reveal, formerly socialist and authoritarian societies are especially vulnerable to catastrophic losses. Although the *Programme of Action* applies to all UN member states, attention naturally focuses on the worst affected. Yet established democracies are not immune from smaller-scale inventory loss and diversion. This chapter illustrates the importance of devoting greater attention to the official small arms of *every* state.

Transparency and anticipatory action are complementary tools with the potential to greatly reduce stockpile dangers. Although they are valuable separately, used together they are particularly effective in minimizing stockpile dangers. Such efforts may not be sufficient to protect individuals against arbitrary or unlawful use of state arsenals. But they can help ensure that people do not suffer from their unintended effects.

The techniques in this chapter reveal the scale and approximate distribution of government small arms inventories, but this only a first step. By providing a glimpse of the lie of the land, of the peaks and valleys in the small arms global map, it can help promote better-informed debate. Such a debate can be sustained, however, only by greater official cooperation and disclosure. Concrete measures require concrete information. ◾

LIST OF ABBREVIATIONS

KPA   Korean People's Army   PAP   People's Armed Police (China)

ENDNOTES

1   Appendices 1 and 2 are available at <http://www.smallarmssurvey.org/publications/yb_2006.htm>

2   Unlike ordinary police, the PAP operates under the joint authority of the Central Military Commission and Ministry of Public Security. Its personnel routinely carry weapons, often seen to include sub-machine guns or assault rifles. The role of heavy armament appears to have been reinforced by expansion in the 1990s. Much of this growth came through the wholesale transfer of units struck from the shrinking PLA and reassigned to the PAP. While ordinary Chinese police are assumed to be armed at the ordinary police rate of 1.4 weapons per officer, the People's Armed Police almost certainly carry the equivalent of many armed forces, approximately 2.3 firearms per person (Tanner, 2002, p. 600).

3   Destructions data is detailed in their respective national reports to the 2003 and 2005 United Nations Biennial Meetings of States.

4   I would like to thank Col. Antulio J. Echevarria II (US Army) for bringing these citations from von Clausewitz to my attention.

5   Private communication with Bruce Stout, August 2005.

6   'There are those who say: "I am a farmer", or, "I am a student", "I can discuss literature but not military arts." This is incorrect. There is no profound difference between the farmer and the soldier. You must have courage. You simply leave your farms and become soldiers. That you are farmers is of no difference, and if you have education, that is so much the better. When you take your arms in hand, you become soldiers; when you are organized, you become military units' (Mao Zedong, 1965, ch. 5).

7   The use of *highest contemporary personnel* levels explains the difference between the estimates in this chapter and lower estimates based on *current* troop levels. An example of the later is Fernandes et al. (2005, pp. 113–16).

8   'Songun Chongch'i [Army First].' <http://www.globalsecurity.org/military/world/dprk/songun-chongchi.htm>

9   'The offense.' <http://www.globalsecurity.org/military/world/dprk/doctrine-offense.htm>

10   Slightly different figures are reported by Minnich (2001, p. 8).

11   'PLA Reserve Forces.' <http://www.globalsecurity.org/military/world/china/pla-reserve.htm> (emphasis added).

12   'PLA Reserve Forces.' <http://www.globalsecurity.org/military/world/china/pla-reserve.htm>, citing Mulvenon and Yang (2002).

13   Extrapolated from data for California and New Jersey (Jacobs, 2002, pp. 150, 162).

14   The general problem of bureaucratic segmentation of information and responsibility is a major theme of Allison and Zelikow (1999). See also Halperin (1985).

BIBLIOGRAPHY

Allison, Graham and Philip Zelikow. 1999. *The Essence of Decision: Explaining the Cuban Missile Crisis,* 2nd ed. New York: Longman.

Alpers, Philip. 2005. *Gun-running in Papua New Guinea: From Arrows to Assault Weapons in the Southern Highlands.* Geneva: Small Arms Survey. June.

—— and Conor Twyford. 2003. *Small Arms in the Pacific.* Geneva: Small Arms Survey. March.

Bermudez, Joseph S., Jr. 2001. *The Armed Forces of North Korea.* London: I. B. Tauris.

Butagira, Tabu and Jamal Abdi Yumbe. 2005. '86 militiamen escape with guns.' *The Monitor* (Kampala). 2 May.

Chivers, Chris J. 2005. 'Ill-secured Soviet Arms Depots Tempting Rebels and Terrorists.' *New York Times.* 16 July.

Clapano, Jose Rodel. 2004 '35,000 of 117,000 Cops Gunless', *Philippine Star.* 17 December.

von Clausewitz, Carl. 1831/1991. *Vom Kriege,* 19th ed. Bonn: Dümmler.

van Crevald, Martin. 1991. *The Transformation of War.* New York: Free Press.

Colt. 2004. 'Colt Challenges Rivals' Illegal Marketing Practices.' *Colt Defense News.* Press release. Hartford, CT: Colt Defense Inc. 21 April.

——. 2005. *Colt Defense Inc., Registration Statement Under the Securities Act of 1933.* Washington, DC: United States Securities and Exchange Commission. 3 June.

Curtis, G. E. 1990. *Yugoslavia: A Country Study.* Washington, DC: Federal Research Division, Library of Congress.

Ezell, Edward C. 1988. *Small Arms Today: Latest Reports on the World's Weapons and Ammunition,* 2nd ed. London: Arms and Armour Press.

Ezell, Virginia. 1995. *Report on International Small Arms Production and Proliferation.* Alexandria, VA: Institute for Research on Small Arms and International Security.

Faltas, Sami and Vera Chrobok, eds. 2004. *Disposal of Surplus Small Arms: A Survey of Policies and Practices in OSCE Countries*. Bonn: Bonn International Center for Conversion, British American Security Information Council, Saferworld, and Small Arms Survey. June.

Fernandes, Ruben Cesar et al. 2005. *Brazil: The Arms and the Victims*. Rio de Janeiro: Viva Rio.

Field, Kimberly C. and Robert M. Perito. 2002–03. 'Creating a Force for Peace Operations: Ensuring Stability with Justice.' *Parameters,* Vol. 32. Winter, pp. 77–87.

Fifield, Anna. 2005. 'Pyongyang is Stepping Up its Nuclear Might, Seoul Claims.' *Financial Times*. 5–6 February, p. 2.

Fitchett, Joseph. 1990. 'Paris Says it Joined NATO "Resistance".' *International Herald Tribune*. 13 November, p. 7.

Florquin, Nicholas and Shelly O'Neill Stoneman. 2004. *'A House isn't a Home Without a Gun': SALW Survey, Republic of Montenegro*. Belgrade: SEESAC and Small Arms Survey.

Galbraith, Peter. 2005. 'Last Chance for Iraq.' *New York Review of Books*. 6 October, p. 22.

Genovese, Mark. 2002. 'Captured SKS' *Small Arms Review,* Vol. 6, No. 1. October, pp. 27–28.

Goldsmith, Paul. 2005. 'Disarming Pastoralists: Kenya Cannot Go It Alone.' *East African* (Nairobi). 23 May.

Gorjanc, Milan. 2000. 'Small Arms and Light Weapons—Possible Contribution to the Stability Pact for Southeastern Europe'. Unpublished conference paper. Ljubljana, 27 January.

Gould, Chandré and Guy Lamb, eds. 2004. *Hide and Seek: Taking Account of Small Arms in Southern Africa*. Pretoria: Institute for Security Studies. October.

Gounev, Philip, Emil Tsenkov, Bernardo Mariani, and Larry Attree. 2004. *Weapons Under Scrutiny: Implementing Arms Export Controls and Combating Small Arms Proliferation in Bulgaria*. London: Saferworld. April.

Grillot, Suzette R., Shelly O. Stoneman, Hans Risser, and Wolf-Christian Paes. 2004. *A Fragile Peace: Guns and Security in Post-conflict Macedonia*. Geneva: Small Arms Survey. June.

Halperin, Morton H. 1985. 'Organizational Interests.' In Daniel J. Kaufman, Jeffrey S. McKitrick, and Thomas J. Leney, eds. *U.S. National Security: A Framework for Analysis*. Lexington, MA: D.C. Heath, pp. 201–32.

Holtom, Paul, Henry Smith, Bernardo Mariani, Simon Rynn, Larry Attree, and Juliana Sokolová. 2005. *Turning the Page: Small Arms and Light Weapons in Albania*. London: Saferworld. December.

IISS (International Institute for Strategic Studies). 2004. *The Military Balance 2004–2005*. Oxford: Oxford University Press.

——. 2005. *The Military Balance 2005–2006*. London: Routledge.

Jacobs, James B. 2002. *Can Gun Control Work?* Oxford: Oxford University Press.

Jahn, George. 2005. 'Iran is Quietly Stockpiling High-Tech Small Arms . . .' *Associated Press*. 25 March.

Jane's Information Group. 2002. *Jane's Infantry Weapons, 2002–2003*. Coulsdon, Surrey: Jane's Infantry Group.

Ji You. 1999. *The Armed Forces of China*. London: I. B. Tauris.

Justiedepartementet. 1999. *Regeringens Proposition 1999/2000:27, En Skaerpt Vapenlagstiftning*. Stockholm.

Kagan, Robert. 2002. 'Power and Weakness.' *Policy Review,* No. 113. June–July, pp. 3–29.

Khakee, Anna and Nicolas Florquin. 2003. *Kosovo and the Gun: A Baseline Assessment of Small Arms and Light Weapons in Kosovo*. Geneva: Small Arms Survey. March.

Kirkup, James. 2005. 'Army Admits Ten Guns Missing.' *The Scotsman*. 5 November.

Kwo'n Kyo'ng-pok. 2005. 'Release of North Korean War Plan Draws Suspicions.' *Choson Iibo* (Seoul). 6 January.

Kytömäki, Elli and Valerie Yankey-Wayne. 2004. *Implementing the United Nations Programme of Action on Small Arms and Light Weapons*. Geneva: UNIDIR.

Macha, Richard. 2003. Presentation before the OSCE Small Arms seminar in Bucharest. 24–26 February.

Mao Zedong. 1965. *Problems of Strategy in Guerrilla War Against Japan*. 3rd edn. Translated from *The Selected Works of Mao Tse-tung, Vol. II*. Peking: Foreign Languages Press.

McNab, Chris. 2001. *The AK47*. St. Paul, MN: MBI Publishing.

Metz, Steven. 1994–95. 'A Wake for Clausewitz: Toward a Philosophy of 21st-Century Warfare.' *Parameters,* Vol. 24, No. 4. Winter, pp. 126–32.

Minnich, James M. 2001. *North Korean Tactics*. Fort Leavenworth, KS: United States Army Command and Staff College. September.

Muggah, Robert. 2005. *Securing Haiti's Transition: Reviewing Human Insecurity and the Prospects for Disarmament, Demobilization, and Reintegration*. Geneva: Small Arms Survey. November.

Mulvenon, James C. and Andrew N.D. Yang, eds. 2002. *People's Liberation Army as Organization*. Santa Monica: RAND.

Nassauer, Otfried. 1995. 'An Army Surplus—The NVA's Heritage,' in Edward J. Lawrence and Herbert Wulf, eds. *Coping with Surplus Weapons: A Priority for Conversion Research and Policy*. Bonn: Bonn International Center for Conversion Research. June.

North Korea Country Handbook. 1997. MCIA-2630-NK-016-97. Quantico, VA: United States Marine Corps Intelligence Activity. May.

Novosti. 2004. 'Over 178,000 Units of Various Arms Registered with Russian Interior Ministry as Missing.' 21 December.

——. 2005. 'Famous Kalashnikov to Unveil Advanced Weapons in Minsk.' 11 May.

OSCE (Organization for Security and Co-operation in Europe). Forum for Security Co-operation. 2000. *OSCE Document on Small Arms and Light Weapons*. 24 November. FSC.DOC/1/00. <http://www.osce.org/docs/english/fsc/2000/decisions/fscew231.htm>

Paret, Peter. 1976. *Clausewitz and the State*. New York: Oxford University Press.

People's Republic of China. 2000. *China's National Defense in 2000*. Beijing: Information Office of the State Council. <http://www.china.org.cn/english/2000/Oct/2791.htm>

——. 2004. *China's National Defense in 2004*. Beijing: Information Office of the State Council. <http://www.china.org.cn/english/2004/Dec/116032.htm>

Pierard, Louis. 2005. 'NZ Staring Down UN's Gun Barrels.' *Hawke's Bay Today*. 28 May.

Polyakov, Leonid. 2005. *Aging Stocks of Ammunition and SALW in Ukraine: Risks and Challenges*. Bonn: Bonn International Conversion Centre. February.

Quinlivan, James T. 1999. 'Coup-proofing: Its Practice and Consequences in the Middle East.' *International Security,* Vol. 24, No. 2. October, pp. 131–65.

Republic of Korea. 1999. *North Korea Military*. Seoul: National Intelligence Service.

Roberts, Adam. 1976. *Nations in Arms: Theory and Practice of Territorial Defence*. New York: Praeger.

Ruggles, Richard and Henry Brodie. 1947. 'An Empirical Approach to Economic Intelligence in World War II.' *Journal of the American Statistical Association,* Vol. 42. March, pp. 72–91.

Rynn, Simon, Philip Gounev, and Thomas Jackson. 2005. *Taming the Arsenal: Small Arms and Light Weapons in Bulgaria*. Belgrade: South Eastern Europe Clearinghouse for the Control of Small Arms and Light Weapons (SEESAC). March.

Small Arms Survey. 2001. *Small Arms Survey 2001: Profiling the Problem*. Oxford: Oxford University Press.

——. 2002. *Small Arms Survey 2002: Counting the Human Cost*. Oxford: Oxford University Press.

——. 2003. *Small Arms Survey 2003: Development Denied*. Oxford: Oxford University Press.

——. 2004. *Small Arms Survey 2004: Rights at Risk*. Oxford: Oxford University Press.

——. 2005. *Small Arms Survey 2005: Weapons at War*. Oxford: Oxford University Press.

Solholm, Rolleiv. 2004. 'The Police Directorate Requests Heavier Arms.' *Norway Post* (Baerum). 26 September.

Stratfor.com. 2001. 'Venezuela – Contributing to Regional Violence.' 4 June.

Swissinfo. 2004. 'Sale of Army Weapons Triggers Heated Debate.' 20 October.

Tanner, Murray Scott. 2002. 'The Institutional Lessons of Disaster: Reorganizing the People's Armed Police after Tiananmen.' In James C. Mulvenon and Andrew N. D. Yang. 2002.

Taylor, Zachary, Charlotte Phillips, and Srdjan Bogosavljevic. 2005. *Living with the Legacy: Small Arms and Light Weapons Survey Republic of Serbia*. London: Saferworld. March.

Tretiakov, Yuri. 2005. 'The Kalashnikov Does Not Miss.' *Trud*. 21 May, pp. 1–2. Translated in *Defense and Security* (Moscow), 25 May 2005.

UN (United Nations). 1997. *Report of the Secretary-General on Small Arms Prepared with the Assistance of the Panel of Governmental Experts on Small Arms*. A/52/298 of 27 August.

——. 1998. *Report of the Evaluation Mission to Albania,* 11–14 June 1998. New York: United Nations.

UNGA (United Nations General Assembly). 2001. *Programme of Action to Prevent, Combat and Eradicate the Illicit Trade in Small Arms and Light Weapons in All Its Aspects ('UN Programme of Action')*. 20 July. Reproduced in UN document A/CONF.192/15 of 9–20 July.

US GAO (United States General Accounting Office). 2003. *Firearms Controls: Federal Agencies Have Firearms Controls, but Could Strengthen Controls in Key Areas*. GAO-03-688. Washington, DC: GAO. June.

Villacres, Edward J. and Christopher Bassford. 1995. 'Reclaiming the Clausewitzian Trinity.' *Parameters*. Vol. 25, No. 3. Autumn, pp. 9–19.

Wilkinson, Adrian. 2005. *South Eastern Europe—Estimates of Weapons Possession (Edition 1)*. Belgrade: SEESAC, 28 February.

Wille, Christina. 2005. *How Many Weapons Are There in Cambodia?* Working paper. Geneva: Small Arms Survey.

Xu Hu. 2005. 'Statement of Mr. Xu Hu, Ministry of Public Security, to the UN Workshop on Small Arms and Light Weapons.' Beijing. 19 April.

Yemen Observer (San'a). 2005. 'Yemen said linked to guns in Saudi attack.' 12 October.

Zivalj, Husein. 2001. Statement by H.E. Mr. Husein Zivalj, Ambassador, Permanent Representative of Bosnia and Herzegovina to the United Nations, to the United Nations Conference on the Illicit Trade in Small Arms and Light Weapons in all its Aspects, New York, 10 July.

ACKNOWLEDGEMENT

Principal author

Aaron Karp

Brazilian police officers check seized weapons believed to be connected to an organized crime group in Sao Paulo in August 2005.
© Alexandre Meneghini/AP Photo

An Uphill Battle

UNDERSTANDING SMALL ARMS TRANSFERS

3

INTRODUCTION

This chapter provides an annual update of the authorized trade in small arms and light weapons. It examines trends in the trade, and provides information on major exporters and importers, whom they trade with, and in what types of weapons. The lack of transparency on the part of many important suppliers and recipients in this global trade makes this effort an uphill battle. In fact, although the issue of small arms and light weapons has been on the international agenda for over a decade, few, if any, states provide full information on their small arms and light weapons exports and imports. So clearly, the Small Arms Trade Transparency Barometer, introduced in *Small Arms Survey 2004,* remains an important tool.

Following up on the analysis of the illicit trade in Europe in *Small Arms Survey 2005,* this chapter provides more systematic information on the illicit trade in South America. The analysis is based on two sources of internationally comparable data partially available for the illicit trade, namely customs and police seizure data.

The main findings of the chapter include the following:

- According to available data and estimates, the top exporters of small arms and light weapons by value (exporting at least USD 100 million of small arms and light weapons, including parts and ammunition, annually) in 2003—the latest year for which data is available—were the Russian Federation, the United States, Italy, Germany, Brazil, and China. Compared to 2002 and 2001, the only change is that Belgium was no longer among the top exporters in 2003.
- The top importers (importing to the value of at least USD 100 million) for 2003, according to customs data, were the United States, Cyprus, and Germany. Top importers tend to vary more than top exporters, but the United States and Cyprus were both on the list for 2001, 2002, and 2003.
- Among the major exporters of small arms and light weapons, the most transparent are the United States and Germany. The least transparent are Bulgaria, Iran, Israel, and North Korea, all scoring zero on the Small Arms Trade Transparency Barometer.
- Customs data in South America has strong limitations in terms of systematization, organization, and comparability. Police data indicates that in this region diversion from military stockpiles from neighbouring countries is as serious a problem as international arms trafficking.

THE AUTHORIZED GLOBAL SMALL ARMS AND LIGHT WEAPONS TRADE: ANNUAL UPDATE

This section provides an update on the authorized global small arms and light weapons trade. It focuses on the major exporters and importers, their top trading partners, and the main products exchanged.[1] It includes information on small arms and light weapons, their parts and accessories, and small arms (as opposed to light weapons) ammunition. Light weapons ammunition is excluded because of reporting limitations. In customs data, light weapons ammunition exports/imports are reported in the same category as ammunition for large conventional weapons. There is no way to single out the former from the latter, hence mixed ammunition categories were excluded from the calculations.[2] However, the trade in military small arms and light weapons is most likely underestimated because of limited transparency on the part of many countries and the lack of reporting for certain types of military weapons.

As in previous years, calculations are provided by the Norwegian Initiative on Small Arms Transfers (NISAT) based on customs data from UN Comtrade. Today, UN Comtrade is the most comprehensive source of comparable information on the trade in small arms and light weapons.[3] Although customs data is compared with national arms export report figures whenever possible, calculations are nevertheless based only on customs data. The main reasons for this are to ensure comparability across countries and to avoid double counting. Figures represent financial values rather than quantities. Another limitation of the data is the time lag in reporting. A number of states do not report promptly to UN Comtrade on their imports and exports, and states can correct their information for more than a year after submission. The following analysis therefore uses 2003 data unless explicitly stated otherwise.

Also as in previous years, 'mirror statistics' are used to get a more complete picture of the trade. Mirror data relies on importers' declarations of their imports to calculate exporters' exports and vice versa. Whoever uses mirror statistics is regularly faced with the necessity of choosing between an importer's declaration of an import and the exporter's reporting on the same transaction when there are discrepancies between the two figures. In such cases, NISAT, following a model developed by the International Trade Centre, has used a reliability measure to select whose reporting to rely on.[4]

Using UN Comtrade has its advantages and disadvantages. While steps have been taken to strengthen its utility, the current approach—analyzing UN Comtrade data in the light of additional information supplied by national reports—has its limitations. Regional reporting mechanisms such as those used in West Africa and in Europe could be more fully mined for data. Lack of transparency, selective and incomplete reporting, and, in the case of the EU Code of Conduct, a focus on *licences granted* and not *actual deliveries* all explain why these sources of information are not currently incorporated into the global assessments presented here. The inescapable element of human error (see Box 3.1) can affect any reporting instrument and is not limited to UN Comtrade.

The documented value of all small arms and light weapons exports in 2003 (as reported to UN Comtrade) is around USD 2 billion. In 2002 the figure was similar (around USD 2.1 billion) (Small Arms Survey, 2005, p. 98). Levels of transparency, in particular by important exporters and on the main categories of small arms and light weapons, have remained roughly the same as in previous years. Little, for example, is still known about Russian and Chinese exports from customs data sources. Moreover, information on certain types of small arms and light weapons (such as light weapons ammunition and certain types of military small arms and light weapons, including the very high-value man-portable air defence systems—MANPADS) is aggregated with information from other weapons types and thus

---

Box 3.1 A CHaotic tale: why export and import data does not always match

UN Comtrade, like any database, suffers from occasional human error. For example, Sudan reported that it had accidentally—and possibly routinely—entered 'CH' to record its imports of weapons from China.[5] CH, however, is not the code for China in UN Comtrade reporting. CN is. CH is the code for Switzerland, and stands for Confoederatio Helvetica.

Box 3.2 Authorized global transfers of small arms ammunition

An examination of authorized small arms ammunition transfers over the period 1999–2003 (the last five years for which data is available from UN Comtrade) shows that the trade in ammunition makes up a large portion of the overall small arms trade. During this period, the share of small arms ammunition exports as part of total small arms and light weapons exports was about one-third (see Figure 3.1).

Ammunition, in contrast to the arms themselves, is a consumable good. This means that users must regularly procure fresh supplies. Trade patterns for ammunition of individual countries could therefore be quite different from those of small arms. Sudden and consistent trends in imports of ammunition and parts of ammunition for military firearms can be detected in countries that are involved in either internal or international conflicts, such as Colombia and the United States.

The period 1999–2003 shows relatively stable trading patterns in small arms ammunition. The top ammunition exporters (defined as those states whose export value was equal to or above USD 150 million for the period as a whole) were the United States, Italy, Belgium, the UK, the Russian Federation, and Germany. For 2003, the top exporters (defined as those states whose export value was equal to or above USD 30 million) were the United States, Italy, Germany, Switzerland, and Spain.[6] The top ammunition importers for the five-year period are the United States, Saudi Arabia, and Germany, and the top importers for 2003 are the United States, Germany, and the United Kingdom.

Data sources on authorized ammunition transfers are more limited than those on small arms transfers as a whole. National arms export reports are as a rule less informative on ammunition than on arms: hence the trends mentioned above are solely based on UN Comtrade data. Customs data, such as that of UN Comtrade, is also far from perfect. The calculations made here only include categories 930621 (shotgun cartridges and parts) and 930630 (small arms ammunition), and mixed categories were excluded.[7] Thus the trade in small arms and light weapons ammunition is most likely underestimated.

Figure 3.1 Worldwide small arms ammunition exports as a share of total small arms and light weapons exports as reported to UN Comtrade, 1999–2003

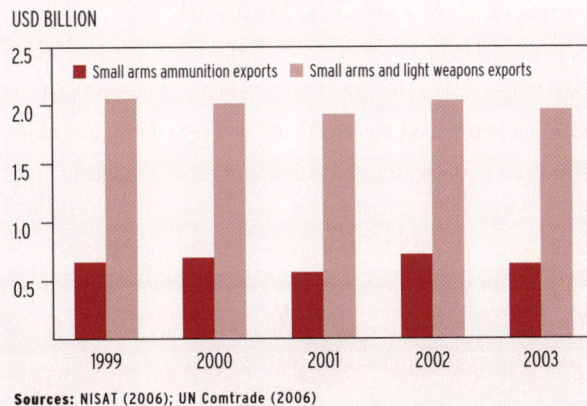

USD BILLION

- Small arms ammunition exports
- Small arms and light weapons exports

Sources: NISAT (2006); UN Comtrade (2006)

Sources: Glatz (2006); NISAT (2006); UN Comtrade (2006)

impossible to quantify. Analysis of UN Comtrade data for 2003 and field research on countries not reporting to UN Comtrade combined suggests there is insufficient reason to challenge the previous estimate for the total global legal trade in small arms and light weapons of USD 4 billion. It is assumed that mirror data captures only a small percentage of actual legal trade from major suppliers of small arms and light weapons that do not report exports to UN Comtrade, or that report only partial information. This list includes Bulgaria, China, Iran, Israel, North Korea, Pakistan, the Russian Federation, and Singapore.

As in previous years, trade in small arms ammunition is an important component of the overall trade in small arms and light weapons (see Box 3.2). Another major component of the trade is military small arms and light weapons, although it is notoriously difficult to estimate its exact share of the trade, as the shotgun and rifle categories may also contain misclassified information about some military weapons that are not necessarily small arms and light weapons, such as heavy artillery systems (Dreyfus, Lessing, and Purcena, 2005, pp. 117–19).[8] Both the ammunition and military weapons components of the total small arms trade are even more important than the figures in Tables 3.1 and 3.2 suggest, given the problem of under- and unclear reporting.

Small arms and light weapons exports

The top small arms and light weapons exporters (exporting at least USD 100 million of small arms and light weapons, including parts and ammunition, annually) according to available data and estimates in 2003 were the Russian Federation, the United States, Italy, Germany, Brazil, and China. In Table 3.1, a complete listing of top and major exporters (major exporters are defined as countries with yearly sales of more than USD 10 million) is presented. Compared to 2002, there are some changes. Belgium was no longer among the top exporters in 2003 after several years in this category. This might be due to a reporting problem: for 2003, there was a very large difference between Belgian and Saudi Arabian declarations of small arms ammunition transfers (Belgium declared an export of USD 37,018, while Saudi Arabia declared an import of USD 47,153,180). The reliability calculator favoured Belgium, and its declaration is included below. This explains, in no small part, the apparent drop in Belgium's exports (and Saudi imports) compared with 2002. Some other countries appear on the list of major exporters for the first time, namely Croatia and Iran, and others, such as Australia and Mexico, reappear after being absent from the 2002 listing. Croatia's main small arms trading partners in 2003 were the United States, Macedonia, Serbia and Montenegro, Afghanistan, and Dominica. For Iran, they were Sudan and Italy. North Korea was newly included as a medium producer. Apart from this, there were no large shifts in major small arms exporters in the period 2001–03.

As in previous years, there are some exporters, presumed to be important in the authorized small arms and light weapons trade, about which relatively little is known. These include Bulgaria, China, Iran, Israel, North Korea, Pakistan, the Russian Federation, and Singapore. Finding information on the small arms and light weapons exports of these countries is often very difficult. Bulgarian policy, for example, is to keep this information classified.[9]

Table 3.1 **Annual authorized small arms and light weapons exports for major exporters (yearly sales of more than USD 10 million), 2003**

Country	USD value (unless otherwise stated)	Main recipients (top five by value)*	Main types of small arms and light weapons exported (top five). NB: types refer to customs codes (see notes)	Remarks
Australia	10 million**	US, Japan, New Zealand, Belgium, UK**	Ammunition, pistols/revolvers, sporting/hunting shotguns, sporting/hunting rifles**	Department of Defence did not report on arms exports in 2003.
Austria	At least 97 million**	US, Germany, Switzerland, Russian Federation, Italy**	Pistols/revolvers, sporting/hunting rifles, parts/accessories sporting/hunting weapons, ammunition, parts/accessories pistols/revolvers**	Reports its exports neither of military weapons nor of pistols/revolvers to UN Comtrade. Hence the value of these categories (based on importers' reports) is likely to be underestimated.
Belgium	At least 75 million**	US, Portugal, France, UK, Italy**	Parts/accessories sporting/hunting weapons, sporting/hunting shotguns, sporting/hunting rifles, ammunition**	Reports its exports neither of military weapons nor of pistols/revolvers to UN Comtrade. Hence the value for these categories (based on importers' reports) is likely to be underestimated. Belgium no longer publishes a national arms export report.

				The Region of Brussels presented arms export reports for the period 1 September 2003–31 December 2004. Although information on licences is disaggregated among 'light',' semi-light', and 'heavy weapons', the report only gives information on granted export licences and not actual deliveries. Values are not broken down by weapons category.
				The Wallonian Region released a report on arms exports for the period 1 September–31 December 2003. This report, however, does not distinguish between heavy conventional weapons and small arms and light weapons and only gives very aggregated information concerning export licences and not actual exports.
				The report from Flanders covers the period 30 August 2003–31 August 2004. It provides information on actual deliveries disaggregated by weapon category, but values for small arms and light weapons are not provided separately from those for other conventional weapons.
Brazil	At least 101 million**	US, Colombia, Saudi Arabia, Germany, Zimbabwe[1]**	Ammunition, pistols/revolvers, sporting/hunting rifles, sporting/hunting shotguns**	Does not report exports of pistols/revolvers to UN Comtrade. However, handguns are the main product and export good of the Brazilian small arms industry. The value for this product is based on importers' reports.[2]
Bulgaria	*Medium producer, but little is reported about its exports.*			
Canada	57 million** Canadian export report for 2003 had not yet been released in January 2006.	US, Belgium, Denmark, UK, Australia**	Ammunition, sporting/hunting rifles, parts/accessories sporting/hunting weapons, pistols/revolvers, military firearms**	Customs data and the national report diverge largely because the latter does not take into account exports to the United States, which, according to the export report, are 'estimated to account for over half of Canada's exports of military goods and technology' (Canada, 2003, p. 7).
China	At least 14 million** Small Arms Survey estimate: USD 100 million (Small Arms Survey, 2004, Annexe 4.1)	US, Bangladesh, Germany, Canada, Malta**	Sporting/hunting rifles, sporting/hunting shotguns, parts/accessories sporting/hunting weapons, parts/accessories pistols/revolvers, pistols/revolvers**	Customs data is likely to underestimate actual exports, as China does not report on many of its exports, and hence figures are based on importers' reporting.

Croatia	13 million**	US, Macedonia, Serbia and Montenegro, Afghanistan, Dominica**	Pistols/revolvers, ammunition, military firearms, rocket/ grenade launchers, parts/ accessories pistols/ revolvers**	
Czech Republic	At least 51 million** EUR 25.5 million (USD 30.8 million)♣	US, Germany, France, Colombia, Slovakia**	Sporting/hunting rifles, pistols/revolvers, ammunition, parts/accessories pistols/ revolvers**	Does not report exports of military weapons to UN Comtrade. Hence the value (based on importers' reports) is likely to be underestimated. Customs and export report data diverge probably largely because small arms ammunition cannot be distinguished from other types of ammunition in the arms export report. The report does not rank recipients for all types of small arms.
Finland	32 million** EUR 2.3 million (USD 2.8 million)♣	US, Sweden, Norway, UK, Italy** Italy, US, Germany, UK, Czech Republic♣	Sporting/hunting rifles, ammunition, parts/accessories sporting/hunting weapons, military firearms, sporting/ hunting shotguns**	Customs and export report data diverge probably largely because civilian weapons are excluded from the export report. Moreover, in the export report, small arms ammunition cannot be distinguished from other types of ammunition.
France	At least 42 million** EUR 47.2 million[3] (USD 57.1 million)♣	Côte d'Ivoire,[4] Turkey, US, Portugal, Canada** Oman, Norway, Tunisia, South Korea, Spain♣	Military firearms, ammunition, military weapons, ammunition, parts/accessories sporting/ hunting weapons**	Does not report exports of military weapons and pistols/revolvers to UN Comtrade. Hence the value (based on importers' reports) is likely to be underestimated. Customs and export report data capture different parts of the small arms and light weapons trade: customs data better captures small arms and civilian weapons, while the arms export report encompasses more high-end light weapons.
Germany	At least 201 million**	US, France, Austria, Switzerland, Japan**	Pistols/revolvers, ammunition, sporting/hunting rifles, parts/ accessories sporting/hunting weapons, parts/accessories pistols/revolvers**	Does not report exports of military weapons to UN Comtrade. Hence the value (based on importers' reports) is likely to be underestimated. Publishes an export report, but it includes information on granted export licences, not actual deliveries of small arms and light weapons, which may be lower.
Iran	At least 16 million**	Sudan, Italy**	Military weapons, parts/ accessories sporting/hunting weapons, ammunition, pistols/ revolvers, parts/accessories pistols/revolvers**	Does not report on its small arms and light weapons exports at all to UN Comtrade. Figures are based on importers' reports. Hence the value is likely to be underestimated.

Israel	At least 15 million**	US, Guatemala, Mexico, Germany, Austria**	Pistols/revolvers, ammunition, parts/accessories pistols/revolvers, military firearms, parts/accessories sporting/hunting weapons**	Does not report on its small arms and light weapons exports at all to UN Comtrade. Figures are based on importers' reports. Hence the value is likely to be underestimated.
Italy	At least 347 million** 529,615♣5	US, France, UK, Spain, Germany**	Sporting/hunting shotguns, ammunition, sporting/hunting rifles, pistols/revolvers, parts/accessories sporting/hunting weapons**	Does not report exports of military weapons to UN Comtrade. Hence the value (based on importers' reports) is likely to be underestimated. Publishes an export report, but it includes information provided by the Ministry of Foreign Affairs on granted licences of military weapons, not actual deliveries of small arms and light weapons, which may be lower. The report includes an annexe with some information from the Customs Agency (Agenzia delle Dogane) on actual exports of military small arms. Information is provided in this annexe by company, product, and value. It is not possible, however, to identify the countries of destination of small arms and light weapons exports (Italy, 2004, pp. 169–96)
Japan	72 million**	US, Belgium, Canada, Germany**	Sporting/hunting rifles, parts/accessories sporting/hunting weapons, sporting/hunting shotguns, shotgun barrels, rocket/grenade launchers**	
Mexico	11 million**	US, Honduras, Peru, Guatemala, Panama**	Shotgun barrels, ammunition, parts/accessories sporting/hunting weapons, military firearms**	Exports of military firearms might actually be returns for repairs or devolution of material to the licensing company, Heckler & Koch. Germany ranks as the sixth importer from Mexico and is the only country that received military firearms from that country.[6]
North Korea	*Medium producer, but little is reported about its exports.*			
Norway	73 million**[7]	Spain, Sweden, Italy, Switzerland, UK**	Military firearms, ammunition, rocket/grenade launchers, sporting/hunting shotguns, sporting/hunting rifles**	Publishes national arms export report, but in the edition covering 2003, it is difficult to distinguish the small arms and light weapons share of arms exports.
Pakistan	*Medium producer, but little is reported about its exports. According to the Stockholm Peace Research Institute (SIPRI), it delivered Anza-2 portable surface-to-air missiles to Malaysia between 2002 and 2003 (SIPRI, 2006).*			
Portugal	17 million**	Belgium, US, Spain, Guinea Bissau, Germany**	Sporting/hunting rifles, sporting/hunting shotguns, parts/accessories sporting/hunting weapons, pistols/revolvers, ammunition**	Publishes an export report, but it does not detail the share of small arms and light weapons in total arms exports.

Romania	*Medium producer, but little is reported about its exports.*		
Russian Federation	At least 43 million** At least 431.8 million (410.3 million for small arms and light weapons and 21.5 million for small arms and light weapons ammunition) (Pyadushkin, 2005, pp. 1–2)♠	US, Cyprus, France, Algeria, Kazakhstan** Jordan, Kyrgyzstan, Kenya (Pyadushkin, 2005, p. 3)♠ India, Indonesia (SIPRI, 2006)♦	Sporting/hunting shotguns, ammunition, sporting/hunting rifles, military firearms, pistols/revolvers** Rocket-propelled grenade launchers, assault rifles (Pyadushkin, 2005, p. 3)[8]♠ MANPADS (SIPRI, 2006)♦	Customs data is likely to underestimate actual exports, as the Russian Federation does not report on many of its exports, and hence figures are based on importers' reporting.
Singapore	*Medium producer, but little is reported about its exports.*		
South Africa	At least 18 million**	Namibia, Colombia, US, UK, Brasil**	Military firearms, ammunition, pistols/revolvers, rocket/grenade launchers, sporting/hunting rifles**	Does not report on its small arms and light weapons exports at all to UN Comtrade. Figures are based on importers' reports. Hence the value is likely to be underestimated. South Africa no longer makes its arms export report public (Honey, 2005).
South Korea	23 million**	US, Venezuela, Indonesia, Australia, Thailand**	Ammunition, parts/accessories pistols/revolvers, parts/accessories sporting/hunting weapons, military firearms**	
Spain	At least 55 million** EUR 5.9 million (USD 7.1 million)♣	US, UK, Portugal, Ghana, France**	Ammunition, sporting/hunting shotguns, parts/accessories sporting/hunting weapons, sporting/hunting rifles**	Does not report exports of military weapons to UN Comtrade. Hence the value (based on importers' reports) is likely to be underestimated. The discrepancy between the arms export report figure and the UN Comtrade figure is most likely due to the fact that civilian weapons are excluded from the export report. Export report does not detail recipients of small arms.
Sweden	At least 29 million** SEK 19 million (USD 2.5 million)♣	US, Norway, Denmark, Germany, Mexico** Pakistan, Thailand, Australia (SIPRI, 2006)♦	Ammunition, parts/accessories sporting/hunting weapons, sporting/hunting rifles, sporting/hunting shotguns** MANPADS (SIPRI, 2006)♦	Does not report exports of military weapons to UN Comtrade. Hence the value (based on importers' reports) is likely to be underestimated. Customs and export report data diverge probably largely because small arms ammunition cannot be distinguished from other types of ammunition in the arms export report. In the export report, it is also difficult to fully distinguish small arms and light weapons from other items. The export report does not detail recipients of small arms.

Switzerland	65 million** CHF 14.1 million (USD 11 million)♣	Germany, US, Austria, Finland, United Arab Emirates** Germany, US, Poland, Italy, Macau (China)♣	Ammunition, pistols/revolvers, military firearms, sporting/hunting rifles, parts/accessories sporting/hunting weapons**	Customs and export report data diverge probably largely because small arms ammunition cannot be distinguished from other types of ammunition in the arms export report. In the export report, it is also difficult to fully distinguish small arms and light weapons from other items. Civilian weapons are excluded from the export report.
Turkey	56 million**	Germany, US, Italy, Austria, Egypt**	Sporting/hunting shotguns, ammunition, parts/accessories sporting/hunting weapons, sporting/hunting rifles, shotgun barrels**	
UK	42 million**	US, Japan, Denmark, Germany, Switzerland**	Sporting/hunting shotguns, ammunition, parts/accessories sporting/hunting weapons, sporting/hunting rifles**	Publishes export report, but does not detail the value of small arms and light weapons exports. Instead it provides numbers of certain types of small arms and light weapons exported and certain destinations.
US	370 million**	Japan, Canada, South Korea, United Arab Emirates, Australia** Germany, Greece, Italy, Egypt (SIPRI, 2006)♦	Ammunition, military firearms, parts/accessories sporting/hunting weapons, rocket/grenade launchers, sporting/hunting rifles** MANPADS (SIPRI, 2006)♦	Publishes export report, but it includes mostly information on granted export licences, not actual deliveries of small arms and light weapons, which may be lower.

* Excluding those contributing less than 1% of the total.

** NISAT (2006); UN Comtrade (2006). Customs codes: 9301 (military weapons), 930120 (rocket and grenade launchers, etc.), 930190 (military firearms), 9302 (revolvers and pistols), 930320 (sporting and hunting shotguns), 930330 (sporting and hunting rifles), 930510 (parts and accessories of revolvers and pistols), 930521 (shotgun barrels), 930529 (parts and accessories of shotguns or rifles), 930621 (shotgun cartridges), 930630 (small arms ammunition).

♣ Export report.

♠ Report by field researcher.

♦ Other academic/research sources such as SIPRI.

Note 1: Customs data and national arms export reports diverge for a number of reasons. Two important causes for discrepancies are: (a) unlike customs data, national arms export reports mostly do not include exports of whatever is categorized as 'civilian' weapons; and (b) many countries use the Wassenaar Arrangement Munitions List (ML) in their arms export reports. The categories of the ML list, with one exception (ML1), do not distinguish small arms and light weapons from other types of weapons. The Small Arms Survey records the ML1 value in the table; hence many types of small arms and light weapons, and all types of small arms ammunition, are excluded. For further details, see Small Arms Survey (2005, pp. 101–02).

Note 2: Some top producers of small arms, e.g. China and the Russian Federation, do not publish any arms export reports and only limited customs information. In order to properly reflect their presumed importance in the small arms trade, the Small Arms Survey goes beyond mirror data. China is still estimated to export USD 100 million worth of small arms and light weapons. For details, see Small Arms Survey (2004, ch. 4, Annexe 4.1). For the Russian Federation, figures were developed on the basis of field research.

Note 3: In this table, 'transfer' can also mean the movement of small arms for repair, and devolution of parts and weapons to licensing companies.

NB 1: 'Ammunition' in the table refers to shotgun cartridges and small arms ammunition combined.

NB 2: Category 9301 (military weapons) is a mixed category, containing both small arms and light weapons and other weapons. It was replaced by four new categories in the newest revision of the UN Comtrade Harmonized System (HS 2002), facilitating differentiation between small arms and light weapons and other weapons, among them the small arms and light weapons categories 930120 and 930190. Some countries still use HS 1996, however, and their reported value for 93010 may therefore include military weapons other than small arms and light weapons. The calculations on which this table is based include data from HS 2002, HS 1996, and HS 1992 to account for all transfers of military small arms and light weapons reported to UN Comtrade. For the older HS codes, the UN Register of Conventional Arms (UN Register) or the *SIPRI Yearbook* were consulted to see whether any of the transfers were likely to concern large-calibre artillery. This was not the case for any transfers during 2003 (otherwise the respective 9301 value would have been taken out of the calculation). For more information on revisions of the UN Comtrade HS, see Marsh (2005).

1 Brazil declares exports to Zimbabwe.

2 Research by Dreyfus, Lessing, and Purcena (2005, p. 116) suggests that Brazil records its firearm exports in a somewhat unorthodox way, filing its pistols and revolvers exports under customs category 930330 (sporting and hunting rifles). The imports declared in category 9302 (revolvers and pistols) by the United States may thus correspond to inflated figures in the category 930330 on the Brazilian side.

3 Calculated from France (2005, Annexe 15).

4 In a phone conversation with the Small Arms Survey on 1 December 2005, an official from the French Mission in Geneva suggested that the small arms and light weapons declared by Côte d'Ivoire may be related to the transfer of French equipment to France's 'Operation Licorne', which was supporting the Economic Community of West Africa States peacekeeping mission in that country at the time.

5 This figure only includes military small arms and light weapons (including parts and ammunition) as reported by the Customs Agency in the national arms export report.

6 On small arms production under Heckler & Koch licence in Mexico, see Small Arms Survey (2004, p. 25).

7 The majority of Norwegian exports to Spain were destined for ships being built in Spain for the Norwegian Navy (e-mail communication with Nicholas Marsh, Peace Research Institute, Oslo, 27 February 2006).

8 Types of small arms and light weapons exported to the top three recipients.

Sources: Brussels (2005); Canada (2003); Czech Republic (2004); Finland (2004); Flanders (n.d.); France (2005); Germany (2004); Italy (2004); NISAT (2006) (UN Comtrade calculations); Norway (2004); Pyadushkin (2005); South Africa (2003); Spain (2004); Sweden (2004); Switzerland (2004); Wallonia (2004); UK (2004)

Small arms and light weapons imports

The top importers (i.e. countries importing more than USD 100 million worth of small arms and light weapons in a given year) for 2003 were the United States, Cyprus, and Germany. In 2002 the list was somewhat different, comprising the United States, Cyprus, Saudi Arabia, and South Korea. Similarly, there are a number of changes in the listing of major importers (i.e. those states that import more than USD 10 million of small arms and light weapons) from 2002 to 2003. Bahrain, Jordan, Kenya, and the Philippines are no longer among the major importers. Newcomers here are Côte d'Ivoire, Egypt, Namibia, New Zealand, Sudan, Thailand, and Venezuela.

For 2003, the top importers were the United States, Cyprus, and Germany.

Listings of top and major small arms and light weapons importers are normally more volatile than a similar list of exporters. Only the United States and Cyprus regularly rate as top importers.[10] While civilian demand (and hence international purchases of civilian weapons) more rarely undergoes drastic changes from one year to the next, procurement decisions of police and the military are more prone to fluctuations, especially in smaller states. For some countries, explanations for sudden drops in imports have to be sought elsewhere, however. Saudi Arabia no longer reports on its imports of military weapons and pistols/revolvers, which probably explains the apparent drastic fall of imports of that country (from USD 132 million in 2002 to only USD 25 million in 2003).

Just as with exporters, information on some countries that are presumed to be major importers remains scarce. This is particularly true for countries in conflict, be it internal or international (although Colombia, Côte d'Ivoire, Israel, the Russian Federation, and Sudan are among the major importers listed in Table 3.2). In some cases, imports to such countries (and especially to insurgents) are illicit and thus do not fit into Table 3.2. Transfers to some war zones are discussed in more detail in the case study chapters of this volume.[11]

Other presumed major importers that are not found in Table 3.2 are post-war states such as Iraq and Afghanistan. Under UN arms embargo until 2004, Iraq's imports gathered pace in that year and 2005. Suppliers of small arms to Iraq include Central and Eastern European countries, e.g. Estonia and Romania, as well as Western countries such as Denmark and the United States (Associated Press, 2005; NATO, 2005; US Central Command, 2004). Afghanistan has also relied on Central and Eastern European suppliers (Croatia, Bulgaria, Hungary, Romania) and the United States in the post-conflict period (Smith & Wesson Holding Corporation, 2005; Ramon, 2004; Associated Press, 2002). Most of the supplies to both countries consist of donations, and hence are not captured by information based on values such as those in Tables 3.1 and 3.2.

Table 3.2 Annual authorized small arms and light weapons imports for major importers (yearly imports of more than USD 10 million), 2003

Country	USD value customs data (UN Comtrade)	Main suppliers (top five)*	Main types of small arms and light weapons imported (top five). NB: types refer to customs codes (see notes)	Remarks
Australia	42 million	US, unspecified countries, Italy, Austria, Finland	Ammunition, military firearms, sporting/hunting rifles, pistols/revolvers, sporting/hunting shotguns	
Austria	At least 27 million	Germany, Switzerland, Turkey, Italy, Belgium	Ammunition, parts/accessories sporting/hunting weapons, sporting/hunting rifles, sporting/hunting shotguns, parts/accessories pistols/revolvers	Does not report on its imports of military weapons and pistols/revolvers to UN Comtrade. Hence the value is possibly underestimated.
Belgium	At least 37 million	Japan, Portugal, US, Italy, Canada	Sporting/hunting shotguns, ammunition, sporting/hunting rifles, pistols/revolvers	Does not report on its imports of military weapons and pistols/revolvers to UN Comtrade. Hence the value (based on exporters' reports) is possibly underestimated. Some imports might actually be returns for repairs.
Canada	58 million	US, Italy, Japan, Germany, UK	Ammunition, parts/accessories sporting/hunting weapons, sporting/hunting rifles, sporting/hunting shotguns	
Colombia	34 million	Brazil, US, South Africa, Czech Republic, Germany	Ammunition, military firearms, pistols/revolvers, rocket/grenade launchers	
Côte d'Ivoire	11 million	France[1]	Military weapons, ammunition, sporting/hunting rifles	Does not report on many of its imports. Hence the value is possibly underestimated.
Cyprus	185 million	Unspecified countries, Russian Federation	Military firearms, sporting/hunting shotguns, ammunition, sporting/hunting rifles	
Czech Republic	11 million	Italy, Germany, US, Turkey, Austria	Pistols/revolvers, parts/accessories pistols/revolvers, ammunition, parts/accessories sporting/hunting weapons, sporting/hunting shotguns	Does not report on its imports of military weapons to UN Comtrade. Hence the value is possibly underestimated.
Denmark	22 million	Germany, Italy, Canada, Sweden, UK	Ammunition, sporting/hunting rifles, sporting/hunting shotguns, parts/accessories sporting/hunting weapons	
Egypt	17 million	US, Turkey, Czech Republic	Parts/accessories sporting/hunting weapons, ammunition, sporting/hunting shotguns, pistols/revolvers, rocket/grenade launchers	

Finland	17 million	Italy, Switzerland, Germany, US, Austria	Ammunition, sporting/hunting shotguns, sporting/ hunting rifles, military firearms	
France	At least 83 million	Italy, Germany, Belgium, US, Spain	Sporting/hunting shotguns, sporting/hunting rifles, ammunition, parts/accessories sporting/hunting weapons	Does not report on its imports of military weapons and pistols/ revolvers to UN Comtrade. Hence the value is possibly underestimated.
Germany	At least 109 million	Switzerland, Turkey, Italy, US, Austria	Ammunition, parts/accessories sporting/hunting weapons, sporting/hunting shotguns, sporting/hunting rifles	Does not report on its imports of military weapons to UN Comtrade. Hence the value is possibly underestimated.
Greece	At least 13 million	Italy, US, Germany, Spain, Russian Federation	Sporting/hunting shotguns, ammunition, parts/accessories sporting/hunting weapons, rocket/grenade launchers, shotgun barrels	Does not report on its imports of military weapons and pistols/ revolvers to UN Comtrade. Hence the value (based on exporters' reports) is possibly underestimated.
Israel	At least 16 million	US, Austria,[2] South Korea, Czech Republic	Ammunition, military firearms, parts/accessories sporting/ hunting weapons, parts/ accessories pistols/revolvers, pistols/revolvers	Does not report any imports to UN Comtrade. Hence the value (based on exporters' reports) is probably underestimated.
Italy	At least 48 million	US, Germany, Belgium, Turkey, Norway	Ammunition, sporting/hunting rifles, parts/accessories sporting/hunting weapons, pistols/revolvers, sporting/ hunting shotguns	Does not report on its imports of military weapons to UN Comtrade. Hence the value is possibly underestimated.
Japan	64 million	US, Germany, Italy, UK, Australia	Military firearms, ammunition, pistols/revolvers, sporting/ hunting shotguns	
Kuwait	At least 14 million	US, Italy, Germany, France, Portugal	Ammunition, sporting/hunting shotguns, military firearms, sporting/hunting rifles	Does not report any imports to UN Comtrade. Hence the value (based on exporters' reports) is probably underestimated.
Malaysia	10 million	US, Switzerland, Italy, Norway, Netherlands	Ammunition, sporting/hunting rifles, rocket/grenade launchers, parts/accessories sporting/ hunting weapons	
Mexico	15 million	Italy, US, Israel, Austria, Sweden	Ammunition, pistols/revolvers, military firearms, rocket/ grenade launchers, sporting/ hunting shotguns	
Namibia	10 million	South Africa, Germany, Czech Republic, Spain, US	Military firearms, pistols/ revolvers, sporting/hunting rifles, sporting/hunting shotguns, ammunition	
Netherlands	At least 23 million	US, Germany, Belgium, Switzerland, Canada	Ammunition, pistols/revolvers, parts/accessories sporting/ hunting weapons, sporting/ hunting shotguns	Does not report on its imports of military weapons and pistols/ revolvers to UN Comtrade. Hence the value (based on exporters' reports) is possibly underestimated.

New Zealand	14 million	US, Italy, Australia, Switzerland, Canada	Ammunition, sporting/hunting rifles, sporting/hunting shotguns, parts/accessories sporting/hunting weapons	
Norway	30 million	Germany, US, Italy, Sweden, Finland	Ammunition, sporting/hunting shotguns, sporting/hunting rifles, military firearms	
Poland	13 million	Germany, Italy, Czech Republic, Russian Federation, Switzerland	Ammunition, pistols/revolvers, parts/accessories pistols/revolvers, military firearms, sporting/hunting rifles	
Portugal	40 million	Belgium, Italy, US, Spain, France	Parts/accessories sporting/hunting weapons, sporting/hunting shotguns, parts/accessories pistols/revolvers, shotgun barrels, ammunition	
Russian Federation	At least 15 million	Germany, Italy, Austria, Belgium, France	Sporting/hunting shotguns, sporting/hunting rifles, parts/accessories sporting/hunting weapons, ammunition	Does not report on its imports of military weapons, revolvers/pistols, and ammunition to UN Comtrade. Hence the value (based on exporters' reports) is possibly underestimated.
Saudi Arabia	At least 25 million	Brazil, US, Germany, Switzerland, UK	Ammunition, parts/accessories sporting/hunting weapons, rocket/grenade launchers, sporting/hunting rifles	Does not report on its imports of military weapons and pistols/revolvers to UN Comtrade. Hence the value (based on exporters' reports) is possibly underestimated.
South Korea	42 million	US, Italy, UK, Germany	Ammunition, rocket/grenade launchers, pistols/revolvers, parts/accessories sporting/hunting weapons	
Spain	At least 99 million	Norway, Italy, Germany, US, Portugal	Military firearms, sporting/hunting shotguns, sporting/hunting rifles, ammunition, pistols/revolvers	Does not report on its imports of military weapons to UN Comtrade. Hence the value (based on exporters' reports) is possibly underestimated.
Sudan	18 million	Iran, China	Military weapons, parts/accessories sporting/hunting weapons, pistols/revolvers, sporting/hunting shotguns, sporting/hunting rifles	
Sweden	At least 24 million	Norway, Germany, Finland, US, Italy	Ammunition, sporting/hunting rifles, sporting/hunting shotguns, pistols/revolvers, parts/accessories sporting/hunting weapons	Does not report on its imports of military weapons to UN Comtrade. Hence the value is possibly underestimated.
Switzerland	24 million	Germany, Austria, Norway, Italy, UK	Ammunition, pistols/revolvers, sporting/hunting rifles, sporting/hunting shotguns, parts/accessories sporting/hunting weapons	

Thailand	At least 10 million	US, Austria, Czech Republic, Germany, Italy	Pistols/revolvers, ammunition, sporting/hunting rifles, sporting/hunting shotguns	Does not report on its imports of military weapons to UN Comtrade. Hence the value is possibly underestimated.
Turkey	32 million	France, US, Italy, Spain, Germany	Military firearms, ammunition, rocket/grenade launchers, pistols/revolvers	
United Arab Emirates	At least 42 million	US, Switzerland, Germany, Brazil, Finland	Military firearms, ammunition, pistols/revolvers, sporting/hunting rifles	Does not report any imports to UN Comtrade. Hence the value (based on exporters' reports) is probably underestimated.
UK	77 million	Italy, US, Germany, Belgium, Spain	Sporting/hunting shotguns, ammunition, sporting/hunting rifles, parts/accessories hunting/sporting weapons	
US	623 million	Italy, Germany, Austria, Japan Brazil	Sporting/hunting shotguns, pistols/revolvers, ammunition, sporting/hunting rifles, parts/accessories hunting/sporting weapons	
Venezuela	10 million	South Korea, Brazil, Italy, Spain, Austria	Ammunition, pistols/revolvers, parts/accessories pistols/revolvers, sporting/hunting shotguns	

* Excluding those contributing less than 1% of the total.

Note 1: Includes customs codes 9301 (military weapons), 930120 (rocket and grenade launchers, etc.), 930190 (military firearms), 9302 (revolvers and pistols), 930320 (sporting and hunting shotguns), 930330 (sporting and hunting rifles), 930510 (parts and accessories of revolvers and pistols), 930521 (shotgun barrels), 930529 (parts and accessories of shotguns or rifles), 930621 (shotgun cartridges), 930630 (small arms ammunition).

Note 2: For the scope of this table, 'transfer' can also mean the movement of small arms for repair, and the devolution of parts and weapons to licensing companies.

NB 1: 'Ammunition' in the table refers to shotgun cartridges and small arms ammunition combined.

NB 2: Category 9301 (military weapons) is a mixed category, containing small arms and light weapons as well as other weapons. It was replaced by four new categories in the newest revision of the UN Comtrade Harmonized System (HS 2002), facilitating differentiation between small arms and light weapons and other weapons, among them the small arms and light weapons categories 930120 and 930190. Some countries still use HS 1996, however, and their reported value for 93010 may therefore include military weapons other than small arms and light weapons. The calculations on which this table is based include data from HS 2002, HS 1996, and HS 1992, to account for all transfers of military small arms and light weapons reported to UN Comtrade. For the older HSs, it was checked whether corresponding conventional arms transfers were reported in the UN Register or the *SIPRI Yearbook*. This was not the case for any country for 2003 (otherwise the respective 93010 value would have been taken out of the calculation). For more information on revisions of the UN Comtrade Harmonized System, see Small Arms Survey (2005, pp. 99–100, Box 4.1) and Marsh (2005).

1 For Côte d'Ivoire's imports from France, which may have been transfers to 'Operation Licorne', see Table 3.1, n. 4.

2 It is registered in UN Comtrade as Austrian exports of category 930510 (parts and accessories of revolvers and pistols). This information is not, however, mirrored by Israel.

Sources: NISAT (2006); UN Comtrade (2006)

DEVELOPMENTS IN TRANSPARENCY: ANNUAL UPDATE

General trends in international transparency

The small arms and light weapons transparency device that received great attention by governments, NGOs, and other actors during 2005—and will probably continue to do so in 2006—is states' reporting on their implementation of the *UN Programme of Action* (*Programme*). These reports can be very useful for understanding legislative developments and practices regarding stockpile security, marking and tracing, collection and destruction of surplus small arms, and other issues.[12] However, they rarely contain any information on small arms imports and exports, as this is outside the scope of the information exchange. The transparency mechanisms on small arms transfers thus remain basically the same as in previous years.[13]

The UN Register (UN, 2006a) is now in its second year of reporting on certain types of light weapons (artillery pieces between 75 and 100 mm, including MANPADS) (Small Arms Survey, 2005, p. 109). The voluntary sharing of information on all types of military small arms and light weapons is also continuing, and that information is now available on the Web site of the UN Department for Disarmament Affairs (UN, 2006b). The *Small Arms Survey 2005* examined mainly the number and geographical spread of countries reporting to the UN Register during the first year (2003). But how useful is the format of the reporting for better understanding light weapons transfers? So far, for export and import figures, few states have provided the kind of detail necessary to single out mortars falling into the category of light weapons from other types of artillery, and MANPADS from other types of missiles. So although some light weapons are now part of the reporting of a number of countries, it is difficult to gain a good understanding of the trade in mortars and MANPADS from the UN Register. As there is no agreed-upon format for the voluntary sharing of information on military small arms and light weapons, the reporting varies widely among different states, and between this reporting and other types of reporting. The case of France is very illustrative here. The French national arms export report is very informative and detailed concerning exports of grenade launchers, light anti-tank weapons, and MANPADS (France, 2005, p. 67 and Annexe 15). However, France does not report these exports to the UN Register.

The annual reports on the implementation of the *EU Code of Conduct on Arms Exports*—which contain data on arms exports from European Union (EU) countries—have become more substantive in recent years. The first report, published in 1999, was of little utility in understanding small arms transfers from EU member states. By the sixth report, published in December 2004, the reporting format had evolved considerably (EU, 2004). It now contains data disaggregated by country and weapon type (using the EU Common Military List categories, very similar to the Wassenaar ML). A major weakness of the reporting (apart from the fact that those countries that do not publish national arms export reports also for the most part do not provide detailed information to the annual EU report) is that the weapons categorization used makes it difficult to fully distinguish small arms and light weapons from other types of weapons (see Table 3.1, Note 1). Furthermore, it only provides data on licences granted, not actual deliveries. Nevertheless,

Box 3.3 Ups and downs in transparency: Belgium and South Africa

In previous years, the Small Arms Survey has noted some positive developments in state transparency, such as new countries starting to publish annual arms export reports. This trend continues (e.g. in February 2005, Bosnia and Herzegovina issued a first report on its arms exports (Bosnia and Herzegovina, 2005)). There are also states moving in the opposite direction, however.

One of the more peculiar cases is Belgium, which regionalized and decentralized its export controls, including reporting on arms exports, in 2003. So far, reporting from the three Belgian regions has been sketchy, at best. Furthermore, there is no federal-level effort to coordinate or assemble the information on arms exports from the regions. All this means that Belgium has effectively discontinued its annual arms export reporting, and the only national governmental source of information on Belgian small arms exports is now customs data.

In South Africa, a number of arms transactions (most of which did not include small arms) have come under intense public scrutiny and debate in the last years. In 2005 the South African government decided to stop the practice of making the national arms export report public (Honey, 2005).

the EU report is very important, as any impetus for improvement in reporting formats, in particular on small arms and light weapons, in this forum would have an impact on many exporting states. This is also one reason why the NGO community carefully scrutinizes the report every year.[14]

Update on the Small Arms Trade Transparency Barometer

One means to encourage states to provide more data on their small arms and light weapons exports is the Small Arms Trade Transparency Barometer. The Barometer assesses information that states provide in two formats: national arms export reports and customs data (as reported to UN Comtrade). The information is analysed from a number of angles, taking into account how accessible, clear, and comprehensive it is. The details states provide on licences and deliveries of small arms and light weapons are also assessed.

This year, the Barometer has been slightly revised to include four new or extended parameters: (a) the timeliness of reporting; (b) whether detailed weapon descriptions are provided (i.e. if there is a differentiation among sub-categories of small arms and light weapons such as pistols/revolvers, sniper rifles, shotguns, machine guns, etc.); (c) whether information on small arms and light weapons ammunition is included; and (d) whether reporting covers both civilian and military small arms (maximum 2 points instead of 1 point previously). The new maximum score is 25.

Details on the scoring system are provided beneath Table 3.3. States are awarded points for information on granted and denied licences only if the *value of the licence* and/or the *quantity of weapons* included in the licence are/is specified. This means that statistics on mere *numbers of licences* are not awarded points. The reason is that such information says little about the scale of the proposed transaction—are a couple of guns or several thousands of them involved? The same is true for deliveries, where, again, information on numbers of shipments (rather than numbers of weapons delivered or their value) does not improve states' scores.

The most transparent states in the 2006 Barometer are the United States and Germany.

The Barometer evaluates the quantity and level of detail of the data made public, but does not assess the accuracy of the information states provide. This is simply beyond the means (in terms of access and resources) of independent, internationally based researchers. It is well known, however, that there are problems of inconsistencies among various reporting formats. Table 3.1 illustrates this extensively and gives some explanations as to why inconsistencies can occur (see Table 3.1, Note 1). Regular revelations in the press of contested and at times secretive arms transfers are also indications that national arms export reports are not always exhaustive.

The Barometer focuses on small arms and light weapons only, and uniquely on those states that are important in the small arms and light weapons trade. It cannot therefore be used as a measure of how transparent states are on their arms exports as a whole (i.e. of all conventional weapons, dual use goods, etc.).

What conclusions can be drawn from the Barometer? At the top of the list are the United States and Germany, and at the bottom, Bulgaria, Iran, Israel, and North Korea (all with zero scores). This state of affairs remains relatively similar to previous years. States are generally better at providing information in certain categories, such as availability on the Internet, availability in a UN language, source information, summaries of export laws and international commitments, and information on values of deliveries. States seem particularly unwilling to provide any data on the intended end-users of small arms, government-sourced transactions (which often, although not always, concern older weapons from state stockpiles), licence denials, and, albeit to a lesser extent, licence approvals. Timeliness of reporting also remains a serious problem in many cases.

Table 3.3 Small Arms Trade Transparency Barometer 2006, covering major exporters

Country and source(s) available	Total points (25 points max.)	Timeliness (2 points max.)	Access (2 points max.)	Clarity (5 points max.)	Comprehensiveness (6 points max.)	Information on deliveries (4 points max.)	Information on licences granted (4 points max.)	Information on licences refused (2 points max.)
Australia[1] E (01–02) C	14.5	0.5	2	2.5	5.5	4	0	0
Austria C	9.5	0	1.5	2.5	3.5	2	0	0
Belgium[2] C	9	0	1.5	2.5	3	2	0	0
Brazil C	10	0	1.5	2.5	4	2	0	0
Bulgaria	0	0	0	0	0	0	0	0
Canada[1] E (02) C	16.5	0.5	2	4	6	4	0	0
China[3] C	10.5	0	1.5	2.5	2.5	4	0	0
Czech Republic E (04) C	16	1	2	3	5	3	2	0
Croatia C	10	0	1.5	2.5	4	2	0	0
Finland E (03) C	16	1	2	3	6	2	2	0
France[4] E (04) C	16.5	1	2	3.5	6	4	0	0
Germany[5] E (04) C	19	1	2	3.5	6	2	4	0.5
Iran	0	0	0	0	0	0	0	0
Israel	0	0	0	0	0	0	0	0
Italy E (04) C	17	2	2	3	5	3	2	0
Japan C	11	0	1.5	2.5	5	2	0	0
Mexico C	9	0	1.5	2.5	3	2	0	0
North Korea	0	0	0	0	0	0	0	0
Norway E (04) C	15	2	2	3	6	2	0	0
Pakistan[3] C	11	0	1.5	2.5	3	4	0	0
Portugal E (03) C	11	0.5	2	2.5	4	2	0	0
Romania E (02)	4.5	0.5	0.5	1.5	1	1	0	0
Russian Federation C	7.5	0	1.5	2.5	1.5	2	0	0
Singapore[3] C	9.5	0	1.5	2.5	1.5	4	0	0
South Africa E (00–02)	1.5	0.5	1	0	0	0	0	0
South Korea C	10	0	1.5	2.5	4	2	0	0
Spain[6] E (04) C	16	1	2	3.5	6	2.5	1	0
Sweden E (04) C	15	2	2	3	6	2	0	0
Switzerland E (04) C	14	1.5	2	3	5	2	0	0.5
Turkey C	10.5	0	1.5	2.5	4.5	2	0	0
UK[7] E (04) C	15	2	2	3	6	2	0	0
US[8] E (04) C	20.5	2	2	3.5	5	4	4	0

E = Export report with year of reporting.

C = Customs data.

Scoring system

(a) *Timeliness (2 points total, score based on national arms export reports data only):* A report has been published within the last 24 months (up to 31 January 2006) (0.5 points); information is available in a timely fashion (1 point if within 6 months of the end of the year in question, 0.5 if within a year).

(b) *Access (2 points total):* Information is: available on Internet through UN Comtrade (1 point); available in a UN language (0.5 points); free of charge (0.5 points).

(c) *Clarity (5 points total):* The reporting includes source information (1 point); small arms and light weapons distinguishable from other types of weapons (1 point); small arms and light weapons ammunition distinguishable from other types of ammunition (1 point); detailed weapons description included (1 point); reporting includes information on types of end-users (military, police, other security forces, civilians, civilian retailers) (1 point).

(d) *Comprehensiveness (6 points total):* The reporting covers: government-sourced as well as industry-sourced transactions (1 point); civilian and military small arms and light weapons (2 points); information on small arms and light weapons parts (1 point); information on small arms and light weapons ammunition (1 point); summaries of export laws and regulations, and international commitments (1 point).

(e) *Information on deliveries (4 points total):* Data disaggregated by weapons type (value of weapons shipped [1 point], quantity of weapons shipped [1 point]), and by country and weapons type (value of weapons shipped [1 point], quantity of weapons shipped [1 point]).

(f) *Information on licences granted (4 points total):* Data disaggregated by weapons type (value of weapons licensed [1 point], quantity of weapons licensed [1 point]), and by country and weapons type (value of weapons licensed [1 point], quantity of weapons licensed [1 point]).

(g) *Information on licences refused (2 points total):* Data disaggregated by weapons type (value of licence refused [0.5 points], quantity of weapons under refused licence [0.5 points]), and by country and weapons type (value of licence refused [0.5 points], quantity of weapons under refused licence [0.5 points]).

Note 1: The Barometer is based on each country's most recent arms exports that were publicly available as of March 2006 and/or on 2003 customs data from UN Comtrade.

Note 2: Under (d), (e), and (f), no points are granted for number of shipments or number of licences granted or denied, as such figures give little information about the magnitude of the trade. The data is disaggregated by weapons type if the share of small arms and light weapons in the country's total arms trade is delineated (x per cent of the total value of the arms exports consisted of small arms and light weapons; x number of small arms and light weapons were exported in total). The data is disaggregated both by country and by weapons type if there is information on the types of weapons that are transferred to individual recipient states (x numbers/x USD worth of small arms was delivered to country y).

Note 3: Under (d), (e), and (f), 'weapons type' means broader weapons categories (that is, 'small arms' as opposed to 'armoured vehicles' or 'air-to-air missiles'), not specific weapons descriptions ('assault rifles' as opposed to 'hunting rifles').

Note 4: The fact that the Barometer is based on two sources—customs data (as reported to UN Comtrade) and national arms export reports—works to the advantage of states that publish data in both forms, since what they do not provide in one form of reporting they might provide in the other. Points achieved from each source of the two sources are added up. However, points are obviously not counted twice (e.g. if a country provides both customs data and export reports in a UN language, it gets 1 point for this under access, not more).

Note 5: The scores of the 2005 and 2006 Barometers are not comparable, due to changes in the scoring system between the two years.

1 Australia and Canada receive full score on deliveries, as they are among the few countries that provide information to UN Comtrade on numbers of small arms transferred in most categories (Australia: except parts/accessories of revolvers/pistols, shotgun barrels, and parts/accessories of sporting/hunting weapons; Canada: except parts/accessories of revolvers/pistols, shotgun barrels, parts/accessories of sporting/hunting weapons, and ammunition).

2 Belgium has not published any national arms export report since 2002 because export control was regionalized in September 2003 (for details, see Wallonia, 2004, pp. 3–12). This means that each Belgian region in principle reports separately on its arms exports, but so far practice in this respect has been sketchy at best (for further details, see Box 3.3). The score is therefore based on customs data submissions only.

3 China, Pakistan, and Singapore all receive full score on deliveries, as they are among the few countries that provide information to UN Comtrade on numbers of small arms transferred. This makes their total scores larger than would otherwise be warranted.

4 France receives full score on deliveries, although it should be stressed that deliveries of quantities (as opposed to values) are provided for a four-year period, rather than yearly (France, 2005, p. 67). France gives details of orders ['*prises de commande*'], which are defined as 'contracts signed and entered into force through a first down-payment during the year under consideration' (France, 2005, p. 54, authors' translation). Orders are not equivalent to licences, and therefore no points are given in the columns pertaining to licences granted and denied.

5 Germany provides more detailed information on licences granted and denied for main trading partners and so-called 'third countries', i.e. countries outside the circle of the EU, the North Atlantic Treaty Organisation (NATO), and NATO-equivalent countries (Australia, Japan, New Zealand, and Switzerland). It has been awarded full points on the relevant criteria nevertheless.

6 Spain makes public its report on small arms and light weapons exports to the Organization for Security and Co-operation in Europe (OSCE) as an annex to its arms export report. The report contains information both on licences granted (volumes by country and weapons type) and on actual deliveries (also volumes by country and weapons type). It covers only the OSCE states, and hence a very limited number of transactions. Spain is hence granted only part of the points on licences and deliveries. Other states make their OSCE reports public, but separately from the arms export reports. These are therefore not taken into account in the Barometer.

7 The score takes into account the UK practice of reporting on export licences on a quarterly basis.

8 The United States receives full score on deliveries, as it is among the few countries that provide information to UN Comtrade on numbers of small arms transferred in all categories except parts/accessories of revolvers/pistols, parts/accessories of sporting/hunting weapons, and sometimes small arms ammunition.

Sources: Australia (2003); Canada (2003); Czech Republic (2005); Finland (2004); France (2005); Germany (2005); Italy (2005); NISAT (2006); Norway (2005); Portugal (2005); Romania (2004); South Africa (2003); Spain (2005); Sweden (2005); Switzerland (2005); UK (2005); US (2005); UN Comtrade (2006)

UNDERSTANDING THE INTERNATIONAL ILLICIT TRADE IN SMALL ARMS: SOUTH AMERICA

How many weapons are trafficked across the globe every year?[15] What types of small arms are most commonly trafficked? Which routes are used, and by which actors are they used?

To answer questions such as these, the Small Arms Survey has started asking governments for information about customs and police seizures of small arms and light weapons. The *Small Arms Survey 2005* concluded that information of this type has been hard to come by, which might seem surprising, given the centrality of the illicit trade to intergovernmental discussions on small arms. Preliminary conclusions for Europe presented last year were that handguns were the type of small arm most commonly smuggled, and that trafficking in the European context has been mostly small-scale (a handful of weapons at most per seizure) (Small Arms Survey, 2005, p. 116). This year, this chapter asks to what extent these findings hold true in another, quite different region of the world: South America.[16]

Unlike most of the European and Western industrialized countries analyzed in the *Small Arms Survey 2005,* South America is composed of weak states whose governments have to cope (to different extents and at various levels) with organized crime; urban violence; institutional corruption; and long, porous, and poorly patrolled borders and/or coastlines.[17] One country in the region is experiencing an ongoing internal armed conflict: Colombia. South America is also the core of the illicit cocaine industry. Cocaine trafficking directly affects three Andean countries (Bolivia, Colombia, and Peru) and indirectly causes violent crime problems in the rest of the region, such as the presence of armed criminal groups in the slums (*favelas*) of Rio de Janeiro that are able to challenge police forces, for example (Dreyfus, 2002). These are all conditions that either provoke or facilitate illicit arms transfers to or across South American countries. South America hosts important small arms producers and exporters, namely Brazil (a major world exporter), Argentina (a relevant exporter at the regional level), and Chile and Colombia (significant producers at the regional level). This contributes to a regional dynamic of licit and illicit small arms transfers, particularly between countries in the south of the region (Dreyfus et al., 2003; Small Arms Survey, 2003, pp. 87–88).

Argentina, Brazil, Chile, and Colombia are the largest small arms producers in South America.

Concerning small arms-related problems in general and illicit small arms transfers in particular, the region is not uniform, however. Brazil and Colombia, for example, have had serious problems with armed groups that exert effective territorial control (criminals in the case of Brazil and insurgent, paramilitary, and criminal groups in the case of Colombia) in either rural (Colombia) or urban (Colombia and Brazil) areas. Both countries at the same time rank among the top five countries globally in terms of firearms-related deaths (Phebo, 2005, p. 16). Bolivia and Peru face the challenge of organized crime in remote coca- and cocaine-producing regions of the country. Countries such as Suriname and Paraguay are used as transhipment routes for the rest of the region (Dreyfus et al., 2003; Cirino, Elizondo, and Wawro, 2004). Venezuela, Ecuador, and Argentina have reported recent cases of diversion of small arms from military and police stockpiles to insurgent or criminal groups in neighbouring countries.[18] On the other hand, countries such as Chile and Uruguay do not register serious small arms-related crime problems (Dreyfus et al., 2003).

The Small Arms Survey sent questionnaires to the National Points of Contact (established according to the *Programme*) and/or small arms control agencies of the 12 countries of the region with questions relating to the type, destination, origin, and place of seizure of small arms seized by either customs or police agencies between 2000 and 2004. Four countries (Argentina, Brazil, Chile, and Peru) replied. Colombian authorities did not directly reply to the questionnaire, so for the case of Colombia, statistics reported by the Colombian Ministry of Foreign Affairs on the implementation of the *Programme* and statistics on arms seizures provided by the Directorate of Judiciary Police (DIJIN) of the Colombian National Police were used.[19]

In the case of Argentina, it was not possible to distinguish among seized, registered, and voluntarily handed over firearms. None of the data reported by these countries actually differentiates between seizures by customs officials and those by police agencies. In the cases of Chile and Peru, the data aggregates seizures made by all law enforcement agencies. Analysis of the cases of Brazil and Colombia, however, is based on disaggregated information regarding seizures made by various government agencies at the local (in the case of Brazil) and national levels. This provided a clearer picture of the places and situations in which these weapons were seized.

As Table 3.5 shows, handguns (revolvers and pistols) dominate among the weapons seized. A closer look at the information provided by specialized agencies such as the Brazilian Federal Police (specialized in the enforcement of federal laws and the repression of interstate crimes) or the Colombian Armed Forces (focusing on counterinsurgency operations), however, reveals a higher proportion of assault weapons. This is true for both statistical information and police reports. This is not surprising: in both countries, assault weapons cannot legally be sold to civilians, with the exception of collectors in Brazil (Brazil, 2000, arts. 9–11) and in very limited, specific, and highly controlled cases in Colombia (COLOMBIA).

Brazil[20]

The Brazilian Federal Police provided a detailed description of the weapons most commonly seized at Brazilian airports, ports, and border controls. Since 1997, Brazil's federal legislation on firearms control has required local police to enter data on seizures into the Federal Police arms database, Sistema Nacional de Armas (Brazil, 1997a, art. 1; 1997b, art. 38). This increased transparency was reinforced by a new firearms control law passed in December 2003 (Brazil, 2003, art. 2; 2004, art. 1). Although the Brazilian government has been implementing the new law, there remains a lack of centralized and systematized data available at the national level (Dreyfus and de Sousa Nascimento, 2005, pp. 126–36).[21] Based on the analysis of information provided by local police, the NGO Viva Rio and its associated research institute Instituto de Estudos da Religião (ISER) have provided additional information on the case of Rio de Janeiro, where there is a serious problem of territorial control by criminal organizations.

In the case of Rio de Janeiro, more than 80 per cent of small arms are Brazilian made. This is not surprising: Brazil is a large producer of small arms and recent research has demonstrated that many weapons that are legally exported by Brazil to neighbouring countries are later smuggled back into Brazil (Dreyfus et al., 2003, p. 34; Delegacia Legal, 2005, pp. 17–18).

Assault weapons are procured in Brazil by illegal non-state armed groups through theft, by diversion from official stockpiles, or via transfers from illicit international markets. According to Brazilian Federal Police sources, assault weapons are trafficked into Brazil primarily via either the ant trade or through a 'pipette' method, involving the disassembly of weapons prior to their shipment in individual part lots. Main identified points of entry for these weapons are the Triple Border Area between Paraguay, Argentina, and Brazil; Santa Ana do Livramento on the border with Uruguay; and Uruguaiana on the Argentinian border.[22] Law enforcement sources from Rio de Janeiro state, however, that in the specific case of that city, the illicit market for assault weapons has essentially been saturated during the 1990s.[23] This is confirmed by the fact that the local police are no longer seizing new weapons, be they of Brazilian or of foreign origin. The illicit market in Rio de Janeiro has been concentrated more recently on the procurement of ammunition for those weapons already acquired (Dreyfus, 2006). Tougher legislation and stricter law enforcement controls and the dismantling of criminal rings specializing in arms trafficking during the mid- to late 1990s also contributed to a decrease in illicit arms flows towards Rio de Janeiro (Dreyfus, 2006).

Most weapons seized in South America are handguns: revolvers and pistols.

Colombia

In the case of Colombia, evidence suggests that large shipments of small arms arrived in the country through trans-national illicit channels, with deliveries to either the Auto Defensas Unidas de Colombia (AUC) or the Fuerzas Armadas Revolucionarias de Colombia (FARC) insurgents. These shipments were part of the trafficking of 10,000 AK-47 assault rifles through Jordan and Peru that took place in 1999, with FARC as the final recipient (COLOMBIA). Additional AK-47s were sent to the AUC in 2001 via Nicaragua and Panama (Small Arms Survey, 2003, p. 87). In addition, ammunition and weapons are diverted from armed groups and security forces of neighbouring countries, particularly Ecuador and Venezuela (COLOMBIA). As Table 3.5 shows, the share of military firearms (assault rifles, machine guns, and sub-machine guns) is higher in the case of seizures made by the Colombian Army and Navy, whose focus is on counterinsurgency operations in rural areas. This is also the case for the Colombian DAS, which specifically targets criminal organizations and illegal armed groups. In the case of the Colombian National Police, who carry out regular policing in urban and rural areas, the number of seized handguns (revolvers and pistols, which are generally used by common criminals or illegally carried by civilians) is overwhelmingly higher than the quantity of seized military firearms.

Chile

When compared to the rest of the region, Chile does not have significant problems related to small arms. The arms market for civilian use is very limited and depends largely on imports. Chile does not share borders with countries that report heavy arms trafficking, such as Paraguay, and its borders are hard to reach and generally well-patrolled. Furthermore, the security forces, especially the Carabineros, have a good reputation in terms of low levels of institutional corruption, which helps to restrain cross-border trafficking (Dreyfus et al., 2003, pp. 39–43). Where it is possible to see a growing increase in the use of small arms is in crime related to the trafficking and sale of illicit drugs. In the last decade, Chile has increasingly been used as a cocaine trafficking route from the Andean countries (primarily Peru and Bolivia), and there has also been a recorded increase in the use of and trafficking in coca paste in large cities. Coca paste distribution occurs mainly in the poor neighbourhoods of the southern metropolitan area (Santiago and Greater Santiago) and in other cities. It is precisely in police operations aimed at suppressing this type of crime in urban areas where more and more firearms are being confiscated (Dreyfus et al., 2003, pp. 39–43).

According to Chilean small arms control authorities, most weapons seized were manufactured in Argentina, Brazil, the United States, Spain, and Italy. The majority are older models.[24] Combined with the fact that 30 per cent of weapons seized in Chile have had previous legal registration, this suggests that in Chile, theft and illicit domestic sales are a greater source for illicit weapons acquisition than international illicit transfers.[25] Furthermore, the production of craft weapons (*hechizas*) remains an important source for robbers and urban youth gang members (Small Arms Survey, 2003, pp. 28–29.)

An incident that received particular media attention in recent years was the diverting of confiscated weapons and small arms to drug trafficking bands. In late 2002 retired members of the military diverted AK-47 rifles captured in the 1980s during an operation against the Manuel Rodríguez Patriotic Front (whose armed branch was active until the mid-1990s) from the storage facilities of state security forces. Those concerned were later tried and removed from their positions (Dreyfus et al., 2003, p. 43). In Valparaíso in late September 2005, Carabineros dismantled a ring of non-commissioned officers and civilian personnel of the Chilean Army who were diverting small arms from military stockpiles to criminal groups (Ramírez, 2005, p. 3). It remains to be seen whether this was an isolated occurrence or whether such cases will come up again in Chile.

Compared to its neighbours, Chile does not have significant small arms-related problems.

Peru

With its insurgent movements defeated and demobilized in the second half of the 1990s, Peru's main small arms problems are related to drug trafficking activities in jungle areas in the eastern and north-eastern parts of the country and common criminality in its big cities. Like Chile, Peru does not have any significant production of commercially traded small arms. Both its licit and illicit markets depend on imports. According to reports by the Peruvian government, small arms trafficking (in small quantities) is concentrated in the northern borders of the country (Peru, 2005, p. 16).

Table 3.4 Reported seizures of small arms in selected South American countries, 2000–04

Country/Agency concerned	2000	2001	2002	2003	2004	Total
Brazil (Civilian Police of the State of Rio de Janeiro)	10,789	10,973	18,232	20,052	5,315	65,361
Colombia (DIJIN–National Police)	42,355	44,656	47,353	47,837	63,037	245,238
Colombian Army	n/a	n/a	n/a	5,924	8,214	14,138
Colombian Navy	n/a	n/a	n/a	680	673	1,353
Colombia (DAS)	n/a	n/a	n/a	315	521	836
Chile	4,909	n/a	2,192	2,537	2,592	12,230
Peru	n/a	c. 2,500	c. 1,500	c. 3,500	c. 2,500	c. 10,000

Table 3.5 Reported seizures of small arms in selected South American countries by weapons type, 2000–04

Country/Agency concerned	Revolvers	Pistols	Rifles and carbines	Shot-guns	Military and assault rifles	Sub-machine guns	Machine guns	Other	Total
Brazil (Civilian Police of the State of Rio de Janeiro)	38,174	13,153	986	8,237	1,300	392	16	3,103	65,361
Colombia (DIJIN–National Police)	118,132	46,302	467	30,268	1,507	n/a	1,209	n/a	197,885
Colombian Army	3,067	2,670	373	2,088	5,263	n/a	308	369	14,138
Colombian Navy	241	266	5	342	484	n/a	15	n/a	1,353
Colombia (DAS)	143	151	11	124	92	n/a	16	299	836
Chile	Handguns: 8,405; long-barrel firearms: 3,389; craft firearms: 1,349. Data on handguns and long-barrel firearms includes 922 small arms that were voluntarily handed over.								
Peru	The information was not digitalized or presented in table format, so it was interpreted from graphs. Most of the seized small arms were revolvers and pistols, followed by craft firearms and shotguns.								

Notes on Brazil (Civilian Police of the State of Rio de Janeiro):

The total quantity of small arms seized in the State of Rio de Janeiro between 2000 and 2004 was actually 66,057; however, 5770 small arms were excluded because their records were not accurate enough to allow them to be classified by type, make, and country. Of the weapons seized within the period, 82% were made in Brazil, predominantly Taurus (43%) and Rossi (24%) revolvers and pistols. Among the foreign-made weapons, 41%

were made in the United States; 13% in Spain; 11% in Argentina; 11% in Belgium; 6% in Germany; 6% in Italy; 4% in Austria; and 8% in other countries. Among the 1,209 seized assault and military rifles, 46% were manufactured in the United States; 21% in Brazil; 11% in Belgium; 10% in Germany; 5% in China; 4% in the Russian Federation; 2% in Argentina; and 1% in other countries. Percentages were calculated by Viva Rio/ISER from a sample of 59,596 weapons that had completed information concerning make, manufacturer and caliber.

Notes on Colombia (DIJIN-National Police):

Information for 2002 is not disaggregated by weapons type, and thus it was not included for the calculation of quantities of seizures by weapons type.

Notes on Colombian Army, Colombian Navy, and Colombia (DAS):

Information was provided only for 2003 and 2004.

Note on all Colombia figures:

The category 'machine guns' includes 'sub-machine guns'.

Sources: Brazil: Civilian Police of Rio de Janeiro, processed and analyzed by Viva Rio/ISER; **Colombia:** Colombia (2005); data from the National Colombian Police, processed by CERAC; Organization of American States, Inter-American Drug Abuse Control Commission (CICAD), 2003–2004 report of the Government of Colombia to the Multilateral Evaluation Mechanism of CICAD; **Chile:** National Directorate of General Mobilization; **Peru:** Peru (2005)

Box 3.4 AKs for Venezuela: reasons for concern

On 18 May 2005 the Venezuelan Ministry of Defence signed a contract with the Russian Federation company Rosoboronexport for the delivery of 100,000 7.62 mm AK-103 assault rifles at a reported price of USD 386 per rifle and the compatible 7.62 x 39 mm ammunition for them. The delivery of the first batch of 28,000 rifles produced by Izhmash, which has to export combat small arms and light weapons through Rosoboronexport, was scheduled for October 2005, while a second batch of 35,000 AK-103s was to be delivered in December 2005. The contract will be completed after the last 37,000 rifles are shipped in March 2006. After that Venezuela will become the first country in the world to launch the licensed manufacturing of AK-103s and ammunition. The deal provoked negative reactions from the United States and Venezuela's neighbour Colombia. Both expressed concerns that these small arms could be diverted into the hands of local insurgents or terrorist organizations (Pyadushkin, 2005).

There is, however, a more worrisome aspect to this purchase and subsequent licensing that is not directly related to the new rifles, but to the old FALs they are going to replace. As mentioned above, there is evidence of the diversion of small arms from the Venezuelan armed forces to insurgent groups in Colombia (Schroeder, 2004, pp. 22–23; Cragin and Hoffman, 2003, pp. 26–32). With the purchase of new rifles to completely equip and train its armed forces and reserves, Venezuela will have a huge surplus of its previous standard assault rifle, the FAL (PRODUCERS).

Another disquieting issue is that CAVIM, the state's arms factory, will start producing 7.62 x 39 mm ammunition, which is exactly the kind of ammunition the FARC in Colombia is currently in desperate need of for 10,000 AK-47s they acquired in 1999 (COLOMBIA).

With plans to purchase large numbers of guns from the Russian Federation, Venezuelan President Hugo Chavez speaks to Russian President Vladimir Putin during a meeting in Moscow in November 2004.
© Sergei Karpukhin/Reuters

CONCLUSION

This chapter, together with those of previous editions of the *Small Arms Survey,* shows that the scale and patterns of the authorized trade in small arms and light weapons and their ammunition, parts, and accessories have been rather stable over the past few years. The trade has been estimated at around USD 4 billion for the last half-decade; the top exporters and—albeit to a lesser degree—the top importers of small arms have largely remained the same. Trading patterns have been stable: e.g. the share of ammunition in the overall small arms trade has remained more or less constant.

It would be wrong, however, to conclude that everything is known about the trade in small arms and light weapons that there is to know. The Small Arms Survey estimates that roughly half of the small arms trade remains unaccounted for. This figure includes exports and imports widely regarded as among the most problematic: trade in certain types of light weapons such as MANPADS and mortars, and (to a lesser degree) other military small arms; trade in light weapons ammunition; the exports of countries such as China, Iran, North Korea, and Pakistan; and other issues.

Although the Small Arms Trade Transparency Barometer is meant to encourage greater transparency, there are generally strong reasons why many non-Western states choose to keep their exports (and imports) classified. Their goods are generally not competitive on Western markets and on some of the markets where Western-made guns are readily available. The incentives for such countries to turn to other markets, some of which would earn them strong and sustained criticism if publicly known, are therefore greater than for many Western states.

Every year, international intergovernmental organizations publish estimates of the scope of the illicit trade in drugs, for example. So far, no such collective intergovernmental effort has been made for guns. The difficulty that a non-governmental research body such as the Small Arms Survey has had to date in obtaining detailed high-quality information in this area may be an indication that a public–private partnership would be needed in this regard.

Lastly, findings for South America indicate that the harmonization of domestic small arms control laws is as important as intelligence, tracing, and other international control measures, as legal flaws in neighbouring countries are used by criminals in order to purchase and smuggle small arms across national borders. This is particularly important in the case of small-calibre handguns, which constitute the predominant group of weapons seized by the police, and are weapons that can be purchased by civilians in gun shops. Although there is evidence of illicit transfers of military firearms (assault rifles and sub-machine guns) from different parts of the world to crime settings and conflict areas in the region, the importance of the diversion of this kind of weapon from police and military stockpiles should not be neglected. Improving surplus disposal practices and stockpile security measures are important steps to prevent the illicit transfers of military firearms. ⬛

LIST OF ABBREVIATIONS

AUC	Auto Defensas Unidas de Colombia	**FARC**	Fuerzas Armadas Revolucionarias de Colombia
Barometer	Small Arms Trade Transparency Barometer		
CERAC	Centro de Recursos para el Análisis de Conflictos (Conflict Analysis Resource Centre)	**HS**	Harmonized System
		ISER	Instituto de Estudos da Religião (Brazil)
		MANPADS	man-portable air defence system(s)

CHF	Swiss franc	ML	Munitions List (of the Wassenaar Arrangement if not otherwise stated)
DARM	Division of Repression of Small Arms Illicit Trafficking (Brazil)	NISAT	Norwegian Initiative on Small Arms Transfers
DAS	Departamento Administrativo de Seguridad (Administrative Security Bureau, Colombia)	*Programme*	*UN Programme of Action*
		SEK	Swedish krona
DIJIN	Directorate of Judiciary Police (Colombia)	SIPRI	Stockholm International Peace Research Institute
EU	European Union		
EUR	euro	UN Register	UN Register of Conventional Arms
		USD	United States dollar

ENDNOTES

1   Data is presented on actual deliveries of small arms rather than licences granted.

2   For details on the issue of mixed ammunition categories, see Glatz (2006).

3   For detailed information on UN Comtrade customs data, see Small Arms Survey (2005, Box 4.1, pp. 99–100). Further information is provided on the UN Comtrade and International Trade Centre Web sites: <http://unstats.un.org/unsd/comtrade/> and <http://www.intracen.org/index.htm>

4   For more information on this reliability measure, which essentially is based on comparing how well each country's records correspond to the mirror reporting of its trading partners, see Marsh (2005) and ITC (2005).

5   Correspondence from the Republic of Sudan to the Embassy of Switzerland in Khartoum, 5 March 2006.

6   In UN Comtrade, China does not rank among the top exporters for either 1999–2003 or 2003 alone, but in 1998 it reported a total export value of USD 36,244,000 in category 930630 (UN Comtrade, 2006). Likely, China changed its reporting to UN Comtrade rather than its actual exports. It can thus be assumed that it ranks close to or among the top exporters of small arms ammunition also for the period 1999–2003.

7   For details on the issue of mixed ammunition categories in UN Comtrade, see Glatz (2006).

8   Dreyfus, Lessing, and Purcena (2005, pp. 117–19) present examples of rocket-propelled artillery saturation systems exported by Brazil to Malaysia, Qatar, and Saudi Arabia being declared as 'sporting and hunting shotguns'.

9   E-mail correspondence with Stefan Raykov, Ministry of Economy, Bulgaria, 26 September 2005.

10   See Small Arms Survey (2003, p. 105; 2004, pp. 109–11, Table 4.3; 2005, pp. 106–09, Table 4.2). Cyprus presents something of an enigma. The value of its imports has been too large to be explained by either local demand or the needs of a relatively small international peacekeeping force. Therefore it seems that Cyprus is the hub of a transit trade about which little is known. However, Cyprus publishes very little information about its exports of military small arms and light weapons, which actually account for the bulk of its imports. Unlike many of the other European Union (EU) member states, it does not publish a national arms export report. Moreover, unlike all other EU member states except Lithuania and Luxemburg, it submitted no information on any arms exports to the 2004 EU report on the implementation of the *EU Code of Conduct on Arms Exports* (EU, 2004, p. 5, fn. 1). To date, Cyprus has not submitted any report on its implementation of the *UN Programme of Action* (*Programme*); neither has it established a point of contact, which is usually considered a first step in implementing the *Programme*.

11   See also Small Arms Survey (2005, CONFLICT SOURCING).

12   For an evaluation of the reporting in 2003, see Kytömäki and Yankey-Wayne (2004). For an overview of information provided in national reports in 2002–2005, see Kytömäki and Yankey-Wayne (2006). For electronic copies of the reports themselves, see <http://disarmament2.un.org/cab/salw-nationalreports.html>.

13   See Small Arms Survey (2005, pp. 109–10) for further details.

14   For a detailed commentary on the reports, see Bauer and Bromley (2004).

15   For the scope of this section, the terms 'illicit international transfers', 'illicit trafficking', and 'trafficking' are used synonymously.

16   South America includes the following countries: Argentina, Bolivia, Brazil, Chile, Colombia, Ecuador, Guyana, Paraguay, Peru, Suriname, Uruguay, and Venezuela.

17   The countries analysed last year were Australia, Canada, Germany, Poland, Romania, Sweden, and the United Kingdom. With the exception of Poland and Romania, which are countries in transition, these countries can be defined as 'strong states' in terms of their socio-political cohesion,

territorial centrality, policy capacity, and socio-economic development. This definition of 'strong' versus 'weak states' is based on Buzan (1991, pp. 96–107, 113–14).

18   See Small Arms Survey (2003, pp. 87–88; 2004, pp. 51–54), Cragin and Hoffman (2003), Cirino, Elizondo, and Wawro (2004), Schroeder (2004), Dreyfus et al. (2003), and Fleitas (2005, p. 15).

19   These statistics were provided directly to the Small Arms Survey and the Centro de Recursos para el Análisis de Conflictos (Conflict Analysis Resource Centre—CERAC).

20   Brazil is a federal state divided into 26 states plus the Federal District.

21   E-mail information sent by an official of the Division of Repression of Small Arms Illicit Trafficking (DARM) of the Brazilian Federal Police Department, January 2006.

22   E-mail information sent by an official of DARM of the Brazilian Federal Police Department, January, 2006.

23   Interview with Delegate Carlos Oliveira, chief of the Firearms Enforcement Division of the Civilian Police of Rio de Janeiro, Rio de Janeiro, August 2005.

24   E-mail communication with Coronel Marcelo Rebolledo, chief of arms and explosives of the National Directorate of General Mobilization of the Chilean Ministry of Defence, October 2005.

25   Percentage calculated from data sent by the National Directorate of General Mobilization.

BIBLIOGRAPHY

Associated Press. 2002. 'Bulgaria Donates Arms, Equipment to Aid Afghan Army.' 21 August.

——. 2005. 'NATO Organizing Shipment of Arms to Iraq.' 12 January.

Australia. 2003. *Annual Report: Exports of Defence and Strategic Goods from Australia 2001/2002.* Canberra: Department of Defence, Defence Trade Control and Compliance, Industry Division. February. <http://www.sipri.org/contents/armstrad/AUS_01-02.pdf/download>

Bauer, Sibylle and Mark Bromley. 2004. *The European Union Code of Conduct on Arms Exports: Improving the Annual Report.* SIPRI Policy Paper No. 8. Stockholm: SIPRI. November. <http://editors.sipri.se/pubs/EUCodeOfConduct.pdf>

Bosnia and Herzegovina. 2005. *Report on Arms Export Control 2005.* Sarajevo: Foreign Trade and Investment Division, Ministry of Foreign Trade and Economic Relations. February. <http://www.seesac.org/export/export_reports.htm>

Brazil. 1997a. Lei N° 9.437, de 20 de Fevereiro de 1997. Institui o Sistema Nacional de Armas—SINARM, estabelece condições para o registro e para o porte de arma de fogo, define crimes e dá outras providências. Brasília: Casa Civil, Subchefia para Assuntos Jurídicos. February. <http://www.mariz.eti.br/Lei%209437.htm>

——. 1997b. Decreto N° 2.222, de 8 de Maio de 1997. Regulamenta a Lei n° 9.437, de 20 de fevereiro de 1997, que institui o Sistema Nacional de Armas—SINARM, estabelece condições para o registro e para o porte de arma de fogo, define crimes e dá outras providências. Brasília: Casa Civil, Subchefia para Assuntos Jurídicos. May. <http://www.mariz.eti.br/Dec_2222.htm>

——. 2000. Portaria No. 024–DMB, de 25 de outubro 2000. Aprova as Normas que regulam as Atividades dos Colecionadores de Armas, Munição, Armamento Pesado e Viaturas Militares. Brasilia: Defense Ministry, Brazilian Armed Forces, Departamento de Material Bélico (Dir G de MB/1952). October.

——. 2003. Lei N° 10.826, de 22 de Dezembro de 2003. Dispõe sobre registro, posse e comercialização de armas de fogo e munição, sobre o Sistema Nacional de Armas—SINARM, define crimes e dá outras providências. Brasília: Casa Civil, Subchefia para Assuntos Jurídicos. December. <http://www.dfpc.eb.mil.br/docs_pdf/lei_10.826_de_22_Dez_2003(Alteradas_pelas_Leis_n%BA_10.826_de_12MAI2004_e_n%BA10884_de_17JUN2000.pdf>

——. 2004. Decreto N° 5.123, de 1° de Julho de 2004. Regulamenta a Lei no 10.826, de 22 de dezembro de 2003, que dispõe sobre registro, posse e comercialização de armas de fogo e munição, sobre o Sistema Nacional de Armas—SINARM e define crimes. Brasília: Casa Civil, Subchefia para Assuntos Jurídicos. July.

Brussels (Belgium). 2005. *'Rapport du Gouvernement au Parlement concernant la loi de 1991 relative à l'importation, à l'exportation, au transit et à la lutte contre le trafic d'armes, de munitions (. . .), période du 01.09.03 au 31.12.04.'* Brussels: Secretariat des commissions. 12 July.

Buzan, Barry. 1991. *People, States, and Fear: An Agenda for International Security Studies in the Post-Cold War Era.* New York/London: Harvester Wheatsheaf, 2nd edn.

Canada. 2003. *Exports of Military Goods from Canada: Annual Report 2002.* Ottawa: Department of Foreign Affairs and International Trade/Export Controls Division of the Export and Import Controls Bureau. December.

Cirino, Julio, Silvana Elizondo, and Geoffrey Wawro. 2004. 'Latin America's Lawless Areas and Failed States.' In Paul Taylor, ed. *Latin American Security Challenges: A Collaborative Inquiry from North and South.* Naval War College Newport Papers No. 21. Newport: Naval War College, pp. 7–48.

Colombia. 2005. *Informe aplicación en Colombia del Programa de Acción de las Naciones Unidas para Prevenir, Combatir y Eliminar el Tráfico Ilícito de Armas Pequeñas y Ligeras en Todos sus Aspectos.* Reporte nacional 2004–2005. Bogotá: Ministerio de Relaciones Exteriores/Dirección de Asuntos Políticos Multilaterales/Coordinación de Desarme. 30 April. <http://disarmament2.un.org/cab/nationalreports/2005/Colombia.pdf>

Cragin, Kim and Bruce Hoffman. 2003. *Arms Trafficking and Colombia.* Santa Monica: Rand Corporation.

Czech Republic. 2004. *Export Controls in the Czech Republic in 2003: Controls of Transfers of Military Equipment; Production, Export and Import of Small Arms and Light Weapons.* Prague: Ministry of Foreign Affairs.

——. 2005. *Annual Report on Export Control of Military Equipment and Small Arms for Civilian Use in the Czech Republic in 2004.* Prague: Ministry of Foreign Affairs.

Delegacia Legal. 2005. *Fontes de abastecimento do mercado criminal de armas.* Rio de Janeiro: Governo do Estado do Rio de Janeiro/Secretaria de Segurança Pública/Polícia Civil do Estado do Rio de Janeiro. September.

Dreyfus, Pablo. 2002. *Border Spillover: Drug Trafficking and National Security in South America.* Doctoral dissertation, Graduate Institute of International Studies (IHUEI), University of Geneva.

——. 2006. 'Crime and Ammunition Procurement: The Case of Brazil.' In Stéphanie Pézard and Holger Anders, eds. *Targeting Ammunition: A Primer.* Geneva: Small Arms Survey.

——, Carolina Iootty de Paiva Dias, Benjamin Lessing, and William Godnick. 2003. *Control de armas pequeñas en el MERCOSUR.* Serie América Latina No. 3. Rio de Janeiro/London: Viva Rio/International Alert. June. <http://desarme.org/publique/media/mercosur_Spanish_24_10_03.pdf>; *Small Arms Control in MERCOSUR.* Latin America Series No. 3. Rio de Janeiro/London: Viva Rio/International Alert. June. <http://www.iansa.org/regions/samerica/mercosur_english_241003.pdf>

——, Benjamin Lessing, and Júlio César Purcena. 2005. 'A Indústria Brasileira de armas leves e de pequeno porte: Produção Legal e Comércio.' In Rubem César Fernandes, ed. *Brasil: as armas e as vítimas.* Rio de Janeiro: 7 Letras/ISER, pp. 126–96. <http://www.desarme.org/publique/media/vitimas_armas_producao_comercio.pdf>; 'The Brazilian Small Arms Industry: Legal Production and Trade.' In Rubem César Fernandes, ed. *Brazil: The Arms and the Victims.* Rio de Janeiro: 7 Letras/ISER, pp. 50–93. <http://www.vivario.org.br/publique/media/The_Brazilian_Small_Arms_Industry_Legal_production_and_Trade_By_Pablo_Dreyfus_Benjamin_Lessing_e_Julio_Cezar_Purcena.pdf>

—— and Marcelo de Sousa Nascimento. 2005. 'Posse de Armas de Fogo no Brasil: Mapeamento das armas e seus proprietários.' In Rubem César Fernandes, ed. *Brasil: as armas e as vítimas.* Rio de Janeiro: 7 Letras/ISER, pp. 94–145. <http://www.desarme.org/publique/media/vitimas_armas_posse.pdf>; 'Small Arms Holdings in Brazil: Toward a Comprehensive Mapping of Guns and Their Owners.' In Rubem César Fernandes, ed. *Brazil: The Arms and the Victims.* Rio de Janeiro: 7 Letras/ISER, pp. 94–145. <http://www.vivario.org.br/publique/media/Small_Arms_Holding_in_Brazil_Toward_a_comprehensive_mapping_of_guns_and_their_owners_By_Pablo_Dreyfus_e_Marcelo_de_Souza_Nascimento.pdf>

EU (European Union). 2004. *Sixth Annual Report according to Operative Provision 8 of the European Union Code of Conduct on Arms Exports.* Official Journal of the European Union. Notice No. 2004/C 316/01. 21 December. <http://europa.eu.int/eur-lex/lex/LexUriServ/site/en/oj/2004/c_316/c_31620041221en00010215.pdf>

——. 2005. *Seventh Annual Report according to Operative Provision 8 of the European Union Code of Conduct on Arms Exports.* Official Journal of the European Union. Notice No. 2005/C 328/01. 23 December. <http://europa.eu.int/eur-lex/lex/LexUriServ/site/en/oj/2005/c_328/c_32820051223en00010288.pdf>

Finland. 2004. *Annual Report According to the EU Code of Conduct on Arms Exports: National Report of Finland for 2003.* Helsinki: Ministry of Defence.

Flanders, Government of. n.d. *Rapport van de Vlaamse Regering: Vergunningen in-, uit- en doorvoer van wapens.* Brussels-Hoofdstad: Government of Flanders. <http://www.nisat.org/Export_Reports/Belgium%202003-2004/Report%20of%20the%20Flemish%20government%20Aug%202003%20-%20Aug%202004.pdf>

Fleitas, Diego. 2005. *Las políticas de control de armas de fuego en la Argentina durante los años 2004 y 2005.* Buenos Aires: Asociación para Políticas Públicas. November. <http://www.app.org.ar/media/LasPoliticasdeControldeArmas.pdf>

France. 2005. *Rapport au Parlement sur les exportations d'armement de la France en 2002 et 2003.* Paris: Ministry of Defence.

Germany. 2004. *Bericht der Bundesregierung über ihre Exportpolitik für konventionelle Rüstungsgüter im Jahre 2003 (Rüstungsexportbericht 2003).* Berlin: Bundestag. December.

——. 2005. *Bericht der Bundesregierung über ihre Exportpolitik für konventionelle Rüstungsgüter im Jahre 2004 (Rüstungsexportbericht 2004).* Berlin: Bundestag. January.

Glatz, Anne-Kathrin. 2006. 'Buying the Bullet: Authorized Small Arms Ammunition Transfers.' In Stéphanie Pézard and Holger Anders, eds. *Targeting Ammunition: A Primer.* Geneva: Small Arms Survey.

Honey, Peter. 2005. 'Arms Sales Kept "Secret".' *Financial Mail* (Johannesburg). 11 August.
<http://www.suntimes.co.za/zones/sundaytimesNEW/business/business1123753234.aspx>

Isbister, Roy and Elizabeth Kirkham. 2005. *An Independent Audit of the UK Government Reports on Strategic Export Controls for 2003 and the First Half of 2004*. London: Saferworld. January. <http://www.smallarmsnet.org/issues/themes/ukaudit.pdf>

Italy. 2004. *Relazione sulle operazioni autorizzate e svolte per il controllo dell'esportazione, importazione e transito dei materiali di armamento nonché dell'esportazione e del transito dei prodotti ad alta tecnologia (Anno 2003)*. Rome: Camera dei Deputati. 29 March.

——. 2005. *Relazione sulle operazioni autorizzate e svolte per il controllo dell'esportazione, importazione e transito dei materiali di armamento nonché dell'esportazione e del transito dei prodotti ad alta tecnologia (Anno 2004)*. Rome: Camera dei Deputati. 30 March.

ITC (International Trade Centre). 2005. 'Reliability of Trade Statistics: Indicators of Consistency between Trade Figures Reported by Countries and Their Corresponding Mirror Estimates, Explanatory Notes, 2003 Data.' Geneva: UNCTAD/WTO/ITC Market Analysis Section. January. <http://www.intracen.org/countries/structural05/reliability03.pdf>

Kytömäki, Elli and Valerie Yankey-Wayne. 2004. *Implementing the United Nations Programme of Action on Small Arms and Light Weapons: Analysis of the Reports Submitted by States in 2003*. Geneva: United Nations. October.

——. 2006 (forthcoming). *Five Years of Implementing the United Nations Programme of Action on Small Arms and Light Weapons: Regional Analysis of National Reports*. Geneva: United Nations.

Marsh, Nicholas. 2005. *Accounting Guns: The Methodology Used in Developing Data Tables for the Small Arms Survey*. Background paper. PRIO/NISAT. 14 November.

NATO (North Atlantic Treaty Organisation). 2005. 'Donated Arms and Equipment from the Government of Romania to Support Iraqi Security Efforts.' NATO Training Mission—Iraq. 26 January. <http://www.afsouth.nato.int/JFCN_Missions/NTM-I/Articles/NTMI_A_02_05.htm>

Netherlands. 2004. *The Netherlands Arms Export Policy in 2003*. The Hague: Ministry of Economic Affairs and Ministry of Foreign Affairs. <http://www.antenna.nl/amokmar/pdf/ArmsExports2003.pdf>

NISAT (Norwegian Initiative on Small Arms Transfers). 2006. *Calculations from the NISAT Database on Authorized Small Arms Transfers*. Unpublished background paper. Geneva: Small Arms Survey.

Norway. 2004. *Eksport av forsvarsmateriell fra Norge i 2003, eksportkontrol og internasjonalt ikke-spredningssamarbeid*. Oslo: Ministry of Foreign Affairs.

——. 2005. *Eksport av forsvarsmateriell fra Norge i 2004, eksportkontrol og internasjonalt ikke-spredningssamarbeid*. Oslo: Ministry of Foreign Affairs.

OAS (Organization of American States). 2004. *Colombia: Evaluation of Progress in Drug Control 2003–2004*. DocumentOEA/Ser.L/XIV.6.1MEM/INF.2004.Add9. Washington, DC: Inter-American Drug Abuse Control Commission (CICAD).

Peru. 2005. *Informe Nacional sobre la Implementación del Programa de Acción 2001 de las Naciones Unidas para Prevenir, Combatir y Eliminar el Tráfico Ilícito de Armas Pequeñas y Ligeras en Todos sus Aspectos del periodo 2001–2004*. Lima: Ministerio de Relaciones Exteriores. <http://disarmament2.un.org/cab/nationalreports/2005/Peru%20informe.pdf>

Pézard, Stéphanie. 2006. 'Sustaining the Conflict: Ammunition for Attack.' In Stéphanie Pézard and Holger Anders, eds. *Targeting Ammunition: A Primer*. Geneva: Small Arms Survey.

Phebo, Luciana. 2005. 'Impacto da arma de fogo na saúde da população no Brasil.' In Rubem César Fernandes, ed. *Brasil: as armas e as vítimas*. Rio de Janeiro: 7 Letras/ISER, pp. 9–36. <http://www.desarme.org/publique/media/vitimas_armas_impacto_saude.pdf>; 'The Impact of Firearms on Public Health in Brazil.' In Rubem César Fernandes, ed. *Brazil: The Arms and the Victims*. Rio de Janeiro: 7 Letras/ISER, pp. 4–25. <http://www.vivario.org.br/publique/media/The_Impact_of_Firearms_on_Public_Health_in_Brazil_By_Luciana_Phebo.pdf>

Portugal. 2005. *Anuário estatístico da defesa nacional 2003*. Lisbon: Ministry of Defence. <http://www.mdn.gov.pt/Publicacoes/anuario_estatistico_2003/anuario_2003.htm>

Pyadushkin, Maxim. 2005. *Research Report on Small Arms Exports from Russia in 2003–2004*. Unpublished background paper. Geneva: Small Arms Survey.

Ramírez, Juan Fernando. 2005. 'Armas, seguridad pública y responsabilidades.' Carta de opinion No. 5. Centro de Estudios Estratégicos. Santiago: Universidad Arcis. 17 October.

Ramon, Robert. 2004. 'Romanians Send Supplies to Equip Growing Afghan Army.' American Forces Information Service, US Department of Defense. 19 March. <http://www.defenselink.mil/news/Mar2004/n03192004_200403195.html>

Romania. 2004. *Raport privind controlulexporturilor de arme în anul 2002*. Bucharest: National Agency for Export Control.

Schroeder, Matthew. 2004. *Small Arms, Terrorism and the OAS Firearms Convention*. Occasional Paper No. 1. Washington, DC: Federation of American Scientists. March. <http://fas.org/asmp/library/OAS/FullReport.pdf>

SIPRI (Stockholm International Peace Research Institute). 2006. Data generated by the SIPRI Arms Transfers Database. January. <http://www.sipri.org>

Small Arms Survey. 2003. *Small Arms Survey 2003: Development Denied*. Oxford: Oxford University Press.

——. 2004. *Small Arms Survey 2004: Rights at Risk*. Oxford: Oxford University Press.

——. 2005. *Small Arms Survey 2005: Weapons at War*. Oxford: Oxford University Press.

Smith & Wesson Holding Corporation. 2005. 'Smith & Wesson Holding Corporation Receives Military Pistol Order: Company Secures $3.4 Million Order from U.S. Army for Afghanistan Border Patrol.' Press release. 21 September.
<http://www.forrelease.com/D20050921/law091.P1.09212005165510.15092.html>

South Africa. 2003. *South African Export Statistics for Conventional Arms 2000–2002*. Pretoria: Directorate of Conventional Arms Control.

Spain. 2004. *Estadísticas españolas de exportación de material de defensa y de doble uso en 2003*. Boletín económico de ICE n° 2827 del 13 al 19 de Diciembre de 2004. Madrid: Subdirección General de Comercio Exterior de Material de Defensa y de Doble Uso.

——. 2005. *Estadísticas españolas de exportación de material de defensa, de otro material y de productos y tecnologías de doble uso en 2004*. Boletín económico de ICE n° 2849 del 11 al 31 de Julio de 2004. Madrid: Subdirección General de Comercio Exterior de Material de Defensa y de Doble Uso. July.

Sweden. 2004. *Strategic Export Controls in 2003: Military Equipment and Dual-Use Goods*. Stockholm: Ministry of Foreign Affairs. March.

——. 2005. *Sweden's Export Control Policy in 2004: Military Equipment and Dual-Use Goods*. Stockholm: Ministry of Foreign Affairs. March.
<http://www.isp.se/documents/public/se/pdf/skr0405eng.pdf>

Switzerland. 2004. *Ausfuhr von Kriegsmaterial im Jahr 2003*. Berne: Staatssekretariat für Wirtschaft (SECO).

——. 2005. *Ausfuhr von Kriegsmaterial im Jahr 2004*. Berne: Staatssekretariat für Wirtschaft (SECO).

UK (United Kingdom). 2004. *United Kingdom Strategic Export Controls: Annual Report 2003*. London: Foreign and Commonwealth Office.

——. 2005. *Strategic Export Controls: Quarterly Report, July–September 2004*. London: Foreign and Commonwealth Office.
<http://www.fco.gov.uk/Files/kfile/strategicexportcontrolsjulsep2004.pdf>

UN (United Nations). 2006a. *United Nations Register of Conventional Arms*. Web site. <http://disarmament.un.org/UN_REGISTER.nsf>

——. 2006b. *Transparency in Armaments: United Nations Instrument for Reporting Military Expenditures*. Web site.
<http://disarmament.un.org/cab/milex.html>

UN Comtrade. Department of Economic and Social Affairs/Statistics Division. 2006. United Nations Commodity Trade Statistics database. Accessed 4 January 2006. <http://unstats.un.org/unsd/comtrade/>

US (United States). 2004. *Fiscal Year 2003 'Section 655' Report*. Washington, DC: US Department of State/US Department of Defense.

US Central Command. 2004. 'MNSTC-1 Continues Iraqi Equipment and Armament Efforts.' CENTCOM News. *The Frontline*. 4 November.
<http://www.stewart.army.mil/frontlineonline/archivedpages/FrontlineOnline11-04-04News.pdf>

Wallonia (Belgium). 2004. *Rapport au parlement wallon sur l'application de la loi du 05 août 1991, modifiée par les lois du 25 et du 26 mars 2003 relatives à l'importation, à l'exportation et au transit d'armes, de munitions et de materiel devant servir spécialement à un usage militaire, et de la technologie y afferente, du 1er septembre 2003 au 31 décembre 2003*.

Wassenaar Arrangement. 2004. *Munitions List*. <http://www.wassenaar.org/list/WA-LIST%20(04)%201%20ML.doc>

Wezeman, Siemon and Mark Bromley. 2005. 'International Arms Transfers.' In SIPRI, *SIPRI Yearbook 2005: Armaments, Disarmament and International Security*. Oxford: Oxford University Press, pp. 417–525.

ACKNOWLEDGEMENTS

Principal authors

Anna Khakee and Pablo Dreyfus with Anne-Kathrin Glatz

Contributors

Nicholas Marsh, NISAT, Júlio César Purcena, Maxim Pyadushkin, and Ruxandra Stoicescu

Iraqi National Police take note of the serial number of an automatic weapon confiscated during a massive house-to-house search in a suburb of Tikrit in November 2003.
© Roberto Schmidt/AFP/Getty Images

Connecting the Dots
THE INTERNATIONAL TRACING INSTRUMENT

4

INTRODUCTION

On the morning of 17 June 2005, the UN tracing negotiations hovered near failure. With its mandate due to expire later that evening, the Open-Ended Working Group on Tracing Illicit Small Arms and Light Weapons (OEWG)[1] lacked agreement on a wide range of outstanding issues, not least of which was whether the new instrument would be legal or political in character. Yet, by the end of the day, UN Member States had made the (often painful) compromises needed to conclude the negotiations. The UN small arms process had advanced another notch.

The *International Tracing Instrument* (UNGA, 2005b), adopted by the UN General Assembly on 8 December 2005 (UNGA, 2005d),[2] represents a modest, but important, step forward in global efforts to address the small arms problem. With limited exceptions, the new *Instrument* consolidates and reinforces key international standards in the areas of marking and record-keeping. In the area of tracing cooperation and, to some extent, implementation, it goes well beyond existing norms. This chapter explores the contents of the *Tracing Instrument* in some detail, highlighting its various strengths and weaknesses. The chapter's major conclusions include the following:

- The failure of UN states to agree on the *mandatory* marking of small arms and light weapons at the time of import is a key weakness of the new *Tracing Instrument*. This is one area where international assistance will be especially important.
- The *Tracing Instrument*'s conflict-tracing functions remain somewhat underdeveloped. Nevertheless, states can use its provisions to extend tracing cooperation to peacekeeping operations if they so wish.
- Following the decision to exclude ammunition from the tracing negotiations, there is a risk that small arms and light weapons will again be separated from their ammunition, even though, in future, there may be no technical justification for doing so.
- In principle, the *International Tracing Instrument* will advance international cooperation in almost all of the areas it covers. The *Instrument*'s real value added will, however, depend on actual implementation as well as the extent to which it spurs further normative development.
- Many states have been slow to prepare for the *Instrument*'s adoption by the General Assembly. In the autumn of 2005, just weeks before they became (politically) bound by the *Instrument,* these countries had yet to determine how to bring national laws and practices into compliance.

The chapter's first sections outline the process leading up to the UN tracing negotiations, and recap the main developments in the negotiations themselves. The chapter then analyses the various provisions of the *Tracing Instrument,* including the OEWG recommendations on ammunition and peacekeeping. The chapter concludes with a review of initial (tentative) efforts at implementation at the national level.

THE PATH TO THE NEGOTIATIONS

Marking and tracing began moving up the list of international priorities in the late 1990s, with the recommendation of the UN Panel of Experts for a study on marking (UNGA, 1997, para. 80(l)(i)) and the adoption of standards for marking, record-keeping, and tracing in the *OAS Firearms Convention* (OAS, 1997). By the end of the decade, governments and civil society had homed in on these issues.

In early 2000, the governments of France and Switzerland launched an initiative to promote the development of new measures for the marking and tracing of small arms and light weapons (France and Switzerland, 2000; 2001a; 2001b). That same year, the Brussels-based research NGO Groupe de recherche et d'information sur la paix et la sécurité (GRIP) released a comprehensive study of the tracing issue, which also argued for such measures (Berkol, 2000). Pro-gun groups, in particular the World Forum on the Future of Sport Shooting Activities (WFSA), likewise weighed in with their views, especially on firearms marking (WFSA, 2000).

The Franco-Swiss initiative and various inputs from civil society fed into the preparatory process for the 2001 UN Small Arms Conference. While there was considerable support at the Conference for recommending that negotiations start on a *legally binding* international tracing instrument, the final *Programme of Action* merely recommended that

The Governmental Experts left ammunition with only a tenuous hold on the future negotiations.

the UN study 'the feasibility of developing' an international tracing instrument. There was no indication that it should be legally binding (UNGA, 2001d, sec. IV.1.c). The politically binding *Programme of Action,* like the legally binding *UN Firearms Protocol* adopted two months earlier (UNGA, 2001b), contained a number of commitments in the areas of marking, record-keeping, and tracing. Such provisions have also featured in many small arms measures adopted at the regional level over the past decade. Yet none of these initiatives met the need for a single, global instrument that would set out comprehensive, detailed, and authoritative standards in the areas of marking, record-keeping, and tracing.

The next step in the UN tracing process was the feasibility study recommended in the *Programme of Action.* The resulting Group of Governmental Experts on Tracing Illicit Small Arms and Light Weapons (GGE) met three times, between July 2002 and June 2003, before issuing its report in July 2003 (UNGA, 2003a). Civil society inputs were 'important' here too, as the GGE itself acknowledged (p. 8). At the GGE's first and second sessions, the Small Arms Survey and the United Nations Institute for Disarmament Research (UNIDIR) presented the findings of a technical study on the scope and implications of an international tracing mechanism (Small Arms Survey and UNIDIR, 2003). GRIP, the Quaker United Nations Office (Geneva), and the WFSA also made presentations to the Group in the course of its work.

In its July 2003 report, the GGE unanimously concluded that it was both 'desirable' and 'feasible' to develop an international tracing instrument. More specifically, the Group recommended that the General Assembly decide at its 58th session to negotiate such an instrument under UN auspices. Yet, as some members of the Group would not support a recommendation for the negotiation of a legal text, the GGE left open the crucial question of instrument character—merely noting that this 'will be determined in the course of negotiations' (UNGA, 2003a, paras. 96–98).

Another contentious issue was that of definitions, especially whether to include ammunition and explosives within the scope of the study. As explained below in the section dealing with ammunition, the GGE kept ammunition (and explosives) on the table by referring to them in the definitions section of its report and noting that they were 'part of' the small arms problem (para. 33). Yet, by failing to provide any guidance on the technical issues specific to ammunition, the GGE left it with only a tenuous hold on the future negotiations. The GGE report contained much language that the OEWG would later use. Especially useful were the provisions articulating 'common minimum

Box 4.1 Civil society and the UN tracing process

Civil society is often seen as a key partner for states in small arms work. While the UN tracing process illustrated the continuing importance of this partnership, it also underlined its limits.

Civil society helped get the tracing issue onto the international agenda in the late 1990s. Initially, a relatively small group of specialists participated in the emerging policy debate, but, as governments singled out tracing for priority attention, NGO interest expanded. The period preceding the UN tracing negotiations saw relatively close collaboration between civil society and governments. In the context of the Franco-Swiss initiative, states actively engaged with civil society in discussing and developing potential tracing standards. The Belgian government also funded important work by GRIP on such standards. As the UN tracing process got under way, civil society remained firmly in the picture. Civil society experts were invited to contribute to meetings convened within the framework of the *UN Programme of Action*, including those that paved the way for the UN tracing negotiations (see text on the GGE).

Pro-control NGOs welcomed the start of the UN tracing negotiations, in June 2004, hoping that these would result in agreement on a legally binding, comprehensive instrument (Anders, 2005b). During the process, these NGOs organized their advocacy efforts under the twin banners of the International Action Network on Small Arms (IANSA) and the Control Arms campaign.[3] The pro-gun groups, meanwhile, coordinated their activities through the WFSA. NGO participation in the tracing negotiations was, however, soon curtailed. After meeting for four days in open session (14–17 June), the OEWG decided to close all of its remaining meetings to NGOs, notwithstanding IANSA's request to allow it and the WFSA to continue to attend as observers.

Both IANSA and the WFSA presented their recommendations on the instrument to OEWG delegates on 17 June, just before the meetings were closed. After June 17, they lobbied delegates outside the meeting room. Throughout the negotiations, pro-control and pro-gun groups distributed position papers and other reports to OEWG participants. Informal lunchtime briefings were also organized at the Working Group's first and second sessions. In a handful of cases, civil society experts, reflecting industry or pro-control perspectives, formed part of national delegations.

There was considerable overlap in the recommendations of the pro-control NGOs and the positions taken by many states during the negotiations. The pro-gun influence was also evident in some cases. The end result, however, seemed quite independent of civil society influence. In particular, the *International Tracing Instrument* failed to satisfy pro-control demands for a legal instrument that included ammunition. The Control Arms campaign also criticized the inclusion of a 'national security' exception in relation to tracing cooperation (Amnesty International, IANSA, and Oxfam International, 2005). Both pro-control and pro-gun groups were also disappointed that they had been denied access to the negotiations.

The UN tracing negotiations left a bitter taste in the mouths of many civil society groups. Having raised the profile of the tracing issue and contributed to its development, their impact on the final *Instrument* appeared minimal. Yet civil society's most important contribution to the tracing process may well lie ahead—in promoting the *Instrument's* effective implementation and encouraging its normative enhancement (see chapter conclusion).

Author: Glenn McDonald

Source: Anders (2005b)

standards' for marking, record-keeping, and cooperation in tracing. Nevertheless, the GGE's less than constructive ambiguity on ammunition, along with its silence on the question of instrument character, would haunt the OEWG for the duration of its mandate.

In December 2003, pursuant to the GGE's recommendation, the UN General Assembly decided 'to establish an open-ended working group . . . to negotiate an international instrument to enable States to identify and trace, in a timely and reliable manner, illicit small arms and light weapons'. The General Assembly did not specify whether ammunition or explosives were in or out of the negotiations. Nor did it take any decision on the question of instrument character. Using the GGE's own language, the General Assembly simply noted that this 'will be determined in the course of negotiations' (UNGA, 2003b, para. 5).

NEGOTIATING THE INSTRUMENT[4]

The OEWG was given three sessions of two weeks each to negotiate the *International Tracing Instrument* (six weeks total). Despite the important basis furnished by the GGE report and the *UN Firearms Protocol*, this was not much time considering the wide range of issues the OEWG needed to resolve—some of which were highly controversial.

On 3 February 2004 the Working Group met to finalize the organizational aspects of its work. Among other things, it elected Ambassador Anton Thalmann of Switzerland as its chairman, set the dates for its three substantive sessions, and decided it would apply to its own work the Rules of Procedure that had been used at the 2001 UN Small Arms Conference (UNGA, 2005a, paras. 3–8; 2001c). In advance of the OEWG's first substantive session, the chairman distributed a non-paper designed to serve as 'a starting point' for the tracing negotiations. Using the GGE report as a 'key reference', the non-paper highlighted common practices and understandings in the areas of marking, record-keeping, and cooperation in tracing. It also identified issues of importance to the negotiations that were controversial or that the GGE had not dealt with (Switzerland, 2004a).

The devil in the details

On 14 June 2004 the Open-ended Working Group began its substantive work. The first substantive session, which ended on 25 June, was not, formally speaking, a negotiating session as no draft text was on the table. States instead exchanged views on the broad range of issues central to the future tracing instrument. The first meetings of the session were devoted to national statements of a general nature. A discussion of instrument character, on Wednesday 16 June, highlighted the sharp division among states on the crucial question of whether the instrument should be legally or politically binding. Representatives of intergovernmental organizations and civil society subsequently took the floor to present their concerns and recommendations.

The last half of the session was taken up with thematic discussions on the topics of marking, record-keeping, and cooperation in tracing. In order to focus the proceedings on the nuts and bolts of the future instrument, the chairman prepared questionnaires for each thematic area. In responding to the chairman's questions, states staked out their initial positions on the content of the instrument and provided the chairman with valuable raw material for the first draft, which the Group agreed he would prepare and distribute in advance of its second substantive session.

States were sharply divided as to whether the instrument should be legally or politically binding.

During the period bridging the first and second substantive sessions, the chairman consulted key stakeholders and partners in order to gain additional inputs on the contents of his first draft instrument. At a meeting at the headquarters of the International Criminal Police Organization (Interpol) in Lyon, France, in early September 2004, the chairman and Interpol officials discussed ways that Interpol could support the implementation of the instrument. In mid-October 2004, Ambassador Thalmann held informal consultations with UN Member States, focusing on several areas where consensus appeared especially elusive. The chairman produced his first draft instrument in December 2004 and the UN Secretariat then distributed it to Member States (Switzerland, 2004b).

The OEWG began wrestling with the specific language of the instrument at its second substantive session (24 January–4 February 2005). Delegates exchanged views on the chairman's draft text during the first part of the session. They also debated the question of instrument character, though without moving any closer to consensus. Drawing upon the OEWG's discussion of the first draft and the written proposals he had received for new language, the chairman presented the Group with a revised draft text during the second week of the session (Switzerland, 2005a). The second session's final meetings were devoted to a consideration of this revised draft. This discussion was to feed into the chairman's third draft text, which he undertook to produce well in advance of the third session.

The chairman's third draft text, based on the inputs he had received from states at the second session, was distributed to UN Member States in mid-April (Switzerland, 2005b). As in previous drafts, Ambassador Thalmann was forced to leave the question of instrument character open. The draft was worded so as to allow, as far as possible, for both legal and political options. In line with his earlier versions, the chairman's third draft text also included provisions relating to small arms and light weapons ammunition. This was the second issue, after instrument character, that had split the OEWG, with some states claiming that ammunition was not part of the Group's mandate and others insisting on its importance to the future instrument.

In general, the chairman's third draft sought to move the Working Group closer to consensus on many substantive issues that remained unresolved. Nonetheless, at the opening of the third substantive session, the chairman indicated—not for the first time—that his draft preserved existing standards in the *UN Firearms Protocol* (UNGA, 2001b), the *UN Programme of Action* (UNGA, 2001d), and the report of the Group of Governmental Experts (UNGA, 2003a).

Make or break

At its third and final session, lasting from 6–17 June 2005, the OEWG continued to negotiate on the basis of the chairman's draft texts. It did not use square brackets or rolling text. Throughout the third session, the Working Group conducted successive readings of the chairman's third draft, adopting paragraphs *ad referendum*[5] whenever the necessary consensus had crystallized. At the end of the first week and the beginning of the second, Ambassador Thalmann asked four delegations to act as facilitators on the most contentious issues:

The role of peacekeeping operations in the instrument became another bone of contention.

- Belgium, marking at the time of import;
- Brazil, definition of small arms and light weapons;
- India, instrument character; and
- South Africa, ammunition.

On the many other issues that could not be agreed during the first week, the chairman asked interested states to work together to develop consensus language on the paragraph at issue. In some cases, the chairman issued his own proposals on such provisions. The chairman structured the process further in the second week, asking specific states to act as 'focal points' on outstanding paragraphs. Here, too, the aim was to develop consensus language for speedy adoption by the Working Group as a whole. During the session, the UN Department for Disarmament Affairs (DDA) circulated a daily list of the paragraphs that had been adopted *ad referendum* to that point.

The facilitators reported on the results of their consultations on Thursday 16 June. While agreement on a definition of small arms and light weapons appeared within reach, states remained far apart on the three other facilitation issues, namely, import marking, ammunition, and instrument character. By this time, the role of peacekeeping operations in the instrument had also emerged as a bone of contention. On Thursday evening, the chairman issued his *Package Proposal* covering all outstanding issues (Switzerland, 2005c). He did not, however, make any proposal on instrument character in light of the sharp divisions that persisted within the Group on this question. On other issues, including that of ammunition, the chairman sought to find compromises that would quickly win the acceptance of the Group as a whole. The package drew heavily on the compromise language developed to that point by the facilitators and focal points.

The Group began its last day of negotiations on Friday 17 June with a discussion of instrument character. Positions remained entrenched, however, with most states continuing to support a legally binding instrument, one state indicating it could not accept a legal instrument, others emphasizing the need for consensus (in essence, on a political instrument), and a few states, for the first time, proposing that the Group leave the question to the General Assembly.

When the OEWG began its last meeting on the afternoon of 17 June, failure seemed a distinct possibility, with much of the future instrument's content—as well as its character—still unresolved. Yet the final hours of negotiations proved decisive. Shortly before the chairman suspended the meeting at the end of the afternoon, the Group reached agreement on most elements of his package. In the course of the evening, the OEWG ironed out its differences on import marking, peacekeeping operations, and ammunition. With time running out, late that evening the Group adopted the last paragraphs of the *Instrument,* agreed it would be political in character, and adopted its report to the General Assembly—by consensus in all cases.

THE INTERNATIONAL TRACING INSTRUMENT

This part of the chapter provides a section-by-section analysis of the *International Tracing Instrument,* including the two recommendations the OEWG made to the UN General Assembly on ammunition and peacekeeping. We start with those aspects, such as character and definitions, that underpin the *Instrument* as a whole.

Instrument character

The negotiations nearly foundered on the question of instrument character. Frustration over the refusal of a few countries to accede to the wishes of the many states that favoured a legal instrument raised the possibility of a backlash that would have upended the negotiations.

The United States expressed its opposition to a legal instrument at the OEWG's first substantive session, repeating the message throughout the rest of the negotiations. Egypt, Israel, and Japan also indicated their preference for a politically binding instrument at various times, though in more hesitant, ambiguous terms. Since no vote was called on the question, it is impossible to assess the real extent of opposition to a legal text within the OEWG. Yet the declared supporters of a legal instrument vastly outnumbered the opponents. They included the EU and ten associated states,[6] the countries of Latin America and the Caribbean, and the countries of sub-Saharan Africa.

Those states that preferred a political text to a legal one gave the following reasons:

- There was insufficient time to negotiate a legal text.
- A legal instrument would enter into force only after a considerable delay (once the ratification threshold had been reached), whereas a political text would become effective as soon as it was adopted by the General Assembly.
- Some states would find it difficult to ratify a legal instrument.
- A political text, agreed by consensus, would apply to all UN Member States, including all producers of small arms and light weapons.
- Legal commitments would be more conservative than their political counterparts.
- A political instrument could be as effective as a legal one provided the commitments it contained were precisely formulated and the political will for implementation was strong.
- States did not need a legal instrument in order to change their laws.
- As the tracing instrument was being negotiated within the framework of the *UN Programme of Action,* it should, like the Programme itself, be political.
- Experience with the *UN Programme of Action* had shown that effective international cooperation did not require a legal instrument.
- A legal instrument would be difficult to revise.

Proponents of a legal instrument offered the following arguments:

- A legal text would have greater normative force.
- The instrument needed to be legal in order to be 'effective'.
- In order to ensure that states complied with their commitments, the text needed to be legal.
- In order to 'enforce' the instrument, it needed to be legal.
- Governments needed a legal instrument in order to make the necessary changes to national laws and practices.
- If it was to build upon the *UN Programme of Action,* it was important that the tracing instrument be legal; a political instrument would set a negative precedent for further *Programme* follow-up.
- A political text would also set a bad example for other UN arms control processes.
- Only a legal instrument would do justice to the small arms problem and demonstrate the international community's determination to address it.
- A political instrument would, conversely, send out a signal of weak commitment to the small arms issue.
- A legal instrument would reinforce local and regional initiatives.

These views appeared to have little, if any, influence over other delegations—at least those delegations that had a position on instrument character. Lacking consensus on a legal text, its proponents agreed to a political instrument in order to secure the substantive gains that had been made in the negotiations and—equally important—preserve the consensus-based approach that had prevailed to that point in UN small arms negotiations.

> Declared supporters of a legal instrument vastly outnumbered the opponents.

As indicated above, the question of precedent was invoked in support of both the legal and political options. Whatever influence the final (political) result may exert on other small arms processes, it is important to note that the *Instrument* itself is not cast in stone. In its last paragraph, UN Member States undertake to 'review the . . . future development of this instrument' as part of *Programme of Action* review conferences (UNGA, 2005b, para. 38). Such 'future development' could involve the transposition of some or all of the *Instrument's* political norms into legally binding form.

Ammunition

Ammunition was the second issue that sharply divided the OEWG. Once again, the United States and EU were at loggerheads, with the United States opposing the inclusion of ammunition in the draft instrument and the EU supporting this. Most other OEWG delegations were in one or the other camp, with only a handful indicating that they were flexible on the issue. Ammunition opponents argued that, as the word had not been included in the Working Group's mandate, it could not form part of the negotiations. Ammunition supporters, by contrast, interpreted the term 'small arms and light weapons' in the same mandate as implicitly covering ammunition (UNGA, 2003b, para. 8).

The agreement reached on ammunition in the final hours of the negotiations contains two elements. The first was the inclusion of a recommendation in the Group's report:

> The Working Group recommends that the issue of small arms and light weapons ammunition be addressed in a comprehensive manner as part of a separate process conducted within the framework of the United Nations. (UNGA, 2005a, para. 27)

The second part of the deal on ammunition was new preambular language: '*Stressing* that all aspects relating to illicit small arms and light weapons should be addressed in a coordinated and comprehensive manner' (UNGA, 2005b, pream. para. 8). This replaced a preambular paragraph that had been proposed at the second session and that the chairman had retained in his second and third draft texts, as well as his *Package Proposal*: 'Stressing the close association between illicit small arms and light weapons and their ammunition, and the need to deal with both issues in a coordinated manner' (Switzerland, 2005c, pream. para. 7).

Ammunition was the second issue that sharply divided the OEWG. The new preambular paragraph broadened the text from ammunition (now deleted) to 'all aspects relating to illicit small arms and light weapons' (UNGA, 2005b, pream. para. 8). Although its normative force is limited, being part of the preamble, the new provision offers additional leverage to those states seeking to ensure that future negotiating mandates incorporate all elements they consider essential to the task at hand, including ammunition. The OEWG recommendation for the creation of a separate UN process on ammunition is rather more tangible (UNGA, 2005a, para. 27), although, as of late 2005, it was unclear where it would lead.

The OEWG's handling of the ammunition question may have implications for future small arms negotiations. Some states that supported the inclusion of ammunition in the tracing negotiations noted that ammunition (and explosives) were part of the 1997 UN Panel definition of 'small arms and light weapons'. They argued that they ought therefore to apply to the tracing negotiations. In its report, the GGE left the question of ammunition's inclusion in the tracing negotiations open—at least in theory. The GGE listed those elements of the Panel definition that related to ammunition and explosives in its own definitions section, indicating that they were 'generally regarded as a part of the problem of small arms and light weapons' (UNGA, 2003a, para. 33). But the Experts Group did not mention ammunition or explosives in the sections of its report dealing with marking, record-keeping, and cooperation in tracing. Nor did it use these terms in its conclusions and recommendations.

Ammunition, particularly in the area of marking, raises technical issues that differ from those relating to weapons alone.[7] The GGE's failure to address such issues made the task of dealing with ammunition in the follow-up negotiations exceedingly difficult. Other small arms measures UN states have expressed an interest in developing, including controls over brokering and international transfers, do not require that weapons and their ammunition be treated differently. Yet, now that a separate UN process has been proposed for ammunition, there is a risk that small arms and light weapons, on the one hand, and their ammunition, on the other, will once again be separated—even though there may be no technical justification for doing so.

Definition of small arms and light weapons

In contrast to the *UN Programme of Action*, the *International Tracing Instrument* contains a detailed definition of 'small arms and light weapons' (UNGA, 2005b, para. 4). There were two main stumbling blocks on the path to consensus on this provision. The first was the choice of template. Taking its cue from the GGE (UNGA, 2003a, para. 32), the chairman's initial draft used a definition derived from the 1997 UN Panel of Experts. Although, in broad outline, it received the support of many delegations at the OEWG's second session, other states expressed a preference for the *UN Firearms Protocol* definition. The final compromise, facilitated in the last stages of the negotiations by Brazil, is in fact a hybrid of the UN Panel and *Firearms Protocol* definitions.

The second controversy surrounding the definition of 'small arms and light weapons' was whether or not to restrict the instrument to weapons 'which are manufactured to military specifications for use as lethal instruments of war' (UNGA, 1997, para. 24). This language, favoured by several states, had appeared in the UN Panel definition and again in the GGE report. Yet several other delegations pointed out that its use by the UN Panel was a function of the mandate it had received 'to consider the types of small arms and light weapons actually being used in conflicts being dealt with by the United Nations' (UNGA, 1997, para. 23). Its insertion in the tracing instrument, these states argued, would remove most civilian weapons from its scope and effectively strip it of any crime-fighting function it might lay claim to.

Negotiations on definitions stumbled over the choice of template.

At the end of the day, these arguments prevailed. The 'military specifications' language became part of a new preambular paragraph, which made the *Instrument*'s application to both crime and conflict settings clear, while skirting any restriction on the definition of 'small arms and light weapons':

> *Noting also* that the tracing of illicit small arms and light weapons, *including but not limited to* [emphasis added] those manufactured to military specifications, may be required in the context of all forms of crime and conflict situations (UNGA, 2005b, pream. para. 2)

The *Tracing Instrument*'s definition of 'small arms and light weapons' borrows liberally from both the *UN Firearms Protocol* (UNGA, 2001b, art. 3(a)) and the UN Panel (UNGA, 1997, paras. 25–26). The first part, drawn in large part from the *Protocol,* uses precise, treaty-like language to define the technical characteristics of small arms and light weapons. While the *Protocol* is the main inspiration for this part of the definition, the *Tracing Instrument* makes several important changes. First, the *Instrument* inserts the term 'man-' before 'portable'. While, at first glance, this seems to make it more restrictive than the *Protocol,* the opposite is true. The states that negotiated the *Firearms Protocol* intended that the word 'portable' limit the definition of 'firearm' to 'firearms that could be moved or carried by one person without mechanical or other assistance' (UNGA, 2001a, para. 3). The *Tracing Instrument,* by contrast, covers light weapons 'designed for use by two or three persons serving as a crew', as well as small arms and light weapons that can be carried by a single person (UNGA, 2005b, para. 4). This leaves aside only a narrow range of light weapons (such as large recoilless rifles) that are not 'man-portable', but may be carried by a light vehicle in accordance with the UN Panel definition (UNGA, 1997, para. 27(a)).

A second important difference between the *Tracing Instrument* and *Firearms Protocol* definitions is the deletion of the word 'barrelled' from the *Instrument*. In the *Protocol,* this term excludes all light weapons that employ a tube or rail, as opposed to a barrel,[8] such as man-portable air defence systems (MANPADS). The *Instrument*'s addition of the word 'launches' after 'expels' confirms its application to a broad range of light weapons, many of which are 'launched' from a tube. Like the *Protocol,* the *Instrument* also excludes antiques and their replicas. These terms are to be 'defined in accordance with domestic law', although '[i]n no case will antique small arms and light weapons include those manufactured after 1899'.

The second part of the *Tracing Instrument*'s definition of 'small arms and light weapons' borrows from the UN Panel. Subparagraph 4(a) of the *Instrument* reproduces, word for word, the UN Panel list of 'small arms'. Subparagraph 4(b) does the same for 'light weapons'. Unlike the UN Panel, the *Tracing Instrument* specifies that these lists are illustrative, not exhaustive, by using the term 'inter alia' in both cases. The *Tracing Instrument* will therefore apply to weapons systems, including future systems, that meet the definition's general criteria, but do not appear in the subparagraph a/b lists.

The *Tracing Instrument*'s definition of 'small arms and light weapons' constitutes one of the OEWG's most important achievements. The new definition is clear, relatively comprehensive, and adaptable. It also combines language used by the arms control (UN Panel and GGE) and law enforcement (*UN Firearms Protocol*) communities. Perhaps most important, it fills a gap left by the definition-deficient *UN Programme of Action*. The *Tracing Instrument* definition thus offers a key reference point for future small arms negotiations, especially within the UN framework. This reference point is limited, however, to weapons; the *Instrument* does not cover ammunition.

The Instrument's definition of small arms and light weapons is clear and adaptable.

Other definitions

The *Tracing Instrument*'s definition of 'tracing' is a simplified version of the definition used by the GGE in its report (UNGA, 2003a, para. 20). It omits, in particular, the notion that a tracing operation will normally end with the identification of 'the person or group that last possessed' the weapons being traced as it was thought that, in practice, this could be the exception, rather than the rule.

Under the terms of the *Instrument,* before a state can request assistance in tracing a weapon, it must first determine that it is 'illicit' (UNGA, 2005b, para. 16).

For the purposes of this instrument, small arms and light weapons are 'illicit' if:

(a) They are considered illicit under the law of the State within whose territorial jurisdiction the small arm or light weapon is found;

(b) They are transferred in violation of arms embargoes decided by the Security Council in accordance with the Charter of the United Nations;

(c) They are not marked in accordance with the provisions of this instrument;

(d) They are manufactured or assembled without a licence or authorization from the competent authority of the State where the manufacture or assembly takes place; or

(e) They are transferred without a licence or authorization by a competent national authority. (UNGA, 2005b, para. 6)

As the *Tracing Instrument* specifies that states make tracing requests based on *their* determination of the illicit nature of a particular weapon (para. 16), the interpretation of these subparagraphs ultimately lies with the requesting state. While the *Instrument* gives the requested state the right to refuse to respond to a tracing request in certain circumstances, it cannot challenge a requesting state's determination that the weapon it wants to trace is illicit (para. 22).

Other general issues

Negotiation of the *Tracing Instrument*'s 'General provisions' section involved many minor disagreements and a few serious ones. Among the latter was the question of whether the drafters should mention 'terrorism' as one of their key concerns, along with crime and conflict. In the final version of paragraph 1, setting out the purpose of the *Instrument,* the OEWG dropped an earlier reference to 'crime and conflict' and moved it to the preamble. Those states that wanted to make the instrument's application to terrorism clear were partly appeased by specifying, in the same preambular paragraph, that '*all forms* of crime and conflict' were at issue (UNGA, 2005b, pream. para. 2; emphasis added). Yet there was no agreement to include the word 'terrorism' in the instrument, so it was left out.[9]

Just as controversial was a proposal to insert in the instrument's 'General provisions' section language the *UN Firearms Protocol* used to exclude many government transfers from its scope (UNGA, 2001b, art. 4(2)). At the end of the day, the Group settled on the following, alternative formulation:

> This instrument does not restrict the right of States to acquire, manufacture, transfer and retain small arms and light weapons for their self-defence and security needs, as well as for their capacity to participate in peace-keeping operations, in a manner consistent with the Charter of the United Nations. (UNGA, 2005b, para. 3)

In essence, this paragraph makes the obvious point that, while the *Instrument* enhances the traceability of small arms and light weapons, it does not affect the underlying right of states to 'acquire, manufacture, transfer and retain' these weapons in conformity with the UN Charter.

The preamble of the *International Tracing Instrument* contains a relatively modest nine paragraphs. This compares with the 22 preambular paragraphs found in the *UN Programme of Action*. Prompted by a warning from some delegations, the OEWG decided to keep most substantive issues out of the preamble in order to avoid a time-consuming 'rebalancing exercise' in which particular principles or concepts (e.g. terrorism) would be 'balanced' with their rivals (e.g. self-determination). By and large, the *Instrument*'s preambular paragraphs merely restate the negotiating mandate and recall the main developments in the UN tracing process. Two exceptions, discussed above, emphasize the *Instrument*'s 'crime and conflict' functions (pream. para. 2) as well as the need for a 'coordinated and comprehensive' approach to the small arms problem (pream. para. 8). In addition, preambular paragraph 9 stresses 'the urgent necessity for international cooperation and assistance' in implementing the *Instrument.*

The three versions of the draft instrument the chairman produced included 'placeholders' for 'technical annexes' covering such things as:

- voluntary best practice: marking;
- voluntary best practice: record-keeping;
- sample terms of reference for the national point(s) of contact;
- standard form of information sought in the context of a tracing request; and
- standard form of information to be provided in response to a tracing request.

In order to avoid a time-consuming 'rebalancing exercise', the preamble was left largely free of substance.

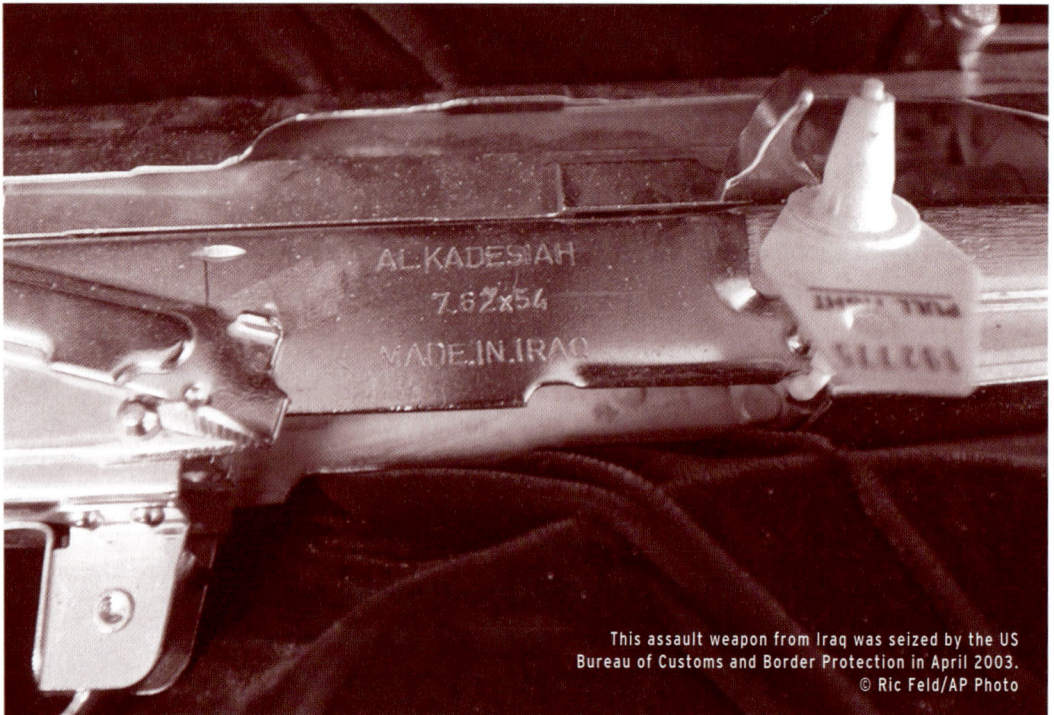

This assault weapon from Iraq was seized by the US Bureau of Customs and Border Protection in April 2003.
© Ric Feld/AP Photo

Several states submitted written proposals for one or more annexes during the negotiations. Nevertheless, as some states indicated that the OEWG would have to agree on the content of such annexes before they could be incorporated in the final instrument—notwithstanding their purely voluntary nature—this initiative was abandoned at the final session

Marking

If small arms and light weapons are to be traced, they need to be marked with basic identifying information. In conjunction with a weapon's physical characteristics, such markings are used to identify the weapon as well as the governments or companies that can assist in reconstructing its history.

Paragraph 8 of the *International Tracing Instrument* sets out the basic rules governing the content of marks. Sub-paragraph (a) largely reproduces the *UN Firearms Protocol* standard for marking at the time of manufacture.[10] Specifically, it allows countries that are using non-alphanumeric marking systems to maintain them. Those countries are, however, obliged to use a numeric and/or alphanumeric code that allows for the universal identification of the country of manufacture. States using alphanumeric marking systems must require weapons to be marked with the name of the manufacturer, the country of manufacture, and a serial number. For both groups of countries, weapons marking must be 'unique'.

Paragraph 8(b) of the *Tracing Instrument* deals with the other essential starting point for weapons traces, namely, the import mark. During negotiations, the OEWG had two standards to choose from: the *UN Firearms Protocol,* which makes such marking mandatory (UNGA, 2001b, art. 8(1)(b)), and the GGE report, which basically makes it optional: 'All small arms and light weapons are marked at the time of manufacture and, *if necessary,* import'

(UNGA, 2003a, para. 62(a); emphasis added).[11] After much argument, the OEWG opted for an intermediate solution. Under the *Tracing Instrument,* import marking is neither mandatory nor entirely optional. After underlining the mandatory nature of import marking under the *UN Firearms Protocol,* paragraph 8(b) directs states to require 'to the extent possible' that imported small arms and light weapons are marked with the country of import and 'where possible' the year of import. Such weapons are also to receive 'a unique marking' (serial number) if they do not already bear such a marking.

Many states were disappointed that the OEWG could not agree to make import marking mandatory, as in the *Firearms Protocol*. Yet the final text, developed through the facilitation efforts of Belgium, is somewhat stronger than the open-ended GGE formulation. In any case, it was more than many countries, concerned about potential costs, were initially willing to accept. This is one area where international assistance will be important.

Paragraph 8(d) of the *Tracing Instrument* addresses the important issue of existing government stocks. The Small Arms Survey has identified the leakage of such stocks as a key source of weapons for war zones and the illicit market generally (Small Arms Survey, 2005, ch. 6; 2002, pp. 137–38). If such weapons have not been marked, as is often the case, they cannot be traced after they leave the government stockpile. The standard articulated in paragraph 8(d) is relatively low. The relevant weapons are to be 'duly marked'; *Instrument* norms for marking at the time of manufacture 'do not necessarily' apply. At the same time, the paragraph introduces a potential loophole in favour of weapons that are held by government forces, but are not 'for their own use'. Nevertheless, the application of paragraph 8(d) to *existing* stocks is clear.

Paragraph 9 covers the confiscation of illicit small arms and light weapons. It builds on *UN Firearms Protocol* article 6(2), specifying that such weapons, if not destroyed, are to be 'uniquely marked' (not merely 'marked', as in the *Protocol*) and recorded in accordance with the *Tracing Instrument*'s record-keeping provisions. States also commit to securely store such arms pending their marking and recording (or destruction)—another element missing from *Protocol* article 6(2).

> Import marking is neither mandatory nor entirely optional.

Countries around the world use a variety of methods to mark small arms and light weapons, including stamping, engraving, and laser etching.[12] The *Tracing Instrument* leaves the choice of such methods to states, while specifying that the resulting marks must be 'easily recognizable, readable, durable and, as far as technically possible, recoverable'. It also insists on the visibility of such marks. They are to be 'on an exposed surface' and 'conspicuous without technical aids or tools' (UNGA, 2005b, para. 7).

The *Tracing Instrument* also prescribes the placement of marks. 'A unique marking should be applied to an essential or structural component of the weapon'. A component is 'essential or structural' if its 'destruction would render the weapon permanently inoperable and incapable of reactivation'. In the case of small arms and some light weapons, such components include the frame or receiver. States are also 'encouraged, where appropriate to the type of weapon,' to mark other weapons parts (para. 10).

Record-keeping

Record-keeping, after marking, is the second prerequisite for effective tracing. Records contain information about the marked weapon itself, along with elements of its history, beginning with manufacture. The *Tracing Instrument*'s record-keeping section has only three paragraphs. This reflects record-keeping's close connection with areas traditionally seen as the preserve of national sovereignty, such as the regulation of civilian weapons possession. Constitutional constraints also led many states to approach the issue with caution.

The *Tracing Instrument* therefore contains few prescriptions or detailed commitments on record-keeping. It also states that the 'choice of methods for record-keeping is a national prerogative' (para. 11). It nevertheless lays down one essential standard:

> States will ensure that accurate and comprehensive records are established for all marked small arms and light weapons within their territory and maintained in accordance with paragraph 12 below in order to enable their competent national authorities to trace illicit small arms and light weapons in a timely and reliable manner. (UNGA, 2005b, para. 11)

By and large, the details of how to accomplish this are left to governments, but the commitment to establish and maintain the records that are needed to ensure 'timely and reliable' tracing is clear.

The *Tracing Instrument* also prescribes the length of time records must be kept. The *UN Firearms Protocol* stipulates a ten-year minimum (UNGA, 2001b, art. 7). The GGE report suggests a longer period of time—'as long as needed for the purposes of tracing'—but allows individual states to decide what this means in practice (UNGA, 2003a, para. 76(d)). Since small arms and light weapons have life spans of several decades if properly maintained, many OEWG delegations argued that records should be kept indefinitely, or at least for much longer than ten years. At the end of the day, the Group agreed on a 30-year minimum for manufacturing records and 20 years for all other records, including those relating to import and export. As a general principle, states undertook, 'to the extent possible', to keep records 'indefinitely' (UNGA, 2005b, para. 12). This is not the only area where the *Instrument* strengthens existing standards, even while extending these to the entire UN membership.

> The *Instrument* establishes detailed modalities for tracing cooperation that have no parallel elsewhere.

Cooperation in tracing

Section V, devoted to 'Cooperation in tracing', forms the operational core of the *International Tracing Instrument*. This is also where it adds greatest value to existing small arms measures, establishing detailed modalities for tracing cooperation that have no parallel elsewhere.[13]

As with marking and record-keeping, the tracing cooperation section leaves the means of compliance with individual states: 'the choice of tracing systems will remain a national prerogative' (UNGA, 2005b, para. 14). It nevertheless sets clear limits to that discretion: 'States will ensure that they are capable of undertaking traces and responding to tracing requests in accordance with the requirements of this instrument' (para. 14). Paragraph 15 underlines the importance of confidentiality. States receiving tracing information agree to 'respect all restrictions placed on its use' and to 'guarantee [its] confidentiality'.

Paragraphs 16–17 set out the threshold requirements for tracing requests. Paragraph 16 reflects the outcome of the OEWG debate concerning the role of peacekeeping operations in the *Tracing Instrument,* mentioning only states with respect to the initiation of tracing requests.[14] In practice, this means that a state is not *required* to respond to a request it receives from a peacekeeping operation or other non-state entity. Yet nothing in the *Instrument* prevents a country from responding to such requests if it wants to. It can apply the provisions of the *Instrument* or develop other modalities for such situations as it sees fit. The Security Council can also order UN Member States to provide tracing assistance to a UN peacekeeping operation or UN sanctions committee if it adopts a binding resolution to this effect.

Under paragraph 16, a state seeking tracing assistance must also determine that the small arm or light weapon is 'illicit' in accordance with the definition of this term discussed above. Further, pursuant to paragraph 17, any request for tracing assistance must contain 'sufficient information' for this purpose. The non-exhaustive list that completes paragraph 17 indicates what this means:

(a) Information describing the illicit nature of the small arm or light weapon, including the legal justification therefor and, to the extent possible, the circumstances under which the small arm or light weapon was found;

(b) Markings, type, calibre and other relevant information to the extent possible;

(c) Intended use of the information being sought.

The information mentioned in paragraph 17(b) is indispensable to any weapons trace. That contained in the second part of paragraph 17(a) may prove useful as well. The fourth element (para. 17(c)) allows the requested state to determine whether it should restrict the use of the information it will provide. By contrast, the first element, concerning the 'illicitness' of the relevant weapon (first part of para. 17(a)), is something of a hindrance to efficient tracing. As noted earlier, the definitions section of the *Tracing Instrument* leaves the determination of 'illicit' firmly in the hands of the requesting state. Yet there is a suggestion in paragraph 17 that the requesting state has to justify that determination when making its tracing request.

Paragraph 21 allows the requested state to 'seek additional information from the requesting State where a tracing request does not contain the information required in paragraph 17'. Yet, crucially, the *Tracing Instrument* does not give the requested state the right to delay or restrict the content of its response, or refuse to respond, simply because it considers the information provided by the requesting state to be insufficient. In particular, it cannot challenge the requesting state's determination that a particular weapon is 'illicit' and thus block the tracing request.

Paragraphs 18–23 of the *Tracing Instrument* deal with responses to tracing requests. Paragraph 18 articulates the general principle of 'prompt, timely and reliable responses'. As for the content of such responses, paragraph 20 specifies that 'the requested State will provide . . . all available information sought by the requesting State that is relevant for the purpose of tracing illicit small arms and light weapons'. The requirement that the information be relevant to tracing stems from a concern, expressed by many OEWG delegations, that the *Instrument* not be used to pursue political goals or other ends not related to weapons tracing. The provision of tracing information is, moreover, subject to a series of exceptions:

> A state asked to help trace a weapon cannot challenge a requesting state's determination that it is 'illicit'.

States may delay or restrict the content of their response to a tracing request, or refuse to provide the information sought, where releasing the information would compromise ongoing criminal investigations or violate legislation providing for the protection of confidential information, where the requesting State cannot guarantee the confidentiality of the information, or for reasons of national security consistent with the Charter of the United Nations. (UNGA, 2005b, para. 22)

These exceptions were part of OEWG discussions from an early stage and, in one form or another, featured in each of the chairman's three draft texts. An earlier reference to a 'breach of domestic law' was subsequently narrowed to refer to 'legislation providing for the protection of confidential information' alone. Throughout the negotiations, many states stressed that the future instrument needed to address the issue of confidentiality. The two exceptions relating to confidentiality in paragraph 22 are complemented by paragraph 15, discussed above.

There was also broad support for the 'national security' exception that appears in paragraph 22, a term that initially appeared in the GGE report (UNGA, 2003a, para. 98) and was borrowed by the General Assembly when formulating the Group's mandate (UNGA, 2003b, para. 7).[15] The reference to the UN Charter limits the scope of this exception somewhat. A more important check on the potential abuse of paragraph 22—as a whole—is the requirement that States explain any use they make of this provision:

> If a State delays or provides a restricted response to a tracing request, or refuses to provide the information sought, on the grounds identified in paragraph 22 above, it will inform the requesting State of the reasons for this. The requesting State may subsequently seek clarification of this explanation. (UNGA, 2005b, para. 23)

Fundamentally, paragraph 22 provides those countries that are intent on evading their commitment to respond to tracing requests with broad cover to do so. Yet paragraph 23 will make it rather more apparent—at least to the requesting state—whether a requested state is applying paragraph 22 in good faith or not.

Implementation

The *Tracing Instrument*'s 'Implementation' section outlines mechanisms and arrangements designed to support its effective functioning. Paragraphs 24–26 lay the foundation for implementation at the national level, including the establishment of the necessary legislative and administrative framework, as well as the designation of one or more national points of contact. Paragraphs 27–29 deal with 'International cooperation and assistance', which many developing countries stressed was essential to their implementation efforts. Despite the express wish of those OEWG delegations for strong language, however, the section retains important qualifiers found in the *UN Programme of Action*. The *Instrument* strongly recommends, but does not make obligatory, the provision of such assistance.[16]

Exchange of information on national points of contact and national marking practices is essential to tracing.

At the first OEWG session, states unanimously stressed the primacy of existing mechanisms and organizations in underpinning *Instrument* implementation at the international level. There was virtually no support for the creation of a new institution, not even one with a modest clearing-house function. The resulting challenge was to fit the future tracing instrument—with its dual crime- and conflict-fighting functions—into the existing web of international organizations and arrangements. Both Interpol and the UN were quickly identified as key partners in this regard.

The UN's role in supporting *Instrument* implementation is set out in paragraphs 30–32. Under paragraph 32, the UN Secretariat is to facilitate the provision of assistance that States request for this purpose (para. 32).[17] The Secretariat (in particular DDA) will also receive from states and issue to them the following information:

(a) Name and contact information for the national point(s) of contact;

(b) National marking practices related to markings used to indicate country of manufacture and/or country of import as applicable. (UNGA, 2005b, para. 31)[18]

This information is essential to tracing. Among other things, the national points of contact are responsible for initiating and receiving tracing requests conducted under the framework of the *Instrument*. Markings used to indicate a weapon's country of manufacture and/or import provide the starting points for the trace.

Interpol's role is no less important to the effective implementation of the *Instrument*. Paragraph 35 outlines several types of assistance the organization may provide 'at the request of the concerned State':

(a) Facilitation of tracing operations conducted within the framework of this instrument;

(b) Investigations to identify and trace illicit small arms and light weapons;

(c) Wherever possible, building national capacity to initiate and respond to tracing requests. (UNGA, 2005b)

Subparagraph (a) covers the typical case where a state takes charge of a tracing operation. Interpol 'facilitation' in this situation could involve the state's use of secure Interpol channels to relay and receive information related to a tracing request. In future, states will also be able to call upon Interpol for automated firearms tracing and help in weapons identification.[19] Interpol assistance in tracing investigations, mentioned in subparagraph (b), could be useful where a tracing operation is especially complex or where the country that wants to trace a weapon lacks the capacity to do so—a situation also addressed in subparagraph (c). Only Interpol members can normally call upon Interpol for the assistance mentioned in paragraph 35. Non-members can also do so, however, if they have a cooperation agreement with the organization.[20]

Tracing small arms and light weapons that have helped fuel a conflict—perhaps in violation of a UN arms embargo—poses several special challenges. In the case of a 'failed state', there may be no governmental authority capable of undertaking a tracing request. A second problem is the longer, more complex lines of supply one usually finds in conflict situations.[21] In principle, there are few, if any, obstacles to Interpol involvement in conflict tracing. Where local law enforcement agencies have ceased to function or instead function under international authority, Interpol can liaise directly with the civilian police component of any peacekeeping operation present on the territory. For example, building on a general *Co-operation Agreement* concluded by the two organizations in 1997 (Interpol and UN, 1997), Interpol is currently supporting UN peace-building efforts in Kosovo. Among other things, this gives the UN (police) point of contact access to Interpol's communications network and databases (Interpol and UN, 2002).

Tracing weapons that have helped fuel a conflict poses special challenges.

The conflict tracing functions of the *International Tracing Instrument* are largely implicit in the provisions reviewed above, in particular paragraph 35. While all references to 'peacekeeping operations' and other non-state entities were ultimately purged from the text, the OEWG did recommend 'that the issue of the applicability of provisions of the draft instrument to United Nations peacekeeping operations be considered further within the framework of the United Nations' (UNGA, 2005a, para. 28). This gives UN Member States an opportunity to further develop the legal and practical framework needed for conflict tracing within the UN system. Paragraph 34 of the *Tracing Instrument* performs a similar function in relation to Interpol, directing states that are members of the organization to 'promote the implementation of this instrument when participating in Interpol's organs' (UNGA, 2005b). The *International Tracing Instrument*, in combination with the OEWG peacekeeping recommendation, thus sets the stage for the further development of conflict tracing, in particular through enhanced UN–Interpol cooperation.

Follow-up

The truc value of the *International Tracing Instrument* will become clear only over time as a function of two factors:

- the extent to which the Instrument is implemented by UN Member States; and
- the extent to which it spurs the further development and strengthening of international norms on marking, record-keeping, and cooperation in tracing.

The *Tracing Instrument*'s 'Follow-up' section provides an important basis for progress on both counts. Paragraph 36 establishes the principle of biennial reporting. In contrast to the *UN Programme of Action* (UNGA, 2001d, para. II:33), such reporting is mandatory:

States *will* report . . . on their implementation of this instrument including, where appropriate, national experiences in tracing illicit small arms and light weapons as well as measures taken in the field of international cooperation and assistance. (UNGA, 2005b, para. 36; emphasis added)

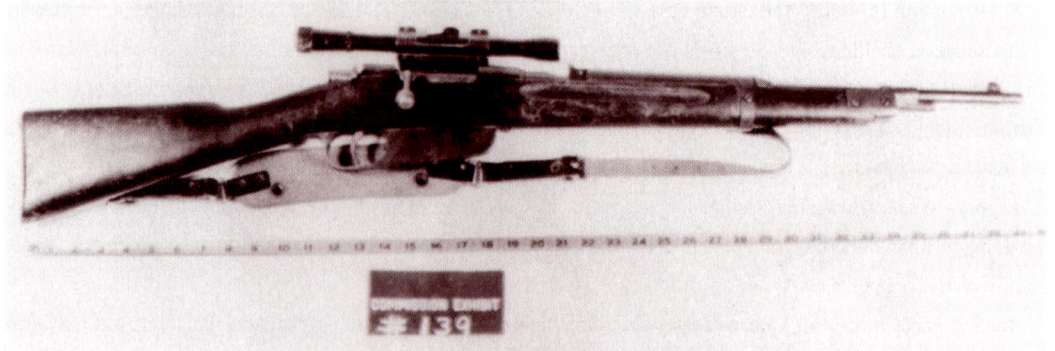

This Mannlicher-Carcano bolt-action, clip-fed rifle, found in the Texas School Book Depository Building in Dallas, was identified by the Warren Commission as the rifle used to shoot President John F. Kennedy. It is inscribed with various markings, including 'MADE ITALY'; 'Cal. 6.5', referring to the gun's calibre; '1940', the year of manufacture; and the serial number C2766. © Warren Commission/AP Photo

The *Tracing Instrument* links these reports and the biennial meetings that will 'consider' them to the UN Conference process, provided that process continues. Reports on the implementation of the *Instrument* can be incorporated into national reports on the implementation of the *UN Programme of Action*. The biennial tracing meetings will be held within the framework of *UN Programme* meetings as long as such meetings 'are in fact convened' (paras. 36–37).

In practice, reporting on *UN Programme* implementation has followed something of a two-year cycle, with reporting activity considerably higher in years that have coincided with a *UN Programme* Biennial Meeting of States (2003 and 2005).[22] Paragraphs 36 and 37 of the *Tracing Instrument* formalize this two-year implementation review process—at least with respect to the *Instrument*. While reporting and meetings held to 'consider' reports do not necessarily translate into actual implementation, they can help (Small Arms Survey, 2004, pp. 257–58). Agreement on *mandatory* reporting is an especially important advance over the *UN Programme of Action* and could conceivably bolster reporting rates for the *Programme* itself.

Last but not least, paragraph 38 of the *Tracing Instrument* addresses the second of the two factors underpinning its long-term effectiveness: normative development. States undertake to 'review the implementation and future development' of the *Instrument* within the framework of conferences held to review the *UN Programme of Action*. Such 'future development' could involve the strengthening of the *Tracing Instrument*'s existing provisions, the formulation of new commitments, and/or the transformation of some or all of the *Instrument*'s norms into legally binding form (see McDonald, 2006). Far from marking the end of the UN tracing process, the *International Tracing Instrument* anticipates its continuation.

IMPLEMENTING THE INSTRUMENT[23]

Following its adoption by the UN General Assembly, in December 2005, the *International Tracing Instrument* applies to all UN Member States (UNGA, 2005b; 2005c; 2005d). In the short term, any initiatives for further normative development are likely to be modest.[24] The key challenge for 2006 is the effective implementation of the current *Instrument*.

In the autumn of 2005, just prior to the *Tracing Instrument*'s adoption by the General Assembly, the Small Arms Survey and three collaborators undertook a preliminary assessment of national preparations for *Instrument* implementation. Government officials in the Americas, Europe, and southern Africa were asked what steps their country had taken or was planning to take for purposes of implementing the *Instrument*, including relevant changes to national laws or regulations, or measures designed to strengthen national capacity.

While, as a rule, states indicated they were aware of the new *Tracing Instrument* and were committed to implementing it, relatively few countries offered details of specific implementation plans or challenges.[25] Thirty of the 37 states that were contacted provided information to researchers.[26] The picture among respondent states was extremely mixed. Simple affirmations of compliance alternated with admissions that the country had still to determine whether existing laws, regulations, and practices met *Instrument* requirements. Several states had already conducted, initiated, or planned to conduct legislative reviews, though in several cases these were to follow the *Tracing Instrument*'s (anticipated) adoption by the General Assembly at the end of 2005. Some countries indicated that these reviews would be conducted at some future date as part of broader reviews of national firearms legislation. Only a handful of countries acknowledged that changes to national laws or practices were needed to meet *Instrument* requirements. A few countries informed researchers of recent changes in national legislation that served this purpose.

Many of the states contacted by researchers had yet to take any decision regarding the designation of the national point(s) of contact required under the *Tracing Instrument* (UNGA, 2005b, para. 25), especially that involved in actual tracing. Several countries indicated that the point of contact responsible for information exchange (see paras. 31, 36) would be the same as that used for *UN Programme of Action* reporting. These points of contact were typically located within the country's ministry of foreign affairs. Those states that had identified a point of contact for tracing operations tended to locate it within the branch of government responsible for law enforcement.

Almost all respondent countries acknowledged that capacity-building would be important to the implementation of the *Tracing Instrument*. Among the capacity-building initiatives in place or under consideration, states cited, in particular, coordination and information exchange among government agencies. This included cooperation between federal and local or regional authorities in several states with a federal structure. One country also referred to plans to link small arms records held by the national armed forces to the national police force, thus enhancing police tracing capacity. Other initiatives mentioned by states included the training of law-enforcement personnel in tracing methods and the development of databases for firearms identification. In preparing to implement the *Instrument*, several states indicated they were cooperating or consulting with industry, NGOs, and other stakeholders, or planned to do so.

The key challenge for 2006 is the effective implementation of the *Instrument*.

While the study found that most countries in the Americas and Europe were relatively focused on the specifics of *Instrument* implementation, southern African states generally seemed more preoccupied with the implementation of the *SADC Firearms Protocol* (SADC, 2001), in force since November 2004. While the two instruments overlap in the areas of marking, record-keeping, and to some extent tracing, they nonetheless differ significantly in their coverage and treatment.[27] Officials in southern Africa indicated that their governments knew of the *Tracing Instrument* and were committed to implementing it. Some of these officials acknowledged, moreover, that relevant national legislation was outdated and needed amending; yet *Tracing Instrument* requirements would be considered only as part of future, general reviews of national firearm legislation.

The specific regional mechanisms now being developed for implementation of the *SADC Protocol*—aimed, above all, at improving the flow of information among SADC states—could be used, at least within the sub-region, for the implementation of the *International Tracing Instrument*. In theory, the two instruments complement each other. In practice, however, there is a risk that *Tracing Instrument* implementation will take a back seat as SADC countries grapple with the implementation of their new *Protocol*. In several cases, it appears that the legislative reviews that are a prerequisite for the full implementation of the *Tracing Instrument* are, in fact, being subordinated to *SADC Protocol* timing. In the context of this research, government officials in southern Africa stressed their commitment to implementing all international—as well as regional—small arms agreements; yet, on the eve of the *Tracing Instrument*'s adoption, their attention seemed to be directed elsewhere.

This problem of competing priorities is not unique to southern Africa. When discussing their implementation of the *Tracing Instrument,* several countries in the Americas and Europe referred to measures they were taking to implement the *UN Firearms Protocol* (UNGA, 2001b). In one or two cases, responding officials appeared to assume that the process of bringing national systems into compliance with the *UN Protocol* would also serve to meet the requirements of the *International Tracing Instrument*—despite the significant differences between these instruments in those areas where they do overlap. Most officials contacted by the research team were, however, aware of the differences, with one official citing the much broader range of light weapons covered by the *Tracing Instrument.* In general, it appeared that states were taking advantage of the *UN Protocol* implementation process to make any changes (sometimes different changes) that might also be needed for purposes of implementing the *Tracing Instrument.*

As of November 2005, it appeared that most countries in the Americas, Europe, and southern Africa were aware of the *International Tracing Instrument* and taking some of the steps needed for its implementation. Overall, however, these states were not attending to this task with the speed and rigour that the *Instrument*'s imminent adoption by the General Assembly warranted.

CONCLUSION

The *International Tracing Instrument* (UNGA, 2005b), now applicable to all UN Member States (UNGA, 2005d), constitutes a modest, but significant, step forward in international efforts to address the small arms problem. Although the final result fell far short of what many had hoped for, the new *Instrument* is nevertheless a useful tool in the fight against illicit small arms.

Overall, the *International Tracing Instrument* consolidates—even strengthens—the standards contained in the *UN Firearms Protocol* (UNGA, 2001b) and the report of the Group of Governmental Experts on Tracing (UNGA, 2003a). Important value added can be found in its sections on definitions, cooperation in tracing, and implementation. Key weaknesses include the failure to make import marking mandatory and to give conflict tracing firmer operational footing within the *Instrument.*

The *Tracing Instrument*'s real value will, however, become apparent only over time as a function of (*a*) its effective implementation by all UN Member States and (*b*) the extent to which it spurs further normative development. As indicated in the previous section, many states have been slow to prepare for the *Instrument*'s adoption by the General Assembly. In the autumn of 2005, just weeks before they became (politically) bound by the *Instrument,* these countries had yet to determine how to bring national laws and practices into compliance. Implementation, in other words, is off to a relatively slow start.

Implementation is one challenge, normative development another. UN Member States have undertaken to review the future development of the *Tracing Instrument* in the context of *Programme of Action* review conferences. This was part of the bargain that allowed the many countries that wanted a strong legal document covering ammunition to accept the final, far-from-perfect result. More diplomatic wrangling is needed before the last word can be written on the UN tracing process. ◼

LIST OF ABBREVIATIONS

DDA	Department for Disarmament Affairs (United Nations)	**MANPADS**	man-portable air defence system
EU	European Union	**OEWG**	Open-ended Working Group to Negotiate an International Instrument to Enable States to Identify and Trace, in a Timely and Reliable Manner, Illicit Small Arms and Light Weapons
GGE	Group of Governmental Experts on Tracing Illicit Small Arms and Light Weapons		
GRIP	Groupe de recherche et d'information sur la paix et la sécurité	**SADC**	Southern African Development Community
IANSA	International Action Network on Small Arms	**UNIDIR**	United Nations Institute for Disarmament Research
Interpol	International Criminal Police Organization	**WFSA**	World Forum for the Future of Sport Shooting Activities
IWeTS	Interpol Weapons Electronic Tracing System		

ENDNOTES

1   Full name: Open-ended Working Group to Negotiate an International Instrument to Enable States to Identify and Trace, in a Timely and Reliable Manner, Illicit Small Arms and Light Weapons.

2   See also UNGA (2005c, para. 2).

3   The Control Arms campaign is jointly run by Amnesty International, IANSA, and Oxfam International.

4   For OEWG-related materials and information, see the UNDDA Web site: <http://disarmament2.un.org/cab/salw-oewg.html>

5   That is, subject to the final approval of the delegate's government.

6   Albania, Bosnia and Herzegovina, Bulgaria, Croatia, Iceland, Macedonia, Norway, Romania, Serbia and Montenegro, and Turkey.

7   See Anders (2006); Small Arms Survey (2005, pp. 25–29); Anders (2005a).

8   In containing the expansion of gases on combustion, a barrel helps impart kinetic energy to a projectile. This results in the phenomenon of recoil transmitted to both weapon and shooter when a barrelled weapon is fired. For this reason, barrels are composed of hard steel alloys. By contrast, launch tubes, such as those used in MANPADS, may guide the projectile's initial direction but do not help drive it forward. The actual thrust comes from the projectile itself. Because a launch tube is open at both ends, little energy is imparted to the tube and consequently to the shooter. Launch tubes are usually made of fibreglass or plastic composites.

9   See also the discussion below of the *Instrument*'s preamble.

10   There are three differences. First, the *Tracing Instrument* applies to manufacture occurring under the 'jurisdiction or control' of the state. The *Firearms Protocol* is silent on this question of territorial scope. Second, under the *Tracing Instrument* states using alphanumeric marking systems must require the marking of 'the country of manufacture'—in contrast to the *Protocol*'s 'country *or place* of manufacture'. Third, unlike the *Protocol*, under the *Tracing Instrument* states agree to 'encourage the marking of such additional information as the year of manufacture, weapon type/model and calibre' (UNGA, 2005b, para. 8(a); 2001b, art. 8(1)(a)).

11   See also para. 62(f).

12   See Small Arms Survey and UNIDIR, 2003, pp. 89–91.

13   The *UN Firearms Protocol* contains one general provision on tracing (UNGA, 2001b, art. 12(4)). The GGE's more extensive 'common minimum standards' served as raw material for the *Tracing Instrument* (UNGA, 2003a, para. 93), but lack its normative force.

14   See also UNGA (2005b, para. 18).

15   This language also found its way into preambular paragraph 6 of the *Tracing Instrument*: 'Noting also that, pursuant to resolution 58/241, this instrument takes into account the national security and legal interests of States.'

16   See, for example, paragraph 27: 'States *in a position to do so* will, upon request, seriously *consider* rendering technical, financial and other assistance . . .' (UNGA, 2005b; emphasis added). Adapted from para. III.3 of the *UN Programme of Action* (UNGA, 2001d).

17   Paragraph 32 does not express this very clearly. In fact, it appears to identify the UN Secretariat itself as the source of such assistance ('providing the assistance . . .'). When this paragraph was agreed on 16 June 2005, States were negotiating on the basis of the chairman's third draft text. The equivalent provision read: 'States request the Secretary-General . . . (c) To facilitate the provision of international cooperation and assistance

designed to enhance the ability of States to effectively implement this instrument.' When paragraph 32 was discussed and adopted *ad referendum,* no delegation said it expected the Secretariat itself to offer such assistance.

18   See also UNGA (2005b, para. 32).

19   As of June 2005, Interpol was upgrading its Interpol Weapons Electronic Tracing System (IWeTS) to include such features. See Interpol (2005).

20   As of 19 September 2005, Interpol had 184 member countries, almost as many as the UN's 191 members. For the latest membership information, see the Interpol Web site: <http://www.interpol.int/Public/Icpo/Members/default.asp>

21   For more information on the specific requirements and challenges of conflict tracing, see Small Arms Survey and UNIDIR (2003, pp. 23–27).

22   See the DDA Web site for a year-by-year listing of the states that have reported on their implementation of the *UN Programme of Action,* together with copies of the actual reports: <http://disarmament2.un.org/cab/salw-nationalreports.html>

23   This section draws on information from: Anders (2005c); Dreyfus (2005); Stott (2005); correspondence with the Government of Canada, 21 November 2005.

24   One possibility would be agreement on a simple mechanism for the voluntary exchange of information on marking, record-keeping, and tracing best practices (McDonald, 2006).

25   Correspondence with the Government of Canada, 21 November 2005.

26   *Responding states:* Angola, Argentina, Austria, Belgium, Botswana, Brazil, Canada, Chile, Colombia, Estonia, Finland, Germany, Latvia, Lesotho, Lithuania, Luxembourg, Malawi, Mexico, Mozambique, Namibia, The Netherlands, Norway, South Africa, Swaziland, Sweden, Tanzania, United Kingdom, Venezuela, Zambia, and Zimbabwe. *Non-responding states:* Czech Republic, Denmark, Ireland, Italy, Mauritius, Spain, and the United States.

27   For a comparison of the two instruments, see Stott (2005, pp. 18–27).

28   The author served as an adviser to Ambassador Anton Thalmann, Chairman of the Open-ended Working Group on Tracing Illicit Small Arms and Light Weapons. The views expressed in this chapter are the author's alone and should not be attributed to Ambassador Thalmann or the Government of Switzerland.

BIBLIOGRAPHY

Amnesty International, International Action Network on Small Arms (IANSA), and Oxfam International. 2005. *UN 'Deal' on Arms Controls Means Business as Usual for the World's Worst Arms Dealers.* Control Arms campaign press release. 14 July. <http://www.controlarms.org/latest_news/un-deal.htm>

Anders, Holger. 2005a. *Scope for International Minimum Standards on Tracing Illicit SALW Ammunition.* Brussels: Groupe de recherche et d'information sur la paix et la sécurité (GRIP). 6 June. <http://www.grip.org/bdg/g4575.html>

——. 2005b. *Role of NGOs in the UN Process on Tracing Illicit SALW.* Background paper. Geneva: Small Arms Survey.

——. 2005c. *National Implementation in Europe of the UN Tracing Instrument: Anticipated Changes in Standards and Practices.* Background paper. Geneva: Small Arms Survey.

——. 2006. *Following the Lethal Trail: Identifying Sources of Illicit Ammunition.* In Stéphanie Pézard and Holger Anders (eds.), *Targeting Ammunition: A Primer.* Geneva: Small Arms Survey.

Berkol, Ilhan. 2000. *Marquage et traçage des armes légères: vers l'amélioration de la transparence et du contrôle.* Brussels: Groupe de recherche et d'information sur la paix et la sécurité (GRIP). <http://www.grip.org/>

Dreyfus, Pablo. 2005. *Assessing National Readiness of Relevant Latin American Small Arms Producer and/or Exporter Countries to Implement the International Tracing Instrument.* Background paper. Geneva: Small Arms Survey.

France and Switzerland. 2000. *Food-for-Thought Paper: Contribution to the Realisation of an International Plan of Action in the Context of the 2001 Conference: Marking, Identification and Control of Small Arms and Light Weapons.* UN doc. A/CONF.192/PC/7 of 17 March.

——. 2001a. *Working Paper by Switzerland and France on Establishing a Tracing Mechanism to Prevent and Reduce Excessive and Destabilizing Accumulation and Transfer of Small Arms and Light Weapons.* UN doc. A/CONF.192/PC/25 of 10 January.

——. 2001b. *Franco-Swiss Workshop on Traceability of Small Arms and Light Weapons: Tracing, Marking and Record-Keeping.* Geneva, 12–13 March 2001. Chair's Summary. UN doc. A/CONF.192/PC/38 of 23 March.

Interpol (International Criminal Police Organization). 2005. *Interpol's Weapons Projects.* 28 June. <http://www.interpol.int/public/weapons/default.asp>

Interpol and UN. 1997. *Co-operation Agreement between the United Nations and the International Criminal Police Organization (Interpol).* 8 July. <http://www.interpol.int/Public/ICPO/LegalMaterials/cooperation/agreements/UN1997.asp>

——. 2002. *Memorandum of Understanding between the International Criminal Police Organization (Interpol) and the United Nations Interim Administration Mission in Kosovo on Co-operation in Crime Prevention and Criminal Justice.* December. <http://www.interpol.int/Public/ICPO/LegalMaterials/cooperation/agreements/UNMIK.asp#>

McDonald, Glenn. 2006. 'The International Tracing Instrument: Challenges and Opportunities.' Paper presented at PrepCom side event (2006 Small Arms Review Conference). United Nations, New York, 13 January.

OAS (Organization of American States). 1997. *Inter-American Convention against the Illicit Manufacturing of and Trafficking in Firearms, Ammunition, Explosives, and Other Related Materials.* Adopted in Washington, DC on 14 November. Entered into force on 1 July 1998. UN doc. A/53/78 of 9 March 1998. <http://www.oas.org/juridico/english/treaties/a-63.html>

SADC (Southern African Development Community). 2001. *Protocol on the Control of Firearms, Ammunition and Other Related Materials in the Southern African Development Community (SADC) Region ('SADC Firearms Protocol')*. Adopted in Blantyre, Malawi, on 14 August. Entered into force on 8 November 2004. <http://www.smallarmssurvey.org/resources/reg_docs.htm>

Small Arms Survey. 2002. *Small Arms Survey 2002: Counting the Human Cost*. Oxford: Oxford University Press.

—. 2004. *Small Arms Survey 2004: Rights at Risk*. Oxford: Oxford University Press.

—. 2005. *Small Arms Survey 2005: Weapons at War*. Oxford: Oxford University Press.

Small Arms Survey and United Nations Institute for Disarmament Research (UNIDIR). 2003. *The Scope and Implications of a Tracing Mechanism for Small Arms and Light Weapons*. Geneva: UNIDIR. <http://www.unidir.ch/html/en/books-reports.php>

Stott, Noel. 2005. *Assessing National Readiness to Implement the International (United Nations) Tracing Instrument: The Case of Southern Africa*. Background paper. Geneva: Small Arms Survey.

Switzerland. 2004a. *Non-Paper of the Chairman: Open-Ended Working Group on an International Instrument for the Identification and Tracing of Illicit Small Arms and Light Weapons*. 7 May.

——. 2004b. *International Instrument for the Timely and Reliable Identification and Tracing of Illicit Small Arms and Light Weapons: Chairman's Draft Text*. December.

——. 2005a. *International Instrument for the Timely and Reliable Identification and Tracing of Illicit Small Arms and Light Weapons: Chairman's Revised Draft Text*. 1 February.

——. 2005b. *International Instrument for the Timely and Reliable Identification and Tracing of Illicit Small Arms and Light Weapons: Chairman's Third Draft Text*. April.

——. 2005c. *Chair's Package Proposal 16 June 2005*.

UNGA (United Nations General Assembly). 1997. *Report of the Panel of Governmental Experts on Small Arms*. A/52/298 of 27 August (annexe). <http://disarmament2.un.org/cab/salw-oewg.html>

——. 2001a. *Interpretative Notes for the Official Records ('Travaux préparatoires') of the Negotiation of the Protocol against the Illicit Manufacturing of and Trafficking in Firearms, Their Parts and Components and Ammunition, Supplementing the United Nations Convention against Transnational Organized Crime*. A/55/383/Add.3 of 21 March.

——. 2001b. *Protocol against the Illicit Manufacturing of and Trafficking in Firearms, Their Parts and Components and Ammunition, Supplementing the United Nations Convention against Transnational Organized Crime ('UN Firearms Protocol')*. Adopted 31 May. Entered into force 3 July 2005. A/RES/55/255 of 8 June (annexe). <http://www.undcp.org/pdf/crime/a_res_55/255e.pdf>

——. 2001c. *Provisional Rules of Procedure of the Conference*. A/CONF.192/L.1 of 14 June. <http://disarmament2.un.org/cab/salw-oewg.html>

——. 2001d. *Programme of Action to Prevent, Combat and Eradicate the Illicit Trade in Small Arms and Light Weapons in All Its Aspects ('UN Programme of Action')*. 20 July. A/CONF.192/15. <http://www.smallarmssurvey.org/resources/2001_un_conf.htm>

——. 2003a. *Report of the Group of Governmental Experts Established Pursuant to General Assembly Resolution 56/24 V of 24 December 2001, Entitled 'The Illicit Trade in Small Arms and Light Weapons in All Its Aspects'*. A/58/138 of 11 July. <http://disarmament2.un.org/cab/salw-oewg.html>

——. 2003b. *The Illicit Trade in Small Arms and Light Weapons in All Its Aspects*. Resolution 58/241 of 23 December. A/RES/58/241 of 9 January 2004. <http://disarmament2.un.org/cab/salw-oewg.html>

——. 2005a. *Report of the Open-ended Working Group to Negotiate an International Instrument to Enable States to Identify and Trace, in a Timely and Reliable Manner, Illicit Small Arms and Light Weapons*. A/60/88 of 27 June. <http://disarmament2.un.org/cab/salw-oewg.html>

——. 2005b. *International Instrument to Enable States to Identify and Trace, in a Timely and Reliable Manner, Illicit Small Arms and Light Weapons ('International Tracing Instrument')*. A/60/88 of 27 June (annexe). <http://disarmament2.un.org/cab/salw-oewg.html>

——. 2005c. *The Illicit Trade in Small Arms and Light Weapons in All Its Aspects*. Resolution 60/81 of 8 December. A/RES/60/81 of 11 January 2006.

——. 2005d. *International Instrument to Enable States to Identify and Trace, in a Timely and Reliable Manner, Illicit Small Arms and Light Weapons*. Decision no. 60/519 of 8 December. A/60/463, para. 95; A/60/PV.61, p. 41.

WFSA (World Forum on the Future of Sport Shooting Activities). 2000. *Firearms Marking: Model Standards and Common Serial Number Codes*. Report of the Workshop held in Baia Sardinia (Olbia), Sardinia, Italy, 22–24 June.

ACKNOWLEDGEMENTS

Principal author

Glenn McDonald[28]

Contributors

Holger Anders, Pablo Dreyfus, and Noel Stott

Army soldiers cross the remains of a dam in 1999, a few months after the Cambodian government initiated country-wide weapons collection programmes.
© Jan Banning/Panos Pictures

Stabilizing Cambodia

SMALL ARMS CONTROL AND SECURITY SECTOR REFORM

5

INTRODUCTION

The international community has supported Cambodia's stabilization since the late 1980s, and much progress has been made. Over the last five years Cambodia has been the beneficiary of support for small arms control programmes through the European Union's Assistance in Curbing Small Arms and Light Weapons (EU ASAC) and the Japan Assistance Team for Small Arms Management in Cambodia (JSAC). Joint efforts in weapons collection and destruction by the government, the EU ASAC, the JSAC, and the Cambodian NGO Working Group for Weapons Reduction (WGWR) have substantially reduced the number of weapons circulating in Cambodia outside of government control. Following the end of the cold war, Cambodia was one of the first conflict-affected countries with a small arms legacy that the international community engaged in a concerted effort. Experiences from Cambodia provide important lessons for other post-conflict countries on what has worked well, and where greater effort is needed.

Since the start of stabilization efforts in Cambodia, the international community has engaged in many more post-conflict operations. Concepts and approaches have developed in the light of experience. Of particular importance is the security sector reform (SSR) model, which emphasizes a broad set of issues in the security services, including professionalism, institutional capacity building, resource allocation, the rule of law, and democratic governance. Until now there has been little sharing of ideas between the small arms and light weapons control community and the SSR community, with opportunities for linking the two agendas on the ground largely unexploited. Using the example of Cambodia, this chapter underlines the utility of incorporating SSR concepts into small arms control planning.

This chapter presents the following findings:

- Small arms and light weapons interventions in Cambodia have been successful in removing large quantities of such arms from circulation outside of government control.
- Reduction of small arms and light weapons has had a positive impact on human security in Cambodia, where guns are less commonly used in violent incidents and homicides.
- Large-scale destruction of surplus military stockpiles and safe storage programmes for government stockpiles have reduced the danger of future leakages and uncontrolled exports.
- Members of the security forces and certain government officials are the only legal small arms users in Cambodia today.
- Problems relating to the competence, professionalism, and integrity of the security forces; rules of engagement for the police in particular; as well as democratic governance of the security sector have not been sufficiently addressed.
- Cambodia is a good example of the significant human security gains that small arms control programmes can yield, yet it also demonstrates the limits of such efforts if broader security sector issues are not addressed.

The chapter begins by providing essential background information on Cambodian politics and the process of post-conflict stabilization. It then considers the impact of small arms and light weapons control programmes on the availability of small arms and overall crime, violence, and gun use. The next section focuses on Cambodia's security services, highlighting continuing reports of official gun misuse. The following section introduces the concept of SSR and notes the limited progress that has been made in Cambodia to date in this area. The chapter's final section discusses how SSR objectives may be taken forward in Cambodia. The conclusion discusses the broader relationships between small arms control and SSR, and the lessons that can be applied to other post-conflict countries.

THE LEGACY OF CONFLICT

The conclusion of the Paris Agreements in October 1991 marked the beginning of the end of more than 20 years of armed conflict in Cambodia. All four protagonists and their international sponsors were parties to the Paris Agreements: the Phnom Penh government (controlled by the Cambodian People's Party—CPP), the Front uni national pour un Cambodge indépendant, neutre, pacifique et coopératif (FUNCINPEC), the Khmer People's National Liberation Front (KPNLF), and the Party of Democratic Kampuchea (PDK). Under the leadership of Hun Sen, the CPP controlled more than 80 per cent of the country, including Phnom Penh, through its army, the Cambodian People's Armed Forces (CPAF). Vietnam had left its CPP allies in power after ending its decade-long occupation of Cambodia in 1989. The three other factions were CPP opponents. Prince Norodom Ranariddh presided over FUNCINPEC and the National Army of Independent Kampuchea (ANKI, formerly the Armée Nationale Sihanoukiste). Ranariddh's father, King Sihanouk, had been deposed in a coup in 1970 by his own prime minister and senior military officials. The KPNLF and its military wing, the Khmer People's National Liberation Armed Forces (KPNLAF), had their roots in the regime that overthrew King Sihanouk, and were, in turn, toppled by the PDK. This last faction, better known as the 'Khmer Rouge', terrorized Cambodia from 1975 until 1978, when it was overthrown by the Vietnamese. The PDK military wing was officially known as the National Army of Democratic Kampuchea.

A UN military observer mission, deployed shortly after the signature of the Paris Agreements, was replaced in March 1992 by the United Nations Transitional Authority in Cambodia (UNTAC), a major peacekeeping operation that administered the country and oversaw the implementation of the peace accords. UNTAC withdrew shortly after it oversaw the May 1993 national elections, which the international community deemed to be substantially free and fair. However,

Map 5.1 **Cambodia**

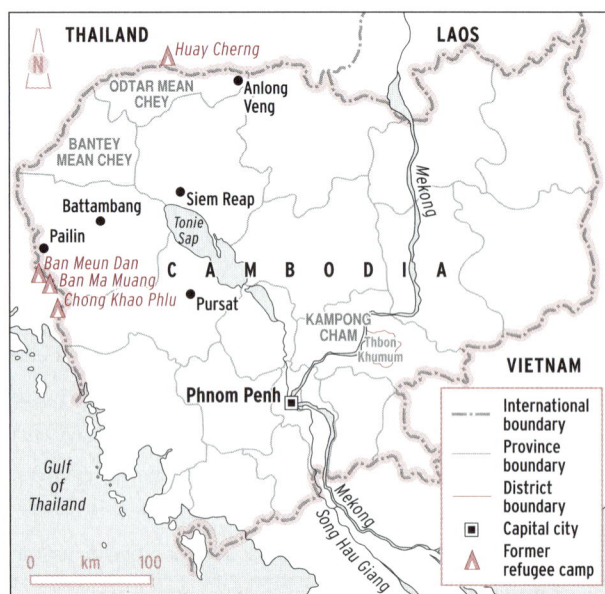

armed conflict continued in parts of the country for several years. While the anti-communist ANKI and KPNLAF complied with the accords and the election results and began integrating their forces within the newly formed Royal Cambodian Armed Forces (RCAF), the Khmer Rouge continued its guerrilla war against the government. The armed conflict was only brought to an end in 1998, when the last Khmer Rouge faction defected to the government.

Three aspects of Cambodia's stabilization process in particular have affected subsequent small arms control and security sector management efforts: Cambodian power politics were gradually transformed from armed conflict to the ballot box through elections, the CPP consolidated its power in the country, and the patronage system underlying government–military relations became entrenched.

The introduction of elections as an alternative, less lethal means of achieving political power proved essential to the stabilization of the country, but this move was also used by Hun Sen to consolidate power. Cambodia's 1993 elections resulted in a coalition government headed by FUNCINPEC and the CPP. Although FUNCINPEC won the election (i.e. received the greatest number of votes), the CPP entered into a power-sharing arrangement with FUNCINPEC. Under this arrangement, all government posts were equally divided between the two parties. The CPP, however, retained control over most of the bureaucracy and the military.

The coalition government came to an end in July 1997, following armed clashes between forces loyal to Prince Ranariddh (FUNCINPEC) and Hun Sen (CPP) (Ledgerwood, n.d.a, p. 3). The confrontation left Hun Sen and the CPP largely in control of the country, and Ranariddh went into exile. The conflict between the coalition partners had built up over several years and illustrates the way in which Hun Sen's CPP had been very successful in weakening its opponents (even FUNCINPEC, which was its coalition partner and ostensible ally). Roberts sees the CPP's refusal to grant positions of authority to Ranariddh's followers at the grass-roots level as one of the root causes of the disagreements (Roberts, 2002, p. 527). The royalist FUNCINPEC was progressively weakened as key members were brought into the CPP through the offer of lucrative positions. As a result, Ranariddh lost power and authority within his own party, while the CPP extended its control over crucial ministries. In an attempt to bolster his position, Ranariddh tried to negotiate an alliance with the outlawed Khmer Rouge against the CPP. He was outmanoeuvred by Hun Sen, who succeeded in breaking up the Khmer Rouge, while integrating those willing to cooperate with him into the RCAF (Roberts, 2002, p. 530).

Beginning in 1996 the CPP encouraged Khmer Rouge leaders to defect by granting them amnesty and providing material incentives. Defectors were given government posts and control over land exempted from central taxation (Faulder, 1999). Of particular importance were the unofficial logging rights granted to defecting Khmer Rouge leaders, for which they did not have to pay taxes and which they could use for their own financial benefit (GW, 1998, p. 44). Ultimately, both the anti-communist opposition and the Khmer Rouge were defeated by bringing those willing to cooperate with Hun Sen and the CPP into the system and offering them attractive rewards in the form of an officially sanctioned share of economic and natural resource rents.

Another key element of the stabilization process was the bringing together of the four armed factions into the RCAF. In order to build loyalty among these disparate forces, Hun Sen pursued a policy of high military spending and granted army commanders control over logging profits and other economic activities (Hendrickson, 2001, p. 72). During the conflict, all factions engaged in natural resource extraction, in particular the timber trade, as a means of financing their military campaigns (TED, 1996). As the fighting declined, many military commanders continued these activities for private gain (GW, 1996, p. 2). The government has allowed them to continue in order to ensure military support and because high-level politicians benefit directly from the timber trade (GW, 1997a; 1997b). Such financial arrangements are not transparent, but informed sources, including the International Monetary Fund (IMF), have highlighted the existence of a parallel budget to the official government budget, with the former being largely financed by the proceeds of logging (GW, 1997b, p. 14). The latest report by Global Witness (GW), entitled *Taking a Cut,* details the complex timber network in operation that links political and commercial interests (GW, 2004). As a result, the security forces remain an important power base for the ruling CPP (ICG, 2000, p. 20). It has been difficult for the prime minister and international donors to introduce policies that run counter to military interests.

The stabilization process entrenched a system based on patronage and clientelism. The result has been a gradual reduction in direct political violence, although violent incidents continued to occur during the 1990s, particularly around the time of elections. While actual voting occurred mostly without incident during the first elections in 1993, international observers recorded widespread political violence against faction supporters and the broader electorate before and after polling day (UNDPKO, 2003; Ledgerwood, n.d.c).

Political rallies in the mid-1990s were often violently disrupted. In the most notorious incident, on 30 March 1997, at least 15 people died and more than 100 were injured when grenades were thrown at a crowd rallying in support of opposition politician Sam Rainsy (Ledgerwood, n.d.a, p. 3). Clashes in July 1997 left several people dead and buildings ruined (Ledgerwood, n.d.b).

Following these events, the CPP consolidated its power. The 1998 elections were largely peaceful, although opposition candidates were subjected to intimidation and targeted violence. Heder asserts that, 'having expelled FUNCINPEC from the state,' Hun Sen 'was able to use its coercive and financial assets to skew those election results in his favour' (Heder, 2003, p. 73). Other commentators have described the process as the creation of 'an ever more monolithic system' (Ledgerwood, n.d.a, p. 1) through which Hun Sen had effectively consolidated almost complete control over the state and electoral machinery (Ashley, 1998).

Khmer Rouge defectors stand behind their weapons during an integration ceremony near the Thai–Cambodian border in June 1998. More than a thousand former rebels returned to the Royal Cambodian Armed Forces that day. © Ou Neakiry/AP Photo

This led to a significant reduction in politically motivated violence, especially that directed against the electorate as a whole. Human rights organizations have observed a general shift towards non-violent methods of influencing the electorate, specifically through vote buying (HRW, 2002, p. 5; 2003a, pp. 8, 10). However, politics remains a risky business for those standing for election. The killing of opposition candidates, while declining in total numbers, continued during the elections of 2002 and 2003. During 2001 and 2002, 15 FUNCINPEC and Sam Rainsy Party members were killed (HRW, 2002, p. 16). In the run-up to the 2003 national election, several high-profile killings occurred and six other political murders were recorded (HRW, 2003b, p. 5). It is difficult to ascertain the scale of the problem country-wide. Nevertheless, there are well-documented cases of killings in particular districts, such as Tbong Khmum in Kompong Cham province. These were often carried out at the victim's home or following an arrest by elements of the police or possibly police impersonators. Victims were usually shot in the head from close range with a handgun. The perpetrators were occasionally also armed with an AK-47 assault rifle (HRW, 2002). The persistence of political violence in Cambodia and the impunity of its perpetrators point to broader governance problems in the country.

Hun Sen's ability to outmanoeuvre rivals by bringing them into his system also continues. Following the communal elections in 2003, two opposition parties, FUNCINPEC and Sam Rainsy, formed the Alliance of Democrats to counter CPP power, but in June 2004 Hun Sen announced that FUNCINPEC had again agreed to enter into a power-sharing arrangement with the CPP. Sam Rainsy fled the country and was sentenced in absentia in December 2005 to 18 months in prison for defaming Hun Sen. However, Sam Rainsy returned in February 2006, announcing that his personal disputes with Hun Sen had been set aside, and was provided with 12 bodyguards from the Ministry of Interior for his protection (*Economist,* 2006).

Politics remains a risky business for those standing for election.

Having outlined the historical and political background, the following sections of this chapter analyse how small arms proliferation has been tackled in Cambodia, and what issues remain relating to the use and misuse of small arms.

TACKLING SMALL ARMS PROLIFERATION IN CAMBODIA

Small arms reduction and control measures

The Cambodian government made control of weapons proliferation one of its first priorities after the elections in 1998. Following the elections, a sub-decree restricted weapons ownership and use. The Ministry of Interior also launched countrywide weapons collection initiatives. In 1999, Prime Minister Hun Sen initiated a programme of public weapons destruction. In the same year, the Cambodian government approached the European Union (EU) for support for small arms control measures, and in April 2000 the EU launched its ASAC programme (Wille, 2005).

The EU ASAC programme was set up as an agency with a single mandate to support small arms and light weapons control processes in Cambodia. One of the first joint actions established under the European Common and Foreign Security Policy, EU ASAC provided a framework for coordinated support by interested EU member states, as well as the European Commission (Wille, 2005). Initially, Canada and Japan also opted to channel their funding through the EU ASAC programme. In 2003, however, Japan established a parallel programme called JSAC, which included activities similar to EU ASAC, but operated in different provinces and focused on the police rather than the RCAF (JSAC, 2006).

The approach of the EU ASAC programme has been to support closely the efforts of the Cambodian government in bringing civil war weapons under government control. It was agreed that the government and EU ASAC would take on separate but complementary roles. While the Cambodian authorities retained legal responsibility for collecting

A policeman stands before a pile of nearly 7,000 guns and other small arms in Kampong Cham province north of Phnom Penh in July 2001. Cambodian authorities lit a massive bonfire to destroy the weapons as part of an ongoing effort to rid Cambodia of illegal arms. © Chor Sokunthea/Reuters

weapons, the EU ASAC programme supported public awareness and funded development schemes. One of the main activities of the programme has been to finance local projects (in particular the digging of wells) in communities where sufficient numbers of weapons had been collected. The EU ASAC programme has also provided financial and technical support for the destruction of surplus military weapons, as well as collected and confiscated weapons. In 2003 an arms cache component was added to the EU ASAC programme. The programme also contained security sector management-related elements, which are discussed later (EU ASAC, 2006).

The EU ASAC programme provided essential support to a government initiative (ensuring local ownership), while allowing for civil society participation (notably the WGWR) and ensuring overall donor coordination. An especially important contribution of the project was its support for awareness raising on the official policies and changes in the law, which disseminated information on government policies to rural Cambodia (de Beer, 2005a).

Estimates of reductions in small arms proliferation

In order to gauge the impact of small arms control programmes in Cambodia, it is important to estimate the number of guns circulating outside of government control before and after the programmes took place. This is difficult to quantify, because estimates of the numbers of weapons circulating outside of government control have been extremely unreliable, e.g. in 1998 the accepted figure was more than 500,000 guns, while others suggested it may have been as high as 900,000 (Fawthrop, 2001). These estimates are now viewed with scepticism, however (Tieng, Long, and Hicks, 2004, pp. 11–12).[1]

In order to assess the impact of the recent weapons collection and destruction programmes, the Small Arms Survey has reviewed available evidence on the number of guns controlled by government and circulating outside of government control as of mid-2005. These estimates are compared with numbers derived from historical analysis of the numbers of soldiers deployed and guns held by each of the four factions and evidence on what happened to these weapons following the conclusion of the Paris Agreements. The main assumptions made and information used in deriving these estimates—discussed in detail in Wille (2006a)—can be summarized here as follows:

- It is assumed that the overwhelming majority of the guns presently held in Cambodia originated from the arsenals of the four factions. There is little evidence that there have been large-scale imports or exports of weapons since 1991. It is also assumed that civilian hunting and sporting firearm ownership at the time of the end of Khmer Rouge rule was negligible.

- For the small anti-communist factions of the KPNLAF and ANKI, a ratio of two guns per soldier is assumed. As there is evidence that the Khmer Rouge were better armed, a ratio of 2.5 guns per soldier is assumed for this group. For the government army we also assume a ratio of 2.5 per soldier. For the government's provincial units and militia forces, two guns per soldier is assumed. The total weapons stockpile of the factions is estimated to have been between 319,500 and 462,500 firearms, which is similar to the figure of 350,000 provided by the factions to the UN in 1991 (UNDPKO, 2003, p. 4).

- Following the conclusion of the Paris Agreements, the CPAF, KPNLAF, and ANKI joined forces to become the RCAF. It is assumed that 70 per cent of their weapons passed into government stockpiles held by the RCAF and police, and that 30 per cent of their weapons leaked into private hands or were stored in arms caches.

- Between 1993 and 1996 around a quarter of the Khmer Rouge defected to the government. Again it is assumed that 70 per cent of their weapons passed into government control. Following the end of the Khmer Rouge insurgency, the remaining fighters and weapons were never properly integrated into the government's security forces. Many of their weapons passed into private hands or were stored in arms caches, and the remainder were held by security forces within the Khmer Rouge enclaves (e.g. Pailin and Anlong Veng).

- Government reports indicate that since 1998 around 120,000 weapons have been collected from private ownership and weapons caches (Ratha, Long, and Vijghen, 2003, p. viii). JSAC reports having collected 11,662 weapons (JSAC, 2006). These weapons were added to government stockpiles. The government, with the support of EU ASAC, has destroyed a total of 180,000 surplus weapons in its stockpiles (collected and military stocks) (EU ASAC, 2005b).

> A substantial proportion of uncontrolled small arms have been collected and destroyed.

Figure 5.1 shows the relevant numbers and flows of weapons; these are discussed more fully in Wille (2006a). The estimates are presented as ranges because of the uncertainties involved. It is estimated that at present there are 107,000 to 188,000 weapons in government stockpiles, and 22,000 to 85,000 weapons remaining outside of government control.

This analysis indicates that government and donor-supported weapons collection and destruction efforts have had a major impact. Prior to collection there were an estimated 154,000 to 216,000 guns circulating outside of government control. Weapons collection programmes have removed 131,000 of these from circulation. The estimates are associated with large uncertainties, but it is reasonable to assert that weapons collection removed a substantial proportion, and perhaps the vast majority, of weapons circulating outside of government control.

Indicators on small arms use in Cambodia today

The available evidence points to a considerable reduction in firearms availability in Cambodia. How has this development affected levels of gun use, violence, and crime? The public health and criminology literature suggests that the proportion of violent acts involving guns is likely to decrease if controls over firearms are strengthened or large

Figure 5.1 **Estimated numbers of guns under and outside of Cambodian government control**

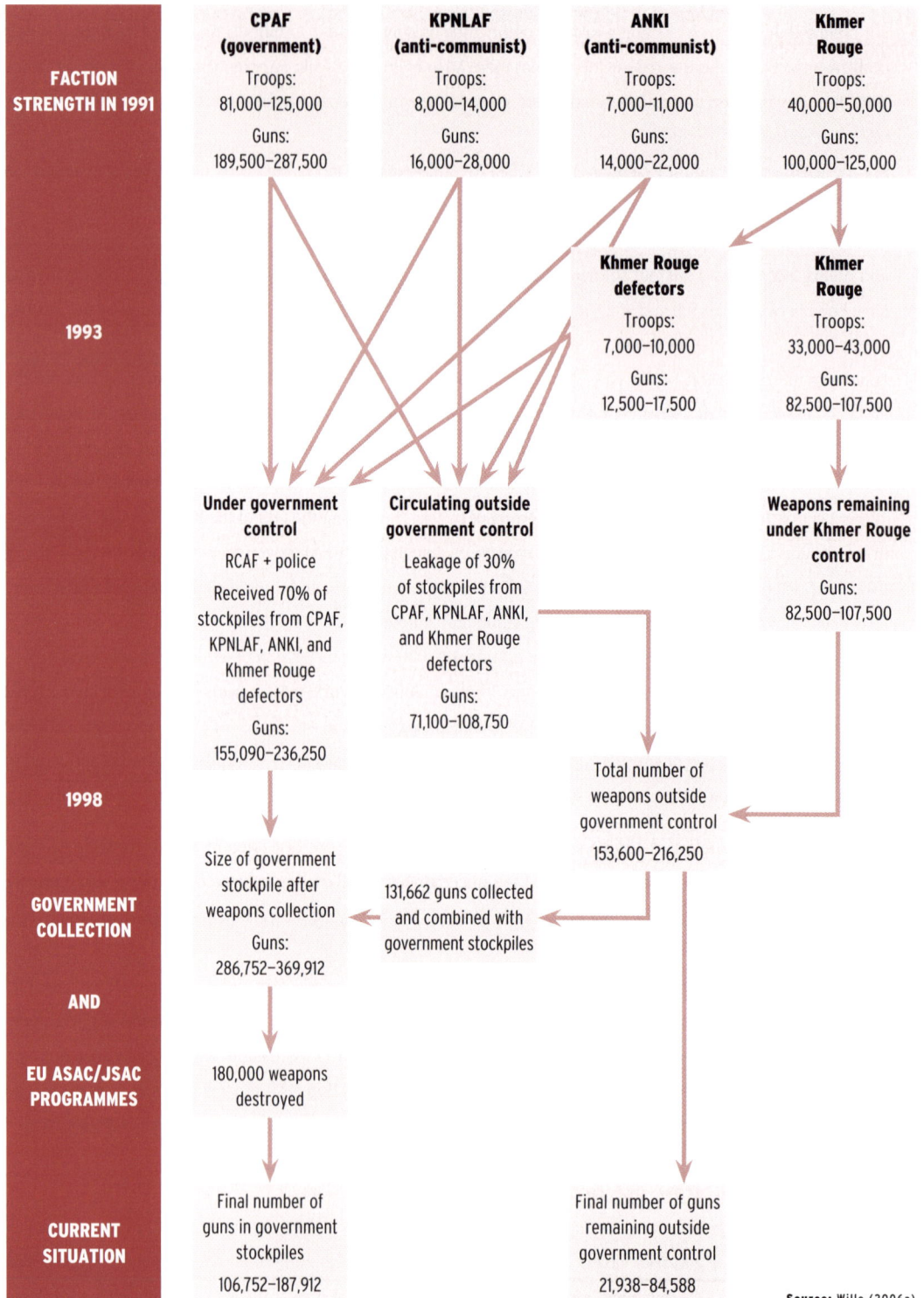

| FACTION STRENGTH IN 1991 | **CPAF (government)** Troops: 81,000–125,000 Guns: 189,500–287,500 | **KPNLAF (anti-communist)** Troops: 8,000–14,000 Guns: 16,000–28,000 | **ANKI (anti-communist)** Troops: 7,000–11,000 Guns: 14,000–22,000 | **Khmer Rouge** Troops: 40,000–50,000 Guns: 100,000–125,000 |

1993

Khmer Rouge defectors
Troops: 7,000–10,000
Guns: 12,500–17,500

Khmer Rouge
Troops: 33,000–43,000
Guns: 82,500–107,500

Under government control

RCAF + police

Received 70% of stockpiles from CPAF, KPNLAF, ANKI, and Khmer Rouge defectors

Guns: 155,090–236,250

Circulating outside government control

Leakage of 30% of stockpiles from CPAF, KPNLAF, ANKI, and Khmer Rouge defectors

Guns: 71,100–108,750

Weapons remaining under Khmer Rouge control

Guns: 82,500–107,500

1998

Size of government stockpile after weapons collection

Guns: 286,752–369,912

Total number of weapons outside government control

153,600–216,250

GOVERNMENT COLLECTION

131,662 guns collected and combined with government stockpiles

AND

EU ASAC/JSAC PROGRAMMES

180,000 weapons destroyed

CURRENT SITUATION

Final number of guns in government stockpiles

106,752–187,912

Final number of guns remaining outside government control

21,938–84,588

Source: Wille (2006a)

numbers of firearms are withdrawn from society.[2] Overall crime and violence may well remain steady or even increase due to other factors, but typically will not involve small arms to the same extent as before such interventions. This section examines the impact of weapons reduction efforts in Cambodia on the proportion of homicides, violent injuries, and robberies that involve small arms, and discusses how these effects relate to overall levels of violence.

It is difficult to derive gun violence indicators for Cambodia because of the paucity of government statistics. While official crime statistics do record homicides, there is likely to be significant under-reporting (Broadhurst, 2002) and they do not indicate the murder weapon. It was therefore necessary to combine press information on the proportion of firearm use reported in assaults and homicides, data on causes for hospital admissions, and official homicide statistics. The Small Arms Survey conducted extensive research to obtain this data. A total of 3,699 incidents reported in the *Phnom Penh Post* between 1994 and 2004 were classified according to whether the homicide or violent incident was committed using a firearm or some other instrument (Flärd, 2005). A Small Arms Survey researcher collected hospital admission data from nine provincial hospitals for the period 1991–2004 that includes categories of bullet injuries and other cases of assault (Son, 2005).

Both these sources point to a considerable reduction in the use of firearms. Among all violent incidents reported by the *Phnom Penh Post*, the use of firearms declined from 80 per cent in 1994 to 30 per cent in 2004. The findings are similar when the analysis is restricted to homicides rather than all reported violent incidents. In 1994, 69 per cent of reported homicides were committed using a gun. By 2004 this figure had dropped to 30 per cent. The hospital admissions data indicates a similar trend. In 1993, 65 per cent of victims of assault admitted to hospital had suffered from bullet wounds. By 2004 this figure had fallen to just 2.6 per cent (Wille, 2006b).[3]

Official crime statistics confirm the conclusions reached in this analysis (Cambodia, 2003). Homicide data indicates a downward trend since 1998. Wille (2006b) discusses interpretation and biases in this data in more detail.

Official crime statistics also report incidents of armed robbery and other types of theft not involving the use of firearms. Again, there is evidence of a shift from the former to the latter. The reduction in gun use appears to be less dramatic

Figure 5.2 **Rate of firearms use in homicides, acts of violence, and causes of hospital admissions (%), 1991–2004**

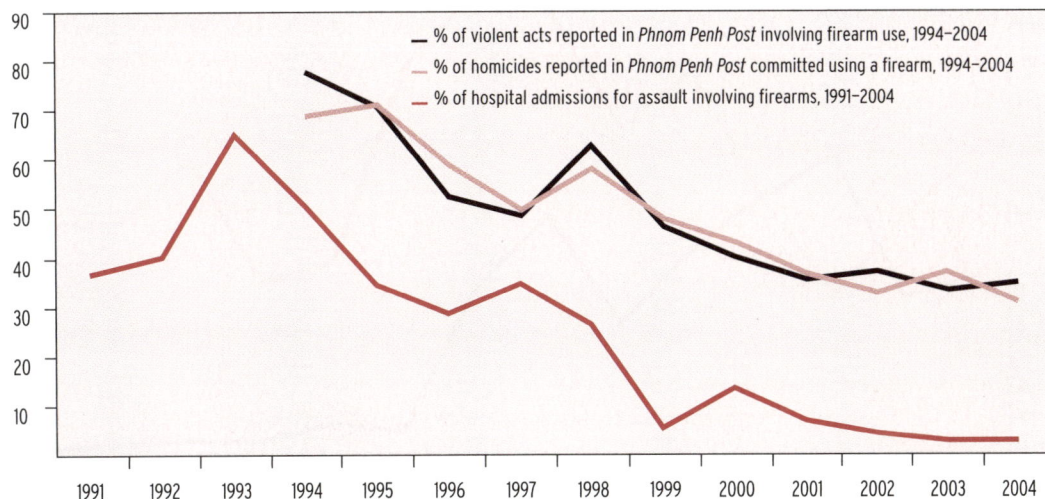

— % of violent acts reported in *Phnom Penh Post* involving firearm use, 1994–2004
— % of homicides reported in *Phnom Penh Post* committed using a firearm, 1994–2004
— % of hospital admissions for assault involving firearms, 1991–2004

Source: Wille (2006b)

for theft than for acts of violence (as indicated by the press reports and hospital admissions data discussed above). As a proportion of all reported offences, armed robbery only fell by 8 per cent between 1996 and 2004. There is also evidence of an increase in armed robbery between 2003 and 2004, but it remains to be seen whether this amounts to a trend.

Figure 5.3   **Total number of acts of violence reported in the Phnom Penh Post by instrument used, 2000–04**

NUMBER OF INCIDENTS PER YEAR

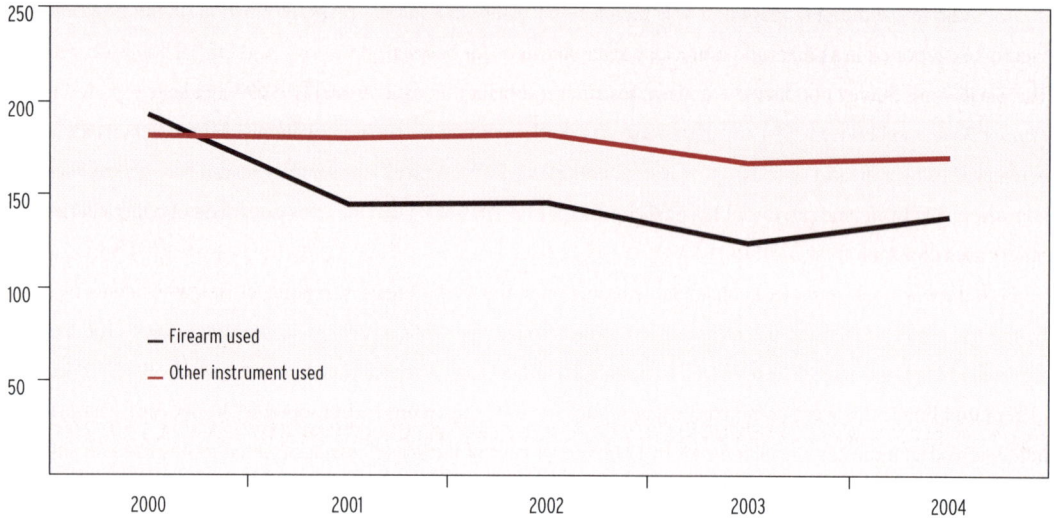

Source: Wille (2006b)

Figure 5.4   **Hospital admissions due to assaults with firearms and with other instruments, 1991–2004**

PER CENT OF TOTAL ADMISSIONS PER YEAR

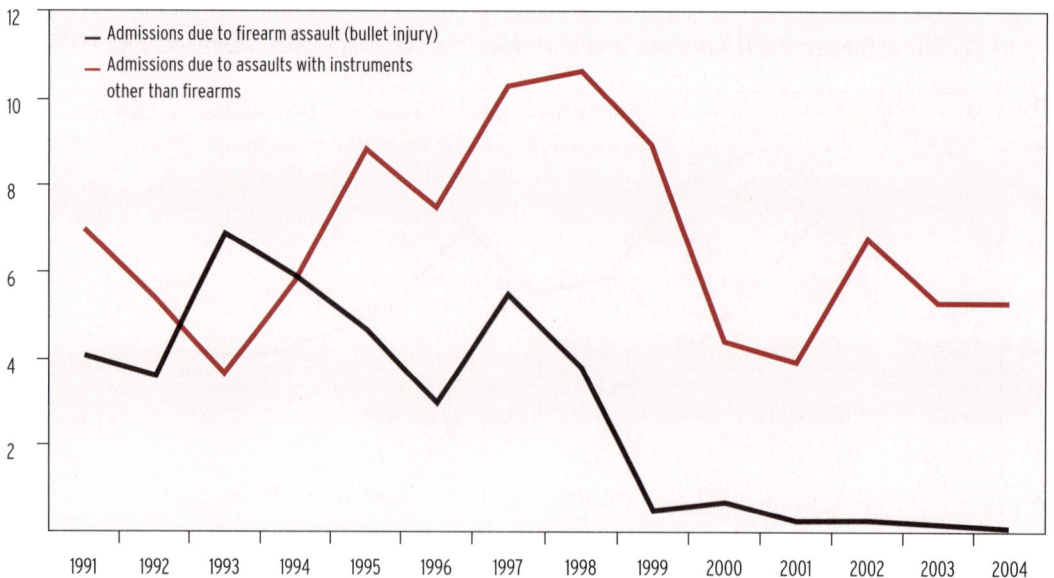

Source: Wille (2006b)

Figure 5.5 **Decline in homicides per 100,000 people, 1998–2003**

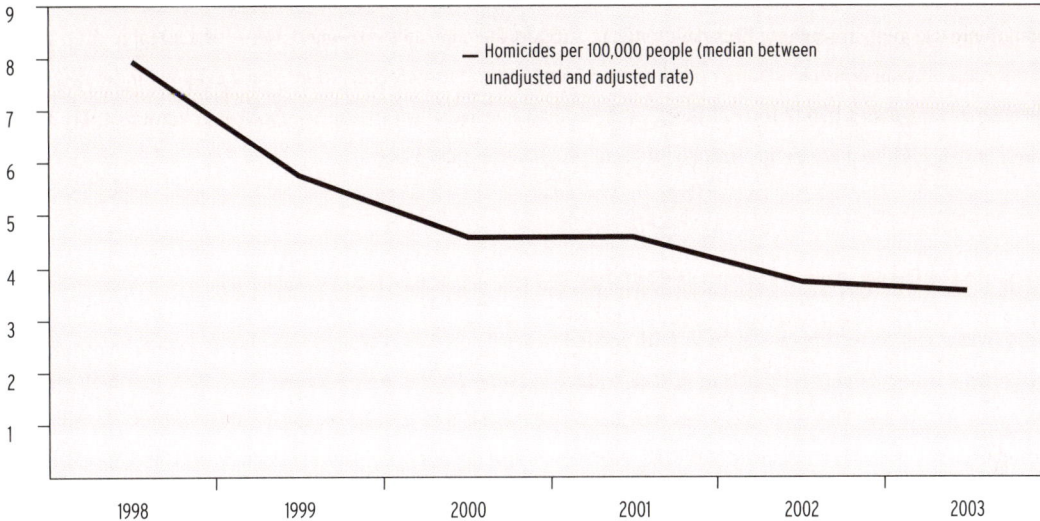

Source: Wille (2006b)

Figure 5.6 **Robbery with and without firearms as percentage of all officially reported offences, 1996–2004**

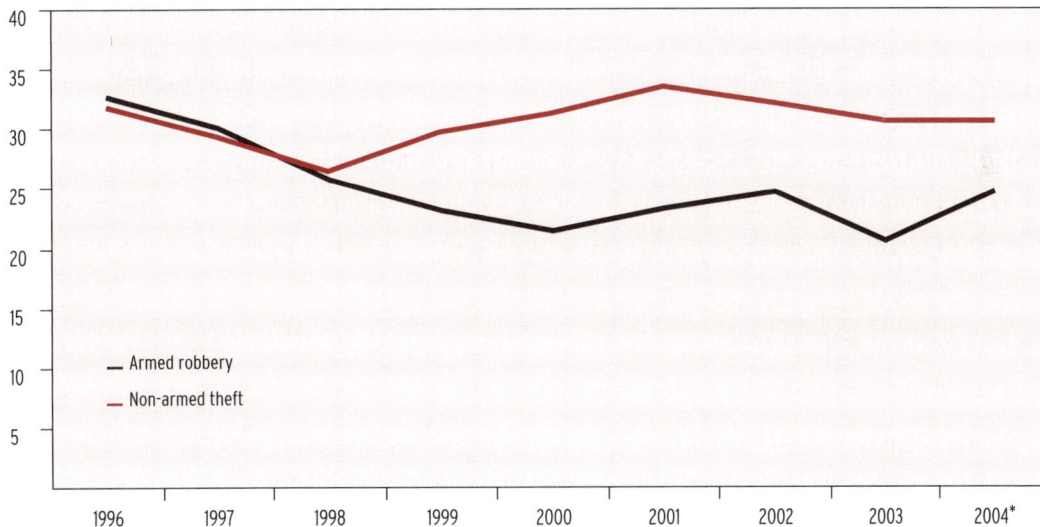

* First nine months.
Source: Wille (2006b)

Recent surveys also indicate that armed robbery is still common. Data gathered in a survey by the WGWR carried out in 2004 found that armed robbery was the most commonly described gun incident. Three-quarters of all interviewed households could describe in detail a gun incident that had occurred during the previous three years (Tieng, Long, and Hicks, 2004, p. 28). Guns remain a popular tool to facilitate theft, although they are most commonly used to intimidate, threaten, and coerce rather than to injure or kill. The survey found that the overwhelming majority of victims of armed robbery did not sustain any physical injuries (Tieng, Long, and Hicks, 2004, p. 27).[4]

Taken together, these sources of data clearly indicate a decline in gun use among the general population. Guns are far less commonly implicated in injury and death than they were ten years ago. Criminals, on the other hand, still commonly use guns in robbery. This data tends to support the estimates presented above of a sizeable drop in the proportion of small arms circulating outside government control. It correlates, more specifically, with the demise of the Khmer Rouge as a bandit force, as well as the comprehensive effort by the Cambodian government, EU ASAC, JSAC, and WGWR to address uncontrolled firearm proliferation post-1998.

GUN USE AMONG THE CAMBODIAN AUTHORITIES AND SECURITY FORCES

With civilian gun ownership mostly prohibited, and following the collection of most unauthorized guns, the possession of firearms is now largely restricted to official bodies, in particular the armed forces, the police, and high-ranking government officials. This section discusses evidence for the use and misuse of guns by these groups. It should be stated from the outset that instances of official weapons misuse are not systematically documented, and are extremely difficult to investigate. However, there is a growing body of evidence of sufficient weight to raise serious concerns about the misuse of guns by public officials. Civil society organizations have been very active in documenting abuses of power, many of which have involved guns (Adhoc, Licadho, and HRW, 1999). United Nations reports by the Special Representative of the Secretary-General for Human Rights in Cambodia have also drawn attention to serious violations of human rights committed by members of the military, the police, the gendarmerie, and other armed forces (Leuprecht, 2004, p. 2).

General evidence on gun use by public officials is provided by a survey carried out by the WGWR (Ratha, Long and Vijghen, 2003). Police, militia, and soldiers topped the list of those 'most likely to use a gun to get their way'. Government officials are also reported to be frequent users of guns (Ratha, Long, and Vijghen, 2003, p. 29). The survey revealed that these groups used guns in different ways: some legitimate, but others clearly indicating an abuse of power. According to the survey, government officials, police, soldiers, militia, or security guards were involved in

---

Box 5.1 Cambodia's law on the management of weapons, explosives, and ammunition

In 1999, Sub-decree 38 categorized all weapons as government property. Yet the decree contained loopholes and was not approved by Parliament. In 2000 EU ASAC hired an international legal adviser to work with the Interior and National Defence Ministries to draft new legislation. The resulting draft law, incorporating most aspects of Sub-decree 38, was completed in 2001 and then circulated to civil society groups for discussion in a series of round tables and national workshops.

The new Law on the Management of Weapons, Explosives and Ammunition (Cambodia, 2005), finalized in June 2005, maintains strict restrictions on gun ownership by civilians. Significantly, it imposes stringent penalties for gun-related offences, including the unauthorized possession, carrying, selling, purchase, lending, hire, production, and repair of weapons.

The law also addresses the responsibilities and duties of police, military, and other government officials, prescribing strict punishments for the misuse of registered government weapons.

Under the new law, the Ministries of Interior and National Defence authorize and monitor arms use, storage, and transport. All weapons must be registered with the government. The law also provides for an amnesty of three months during which illegally held arms can be handed in to the authorities without punishment.

As of January 2006, the Cambodian government had yet to introduce an implementation plan for the law. It also needs to be disseminated and explained to the general public, civil society organizations, the police, and the military. A key, longer-term challenge is the law's effective, consistent, and impartial application.

Source: Eileen Kilgour, WGWR

Table 5.1 Persons 'most likely to use a gun to get their way' in Cambodia, 2002	
	%
Police, militia, and soldiers	29
Robbers and criminals	22
Government officials	14
Security guards	6
Powerful rich men	6
Other (including political party members and civilians)	11
Unspecified	12

Source: Ratha, Long, and Vijghen (2003, p. 29)

16–39 per cent of reported incidents in which a gun was drawn for the purpose of extorting money (Ratha, Long, and Vijghen, 2003, p. 33).

Concerns about the use and misuse of guns by the security forces indicate two sets of problems. Firstly, there are concerns that the security forces do not perform effectively in upholding the rule of law and tackling crime. Secondly, there are concerns about the abuse of power by security forces, and their ability to flout the law with impunity.

This section illustrates these issues using two examples, which together cover some of the most serious governance and security problems in Cambodia: (i) gun misuse/abuse by the police force; and (ii) coercion in the logging industry.

The police force

Various international instruments define practices for firearm use by the police. The two most important are the *UN Code of Conduct for Law Enforcement Officials* (UNGA, 1979) and the *Basic Principles on the Use of Force and Firearms by Law Enforcement Officials (UN Basic Principles),* adopted in September 1990 (UN, 1990; see Small Arms Survey, 2004, ch. 7).

Although these instruments are not legally binding, they do establish a set of international norms that are widely accepted. In particular, they subject firearms use by police officers to the principles of *necessity* and *proportionality*. They also address the problem of corruption. A complete assessment of the extent to which these principles have been adhered to in Cambodia has never been attempted, and is beyond the scope of this chapter. However, newspaper reports, civil society organizations such as the WGWR, and reports by the Special Representative of the UN Secretary-General clearly document a number of problems.

A review of reports on crime incidents contained in the *Phnom Penh Post* suggests that police officers struggle to adopt effective tactics to deal with armed robbery without unnecessary firing of arms. Many reports describe officers intervening in a crime scene by shooting the suspected criminal. Such conduct has put the police officers' own safety at serious risk, has killed innocent bystanders, and ultimately pre-empts the judicial system (Flärd, 2005). It also contravene Principles 4 and 5 of the *UN Basic Principles,* which stipulate that firearms must only be used as a last resort, and in proportion to the seriousness of the offence (UN, 1990).

There does appear to have been some improvement in police tactics over the years. Analysis of reported deaths in the *Phnom Penh Post* shows that in 1997, 12.6 per cent of all reports of firearms deaths in the newspaper referred to circumstances where the police had killed a suspect. By 2002 such reports accounted for only 3.7 per cent of firearm deaths. Such events are still far from being rare or unusual, however (Flärd, 2005).

An effective police force will not only handle firearms in accordance with international standards, but will provide effective and humane protection against crime, violence, and civic strife (Marenin, 2005, p. 1). The lack of effective policing has hampered efforts to tackle firearm crime in Cambodia. Surveys of crime victims by the WGWR report very low arrest rates following gun crime incidents: 11 per cent in the 2002 survey and 20 per cent in the 2003 survey (Ratha, Long, and Vijghen, 2003, p. 36; Tieng, Long, and Hicks, 2004, p. 44, respectively).

There are several reasons why the police are ineffective in tackling crime, and gun crime in particular. Inadequate training, resource constraints, low pay, lack of equipment, and weakness of forensic and investigative capacity all clearly hamper the ability of the police to combat crime. Furthermore, the core values and norms that govern the behaviour of the police play a major role. In the Cambodian context, informal relationships often override formal rules, providing the dominant set of incentives that govern individual behaviour. Anthropologists and sociologists have pointed to the importance of patron–client relationships in Cambodia. Ledgerwood, for example, states that Khmer people tend to 'organize their daily interactions with others by attaching themselves to someone higher in the social hierarchy' (Ledgerwood, n.d.d, p. 5). These complex webs of social relations and obligations provide powerful individuals with a support base, and offer a means of social advancement for their supporters.

The behaviour of the police in Cambodia is clearly affected by such social relations. Investigations by human rights organizations of homicides linked to the security forces over a 22-month period in 1997 and 1998 find that none of the 209 alleged perpetrators had been brought to justice (Adhoc, Licadho, and HRW, 1999, pp. 37–38). The report concludes that both the courts and the police are vulnerable to intervention, pressure, and directives from high-ranking political leaders or other powerful people protecting their subordinates (Adhoc, Licadho, and HRW, 1999, p. 3).

Recent public surveys also confirm that there is a commonly held perception that powerful people can behave with impunity (Ratha, Long, and Vijghen, 2003, p. 36). The Special Representative of the UN Secretary-General in Cambodia highlights a 'continuing problem of impunity' where 'individuals and groups responsible for committing serious violations of human rights, in particular members of the military, police, the gendarmerie and other armed forces, have not been arrested or prosecuted' (Leuprecht, 2004, p. 2).

Formal rules governing police behaviour are not well respected. Police corruption is common, and police often sell their services, including the threat of the use of official force, on the private market (Hendrickson, 2001, p. 74). The WGWR reports that police officers sometimes demand payment from victims before investigating a case (Tieng, Long, and Hicks, 2004, p. 45). Police work may also overlap with private business, sometimes on a large scale. For example, the Ministry of Interior part owns the private company MPA Security Services. Half of the company's personnel come from the Ministry of Interior, and are mostly serving police officers (Hendrickson, 2001, p. 74).

Another aspect of police behaviour in Cambodia has been the mixing of law enforcement and judicial functions. Police officers reportedly perform quasi-judicial functions by offering mediation services to both victims and perpetrators (Tieng, Long, and Hicks, 2004, p. 45). Some observers claim that as many as a quarter of criminal cases are mediated by the police directly, and are never referred to a court of law (Ratha, Long, and Vijghen, 2003, p. 36).

All of these examples raise serious concerns about the role of the police, both in the way that they use guns and in their ability to tackle gun crime. In many ways the police have become part of the gun problem in Cambodia

Recent surveys confirm that powerful people are perceived to act with impunity.

rather than the solution. Transforming their role will require major training efforts and institutional reform, including of the judicial system—issues reviewed in the discussion of SSR below.

Coercion in the logging industry

Concessions for fisheries, plantations, and forest exploitation have been granted since the early 1990s. In 2005 it was reported by the *Phnom Penh Post* that 14 per cent of Cambodia's total landmass was administered under concession (Hamilton, 2005), which suggests a decline from 35 per cent of Cambodia's territory reportedly given out under concession in 1995 (GW, 2002, p. 3). According to GW, the armed forces (i.e. the RCAF) are effectively the largest land concession holder in Cambodia. The army was granted over 700,000 hectares of so-called Military Development Zone land in the early 1990s as part of the peace dividends (GW, 2004, p. 26).[5] Officially all concession logging activities were suspended in December 2002, but there is evidence that illegal logging continues (World Bank, 2004, p. 2; UNCOHCHR, 2004; EIC, 2004, ch. 9.2).

In a series of reports, GW has documented the extent of logging and the processes involved. Private companies carry out most of the logging, while the military is closely involved with providing security services in return for part of the income. Many of the security personnel are detached or 'inactive' soldiers, who remain on the RCAF payroll (UNCOHCHR, 2004, p. 28). There are also reports that armed RCAF soldiers have been dispatched to protect company property (UNCOHCHR, 2004, p. 29).

Firearms are used as a tool to protect logging concessions and to extract unofficial 'taxes' from commercial logging companies and local people. Junior officials within the RCAF, the police, and the forest administration system reportedly collect payments from wood traders at every stage of the timber extraction process. This is facilitated by the fact that most logging operations contravene forestry law (GW, 2004, p. 19).[6] Payments provide protection against law enforcement, and are passed up the chain of command from junior officers to superiors, and ultimately to senior officials in the capital (GW, 2004, p. 19). Business people involved in the logging industry do not perceive extortion by the military or other groups as a threat to their business, because the trade is very profitable, in spite of the heavy 'taxes' levied on it (GW, 2004, p. 20).

> Firearms are used to extract 'taxes' from commercial logging companies and local people.

The logging industry—and other concessions connected with agricultural land use and fishing rights—has created conflicts with the resident rural population. In many cases commercial activities directly threaten traditional livelihoods. There are several documented cases of land grabbing and forced displacement of local people. Firearms have commonly been used as a tool of intimidation and coercion in such cases (UNCOHCHR, 2004, pp. 26, 28, 29). In some instances, security guards protecting forestry concessions have taken farmland from local people for their own use (Ratha, Long, and Vijghen, 2003, pp. 43–44). Such practices, underpinned by the coercive power of official weapons, undermine human security in Cambodia and point to the urgency of SSR.

THE SSR AGENDA IN CAMBODIA

SSR is a broad-ranging concept that covers the reform of the entire security system, including the state security forces (military, police), the civilian bodies that manage the security apparatus, and non-statutory security providers. The SSR policy agenda covers three interrelated aspects: the institutional framework for security provision, governance of the security institutions, and accountability of the security forces (OECD, 2005).

The Organisation for Economic Co-operation and Development–Development Assistance Committee (OECD–DAC) guidelines (OECD, 2005) identify the following actors that comprise the security sector: (a) *core security actors* (such as the armed forces, the police, etc.); (b) *security management and oversight bodies* (such as legislative select committees, financial management bodies, and civil society organizations); (c) *justice and law enforcement institutions* (such as the judiciary, prosecution services, and human rights commissions); and (d) *non-statutory security forces* (such as liberation armies, private bodyguard units, private security companies, and political party militias).

SSR has come to be accepted as a core pillar of efforts to reconstruct post-conflict and collapsed states, and to facilitate the democratic transition of post-authoritarian states. However, the concept is less than a decade old, and did not inform the design of post-conflict measures implemented in Cambodia in the early 1990s. The links between small arms control and the broader SSR agenda are now widely recognized (Sedra, 2006). Most obviously, small arms control programmes that remove civil war guns from uncontrolled circulation can be regarded as a crucial precondition for SSR. SSR only makes sense when the state security forces exercise a monopoly over the legitimate use of force. By supporting the Cambodian government's efforts to bring all weapons within the country under its control, small arms collection and destruction programmes have contributed to this important precondition for SSR.

Security sector efforts in Cambodia have to a large extent focused on strengthening the core state security forces. A number of countries, including Australia, France, Indonesia, and the United States, provided military assistance to the RCAF to increase its fighting capabilities against the Khmer Rouge. However, Australia and the United States suspended defence cooperation following the events of July 1997. Hendrickson reports that the Chinese filled the gap at this point (Hendrickson, 2001, pp. 70–71). Australia is presently helping build police capacity to combat transnational crime (Australia, 2002). France maintains attachés for the police, military, and gendarmerie at its embassy. The Chinese and Vietnamese have worked with the military and have provided police training, but little is known about the specifics of this support.

> **Programmes that remove civil war guns from uncontrolled circulation are a precondition for SSR.**

Some of this work was also carried out under the umbrella of small arms projects. Within their remits, the EU ASAC and JSAC programmes have been active in building the institutional capacity of the armed forces (including the navy and air force) and the police. The main activities have been providing the armed forces with safe storage facilities for weapon stocks and introducing a central weapons register (EU ASAC, 2005b). In addition, other equipment, such as bicycles and radios, was also provided to the police (EU ASAC 2003c, p.12). The JSAC programme built safe storage facilities for the police in Otdar Mean Chey, Bantey Mean Chey, and Siem Reap provinces (JSAC, 2006).

These interventions have addressed a serious institutional shortcoming in stockpile management. The accompanying small arms destruction process has also helped to rationalize stockpiles and reduce the risk of future leakage. Important legislative changes have also strengthened governmental control over small arms throughout the country. These programmes have helped to stabilize Cambodia and restore the role of core security forces. But the rest of the SSR agenda in Cambodia remains largely unfinished. Perhaps most importantly, efforts to reform government institutions responsible for overseeing the security forces have not advanced far. Nor has Cambodia's defence sector been revamped. There is a need to restructure and professionalize the army in the light of the modest level of threat posed to Cambodia's external security.

During the 1990s there were several largely unsuccessful attempts by donors to promote army downsizing and demobilization, with economic considerations the major driving force. Donors expressed concern at the size of the military budget, estimated at 3.3–5 per cent of gross domestic product (World Bank, 1999, p. viii), and argued that

A riot police officer fires at anti-government demonstrators during a clash in Phnom Penh in September 1998. © Sakchai Lalit/AP Photo

these resources could be used more productively to invest in health and education. Yet in 1997 the Cambodia Veteran Assistance Programme was suspended following the election-related violence in July. Its successor programme initially achieved some results in reducing the payroll. In 1999, 15,551 irregulars were removed from the payroll, and in the following two years 1,500 soldiers and 15,000 ex-combatants were demobilized (EIC, 2005). However, the second phase of this programme was suspended in 2003 on the discovery of corrupt procurement practices within the Cambodian government.[7] As of the end of 2005 the size of Cambodia's army and military budget remained a cause of concern. The government has expressed its commitment to continued reform in the *White Book for National Defence Reform* (EIC, 2005, pp. 66–67). It remains to be seen whether more progress will be made over the next few years.

This history highlights an important consideration for both SSR and small arms management. Richardson and Sainsbury (2005) argue that the RCAF demobilization programme has been unsuccessful because donors misjudged the root of the problem. They believed that the excess military spending was the result of an excess of soldiers and failed to recognize the role of patronage and corruption in siphoning off resources, which is a key feature of the military–state relationship, as outlined earlier (Richardson and Sainsbury, 2005, p. 291).

Small arms programmes have not been ignorant of SSR issues, and there have been some attempts at addressing them. For example, the EU ASAC programme has also undertaken small-scale activities aimed at improving relations between the security forces and the general public (de Beer, 2005b; EU ASAC, 2003a). While these programmes appear to have resulted in an improved perception of security among the population (EU ASAC, 2002; 2003b), they have not addressed the overall institutional culture and structural weaknesses in the police force.

The fundamental problem with the security sector in Cambodia is the weakness of democratic oversight and the inability of other state and civil society institutions to provide checks and balances. In the absence of greater accountability of the Cambodian security forces to the people of Cambodia, the benefits of other donor-supported elements of SSR will be limited. Hendrickson concludes that unless civilian control over the security forces is strengthened, it is unlikely that the abuse of firearms by official bodies can be effectively tackled (Hendrickson, 2001, p. 79).

CONCLUSION

Great progress has been made in bringing about stabilization in Cambodia since 1993 due to a range of factors, including the introduction of elections. Small arms control programmes have made an important contribution to the stabilization process. The joint efforts of the EU ASAC/JSAC programmes and the Cambodian government have success-fully removed a substantial proportion—perhaps the vast majority—of weapons from circulation outside government control. These programmes have also helped improve security sector management, with new laws and practices governing the safe storage of government stockpiles. Large-scale destruction of surplus military stockpiles has also reduced the danger of future leakages and uncontrolled exports.

This chapter has presented new evidence of the impact of these changes on Cambodia. It is clear that guns are far less commonly used in violent incidents and, to some extent, in ordinary crime. Cambodia thus demonstrates the significant human security gains small arms control measures can yield with a relatively modest investment over the space of a few years. The EU ASAC approach of working closely with the government and civil society actors similarly provides a useful model that could be replicated elsewhere.

Small arms control programmes have contributed to important changes in Cambodia, but have not addressed all of the human security problems arising from gun use in Cambodia. With the removal of most guns from civilian hands, the possession of guns is now largely restricted to public officials. Yet many of these officials misuse their weapons. To a large extent, these problems stem from the entrenchment of a system of patronage and clientelism during the period of Cambodia's post-conflict stabilization. Hence they can only be addressed through fundamental change in the political, economic, and social systems in the country. There is a pressing need to address such issues as the role and size of security forces, and their competence, professionalism and integrity, rules of engagement, democratic governance, and oversight.

These issues will be more difficult to address in Cambodia than illegal civilian small arms possession and use. Many are highly sensitive politically, because they are fundamental to the power relations and patron–client networks that keep the regime in power. While broader-ranging reforms to the security sector will be difficult to achieve, it is important that the international community not lose sight of this objective. A useful first step would be for donors to recognize more openly the deep-seated governance problems in Cambodia, and the failings of the security forces in particular. Donor pressure on the Cambodian government has often been insufficient in this respect, though donors have occasionally lost patience with the government.[8]

As experience from other South-East Asian countries indicates, SSR is a long-term process that is highly political in nature (Huxley, 2001). Although external actors have a legitimate role to play in stressing key principles and goals, SSR cannot rest on external pressure alone. The involvement of civil society will be crucial, as will the genuine buy-in of Cambodia's top leadership and the gradual strengthening of the powers of oversight bodies, in particular the judiciary. Democratization processes in Cambodia provide some grounds for optimism, although SSR has not yet been a major topic of social debate. Civil society organizations, while still limited in their capacity and influence, are beginning to address security sector issues, and are gradually emerging as sources of pressure on government. The media are also relatively free and have highlighted numerous cases of abuse by the security services. If SSR is to take root, it will also have to reflect local social norms and concerns (Clegg, Hunt, and Whetton, 2000, p. 88).

Cambodia provides important lessons for other countries with ongoing post-conflict stabilization programmes, such as Afghanistan and Iraq. The experience of Cambodia demonstrates the gains that can be made through small arms control programmes. However, it warns that improvements in security will only be partial unless such interventions address all small arms users, both governmental and non-governmental. The case of Cambodia also illustrates that the framework and phasing of international support has a strong influence on the process of stabilization, and the political and security arrangements that follow. In hindsight, important opportunities may have been missed to take up security sector issues in the first few years after the Paris Agreements, when international engagement was at its height. ◾

LIST OF ABBREVIATIONS

ANKI	National Army of Independent Kampuchea (formerly Armée Nationale Sihanoukiste)	IMF	International Monetary Fund
		KPNLAF	Khmer People's National Liberation Armed Forces
CPAF	Cambodian People's Armed Forces		
CPP	Cambodian People's Party	KPNLF	Khmer People's National Liberation Front
EU	European Union	PDK	Party of Democratic Kampuchea
EU ASAC	European Union's Assistance in Curbing Small Arms and Light Weapons	RCAF	Royal Cambodian Armed Forces
		SSR	Security sector reform
FUNCINPEC	Front uni national pour un Cambodge indépendant, neutre, pacifique et coopératif	*UN Basic Principles*	*Basic Principles on the Use of Force and Firearms by Law Enforcement Officials*
GW	Global Witness		
JSAC	Japan Assistance Team for Small Arms Management in Cambodia	WGWR	Working Group for Weapons Reduction

ENDNOTES

1   This report by the WGWR (Tieng, Long, and Hicks, 2004) cites the often 'quoted but never sourced calculation of gun ownership per three households or one gun per twelve people' as being 'of particular concern' (pp. 11–12).

2   For a review of the literature on the impact of gun availability on violence levels, see Small Arms Survey (2004, pp. 183–90).

3   Anecdotal evidence suggests that there has been a particular decline in the use of military-style weapons (such as AK-47s and M16s) in homicides, assaults, and other crimes. Such a trend might well imply that the emphasis on removing civil war weapons has had a significant impact on crime patterns and effects. The connection cannot be definitively established, however, since available reports do not systematically record the types of guns used.

4   Eighty-five per cent of interviewed victims said that they incurred no physical harm in the incidents, but this figure should be treated with caution. As only 1.7 per cent of interviewed households reported being victims of armed robbery, the total sample on which it is based is very small.

5   Neither the nature of the agreement nor the precise location of the different areas that comprise the 700,000 hectares has ever been made public (GW, 2004, p. 26).

6   Article 100 of the Law on Forestry of 2002 stipulates between one and five years of imprisonment and fines for 'any activities carried out by local authority officials, police officers, Royal Cambodian Armed Forces or other authorities that directly or indirectly allow forest exploitation' (quoted in GW, 2004, p. 19).

7   The World Bank requested that the Cambodian government repay the USD 2.8 million related to the misprocurement of a contract for motorbikes.

8   The IMF, for example, temporarily froze financial support to Cambodia because of concerns about the lack of budgetary transparency in that country, including the failure to disclose the existence of logging revenue and the parallel budget in 1996 (GW, 1997b, p. 14).

BIBLIOGRAPHY

Adhoc, Licadho and HRW (Human Rights Watch). 1999. *Impunity in Cambodia: How Human Rights Offenders Escape Justice.* Vol. 11, No. 3. Report. HRW. June. <http://www.licadho.org/reports/files/21Impunity%20in%20Cambodia.pdf>

Ashley, David. 1998. 'Between War and Peace: Cambodia 1991–November 1998.' In Dylan Hendrickson, ed. *Cambodia's Constitutional Challenge.* London: Accord. <http://www.c-r.org/accord/cam/accord5/index.shtml>

Australia, Government of. 2002. 'Australian and Cambodian Law Enforcement Agencies Working Together in the Region.' Press release. 5 June.
    <http://www.ag.gov.au/agd/WWW/justiceministerHome.nsf/Page/Media_Releases_2002_2nd_Quarter_5_June_2002_-_Australian_and_
    Cambodian_law_enforcement_agencies_working_together_in_the_region>

Broadhurst, Roderic. 2002. 'Lethal Violence, Crime and State Formation in Cambodia.' *Australian and New Zealand Journal of Criminology*. Vol. 35.
    April, pp. 1–26.

Cambodia, Kingdom of. National Police. 2003. *Cambodia National Police Crime Statistics, 1996–2003.*

——. 2005. Law on the Management of Weapons, Explosives and Ammunition. Adopted 26 April 2005.
    <http://www.eu-asac.org/programme/arms_law/arms_law_april_2005.pdf>

Clegg, Ian, Robert Hunt, and Jim Whetton. 2000. *Police Guidance on Support to Policing in Developing Countries.* Centre for Development Studies,
    University of Wales, Swansea. <http://www.swan.ac.uk/cds/pdffiles/POLICE%20GUIDANCE.pdf>

de Beer, David. 2005a. 'Disarmament Education in the SALW Programme of EU ASAC.' Paper presented at the Workshop on the Promotion of the
    United Nations Study on Non-proliferation and Disarmament Education. Bali, Indonesia, 21–22 December.
    <http://www.eu-asac.org/media_library/reports/051222.pdf>

——. 2005b. *Small Arms Control in Cambodia: Lessons Learned from the EU ASAC Programme.* Eschborn: GTZ.
    <http://www2.gtz.de/dokumente/bib/05-0774.pdf>

EIC (Economic Institute of Cambodia). 2004. *Cambodia Economic Watch.* October.
    <http://www.eicambodia.org/publications/cew/forestrymanagement.php>

——. 2005. *Cambodia Economic Watch II.* April. <http://www.eicambodia.org/publications/cewII/chapterix_civiladministration.php>

Economist, The. 2006. 'Cambodia: A Sudden Outbreak of Niceness; Hun Sen's Change of Heart.' 25 February, p. 56.

EU ASAC (European Union's Assistance in Curbing Small Arms and Light Weapons). 2002. *EU ASAC Preliminary Summary Report 2002.* Phnom Penh.
    <http://www.eu-asac.org/programme/vwc/report2002.pdf>

——. 2003a. *Final Report on the Implementation of the Capacity Building Project for Community Relations and Code of Conduct of the Cambodian
    National Police Force in Selected Provinces.* Phnom Penh. December.
    <http://www.eu-asac.org/media_library/reports/Reporton2003PoliceTrainingproject.pdf>

——. 2003b. *EU ASAC Preliminary Summary Report (NEW).* <http://www.eu-asac.org/programme/vwc/report2003.pdf>

——. 2005a. 'EU ASAC to Assist in Destruction of Over 4,000 Weapons in Sihanoukville.' Press release. 23 September.

——. 2005b. Web site. <http://www.eu-asac.org/>

——. 2005c. *EU ASAC Annual Report 2004.* Reference No. 117/2005/Manager EU ASAC. Phnom Penh.

Evans, Roderick. 2003. 'Security Sector Transformation from a Development Donor Perspective.' In United Nations Office at Geneva (UNOG) and the
    Geneva Centre for the Democratic Control of Armed Forces (DCAF), *Security Sector Reform: Its Relevance for Conflict Prevention, Peace Building
    and Development.* Presentations made at the first joint seminar of UNOG and DCAF. Geneva, 21 January.
    <http://www.dcaf.ch/publications/SSR_DCAF-UNOG/Evans.pdf>

Faulder, Dominic. 1999. 'The Town of Last Resort. Pailin.' *Asia Week.* 12 March. <http://www.asiaweek.com/asiaweek/99/0312/nat7.html>

Fawthrop, Tom. 2001. '900 000 Small Arms Plague Cambodia.' *Jane's Intelligence Review.* April.

Flärd, Helge. 2005. *Survey Report:* Phnom Penh Post. Background paper. Geneva: Small Arms Survey.

GW (Global Witness). 1996. *Corruption, War and Forest Policy: The Unsustainable Exploitation of Cambodia's Forests.* February.
    <http://www.globalwitness.org/reports/download.php/00087.doc>

——. 1997a. *Just Deserts for Cambodia? Deforestation and the Co-Prime Minister's Legacy to the Country.* Briefing. June.
    <http://www.globalwitness.org/campaigns/forests/cambodia/deserts/contents.htm>

——. 1997b. *A Tug of War: The Struggle to Protect Cambodia's Forests.* Briefing. March.
    <http://www.globalwitness.org/campaigns/forests/cambodia/tugowar/tugofwar.html>

——. 1998. *Going Places . . . Cambodia's Future on the Move.* Briefing. March 1998.
    <http://www.globalwitness.org/campaigns/forests/cambodia/goingplaces/index.htm>

——. 2002. *Deforestation without Limits.* Report. July 2002. <http://www.globalwitness.org/reports/index.php?section=cambodia>

——. 2004. *Taking a Cut: Institutionalised Corruption and Illegal Logging in Cambodia's Aural Wildlife Sanctuary.* Report. November.
    <http://www.globalwitness.org/reports/download.php/00208.pdf>

Hamilton, Janna. 2005. 'UN Decries Gov't Default on Land Concession Commitment.' *Phnom Penh Post.* 25 March–7 April.
    <http://www.cambodiamorning.blogspot.com/2005_04_01cambodiamorning_archive.html>

Heder, Steve. 2003. 'Cambodia (1990–98): The Regime Didn't Change.' In Roger Gough, ed. *Regime Change: It's Been Done Before.* London:
    Policy Exchange, pp. 66–75. <http://www.haitipolicy.org/regimeChange.pdf?PHPSESSID=4e60885ada1a630beba0c69a62a61065>

Hendrickson, Dylan. 2001. 'Cambodia's Security-sector Reforms: Limits of a Downsizing Strategy.' *Journal of Conflict, Security and Development.* Vol.
    1, No. 1, pp. 67–82.

HRW (Human Rights Watch). 2002. *Cambodia's Commune Elections: Setting the Stage for the 2003 National Elections.* Vol. 14, No. 4. HRW. April.
    <http://www.hrw.org/reports/2002/cambo0402/>

——. 2003a. *Don't Bite the Hand that Feeds You: Coercion, Threats and Vote-buying in Cambodia's National Elections.* Briefing paper. HRW. July.
    <http://www.hrw.org/backgrounder/asia/cambodia/elections.htm>

——. 2003b. *The Run-up to Cambodia's 2003 National Assembly Election: Political Expression and Freedom of Assembly under Assault.* Briefing paper.
    HRW. June. <http://www.hrw.org/backgrounder/asia/cambodia/>

Huxley, Tim. 2001. *Reforming Southeast Asia's Security Sectors.* Working Paper No. 4. London: King's College, London, Centre for Defence Studies, Conflict,
    Security and Development Group.

International Crisis Group (ICG). 2000. *Cambodia: The Elusive Peace Dividend.* ICG Asia Report No. 8. Phnom Penh/Brussels: ICG. 11 August.

JSAC (Japan Assistance Team for Small Arms Management in Cambodia). 2006. Web site. <http://www.online.com.kh/~admin.jsac/newsENG.html>

Ledgerwood, Judy. n.d.a. 'Cambodia since April 1975.' <http://www.seasite.niu.edu/khmer/Ledgerwood/Part6.htm>

——. n.d.b. 'The July 5–6 1997 Events: When is a Coup not a Coup?' <http://www.seasite.niu.edu/khmer/Ledgerwood/july_56_1997_events.htm>

——. n.d.c. 'The UN Sponsored Elections of 1993: Were They "Free and Fair"?' <http://www.seasite.niu.edu/khmer/Ledgerwood/free_and_fair.htm>

——. n.d.d. 'Understanding Cambodia: Social Hierarchy, Patron–Client Relationship and Power.'
    <http://www.seasite.niu.edu/khmer/Ledgerwood/patrons.htm>

Leuprecht, Peter, Special Representative of the Secretary-General for Human Rights in Cambodia. 2004. *Advisory Services and Technical Co-operation in the Field of Human Rights: Situation of Human Rights in Cambodia*. Report. New York: UNECOSOC. 20 December.
    <http://cambodia.ohchr.org/download.aspx?ep_id=218>

Marenin, Otwin. 2005. *Restoring Policing Systems in Conflict Torn Nations: Process, Problems, Prospects*. Occasional Paper No. 7. Geneva Centre for the Democratic Control of Armed Forces (DCAF). June.
    <http://kms.isn.ch/serviceengine/FileContent?serviceID=PublishingHouse&fileid=365DAB3D-DCA7-E969-9A93-84692CD0F4E2&lng=en>

OECD (Organisation for Economic Co-operation and Development). 2005. *Security System Reform and Governance*. Reference document. DAC Guidelines and Reference Series. Paris: OECD. <http://www.oecd.org/dataoecd/8/39/31785288.pdf>

Ratha, Sourn, Dianna Long, and John Vijghen. 2003. *Gun and Livelihood: The Use of Small Arms and Their Impact on People's Livelihood*. Report No. 5. Research Series on SALW Issue in Cambodia. Phnom Penh: WGWR. December.

Richardson, Sophie and Peter Sainsbury. 2005. 'Security Sector Reform in Cambodia.' In Albrecht Schnabel and Hans-Georg Ehrhart, eds. *Security Sector Reform and Post-conflict Peace-building*. Tokyo: United Nations University Press, pp. 283–96.

Roberts, David. 2002. 'Democratization, Elite Transition, and Violence in Cambodia, 1991–1999.' *Critical Asian Studies*. Vol. 34, No. 4, pp. 520–38.

Sedra, Mark. 2006. *Linking Security Sector Reform and Small Arms and Light Weapons Control and Reduction Programming*. Unpublished background paper. Geneva: Small Arms Survey.

Small Arms Survey. 2004. *Small Arms Survey 2004: Rights at Risk*. Oxford: Oxford University Press.

——. 2005. *Small Arms Survey 2005: Weapons at War*. Oxford: Oxford University Press.

Son, Seng. 2005. Unpublished collected Cambodian hospital data. Geneva: Small Arms Survey.

Tieng, Saman, Dianna Long, and Natalie Hicks. 2004. *Security Promotion Efforts: Existing State Mechanisms to Control and Prevent Urban Gun Crime in Cambodia*. Report No. 6. Research Series on SALW Issues in Cambodia. Phnom Penh: WGWR. December.

TED (Trade Environmental Database). 1996. *Cambodia Timber Export Ban (CAMWOOD Case)*. Case No. 285.
    <http://www.american.edu/TED/camwood.htm>

UN (United Nations). 1990. *Basic Principles on the Use of Force and Firearms by Law Enforcement Officials ('UN Basic Principles')*. Adopted by the Eighth United Nations Congress on the Prevention of Crime and Treatment of Offenders. Havana, 27 August–7 September.
    <http://www.unhchr.ch/html/menu3/b/h_comp43.htm>

UNCOHCHR (United Nations Cambodia Office of the High Commissioner for Human Rights). Special Representative of the Secretary-General for Human Rights in Cambodia. 2004. *Land Concessions for Economic Purposes in Cambodia: A Human Rights Perspective*. Phnom Penh: UNCOHCHR. November. <http://www.cambodia.ohchr.org/report_subject.aspx?sg_id=5>

UNDPKO (United Nations Department of Peacekeeping Operations). 2003. *Cambodia: United Nations Transitional Authority in Cambodia (UNTAC)*.
    <http://www.un.org/Depts/dpko/dpko/co_mission/untacbackgr2.html>

UNGA (United Nations General Assembly). 1979. *Code of Conduct for Law Enforcement Officials*. Adopted by UNGA Resolution 34/169 of 17 December. <http://www.unhchr.ch/html/menu3/b/h_comp42.htm>

Wille, Christina. 2005. 'European Union Assistance on Curbing Small Arms and Light Weapons in the Kingdom of Cambodia: A Case Study on European Action on SALW under the CFSP.' In *The European Union in Small Arms Action*. Geneva: UNIDIR. November.
    <http://www.unidir.org/pdf/1-SALW-ActionForView25-11-05.pdf>

——. 2006a. *How Many Weapons Are There in Cambodia?* Background paper. Geneva: Small Arms Survey.

——. 2006b. 'Finding the Evidence: The Links between Weapons Collection Programmes and Gun Use in Cambodia.' African Security Review, Vol. 15, No. 2. Forthcoming. <http://www.iss.org.za/Publications/Asrindex.html>

World Bank. 1999. *Cambodia Public Expenditure Review: Enhancing the Effectiveness of Public Expenditures*. World Bank Document. Report No. 18791-KH. 8 January. <http://www.wdsbeta.worldbank.org/external/default/WDSContentServer/IW3P/IB/1999/09/10/000094946_99031910564134/Rendered/PDF/multi_page.pdf>

——. 2004. *Seizing the Global Opportunity: Investment Climate Assessment and Reform Strategy for Cambodia*. World Bank Document. Report No. 27925-KH. 12 August. <http://www.siteresources.worldbank.org/INTCAMBODIA/Resources/Global-opportunity.pdf>

ACKNOWLEDGEMENTS

Principal author

Christina Wille

Contributors

Christine Beeck, Helge Flärd, Eileen Kilgour, Seng Son, and WGWR

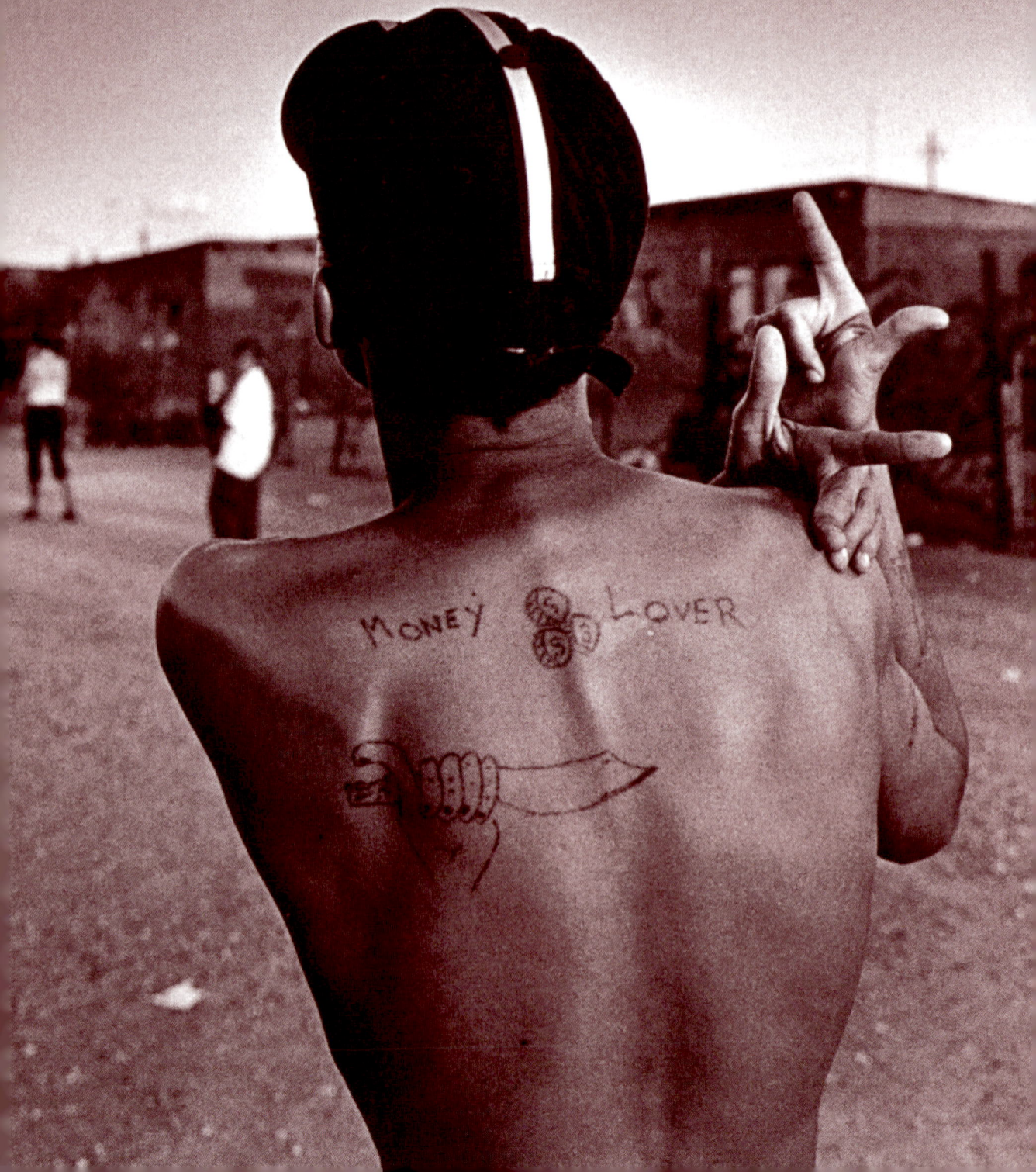

A gang member gestures as he shows off his tattoos on the streets of a township near Upington, South Africa. February 2002.
© Per-Anders Pettersson/Getty Images

The Other Side of the Coin

DEMAND FOR SMALL ARMS

6

INTRODUCTION

When the State loses control over its security functions and fails to maintain the security of its citizens, the subsequent growth of armed violence, banditry and organized crime increases the demand for weapons by citizens seeking to protect themselves and their property.

—UNGA (1997, para. 42)

In general the 'demand side' approach is promising because for many criminals the choice of whether to carry and use a gun depends in part on the consequences. Law enforcement should seek to tip the balance by making the legal consequences more severe and salient.

—Cook and Ludwig (2004, p. 605)

Between 2003 and 2004, the number of firearms deaths registered in Brazil dropped by eight per cent. This substantial fall in gun violence coincided with a gun buy-back programme that was conducted under a new disarmament statute, in force since July 2004. Greatly exceeding expectations, some 200,000 firearms had been turned in within three months of the launch of the programme. By mid-2005, this number had doubled (Kingstone, 2005; Instituto Sou da Paz, 2005; Logan, 2004, p. 1). Neighbouring Colombia meanwhile undertook a series of violence reduction and arms control programmes in the country's largest municipalities, including Bogotá and Medellín, to reduce epidemic levels of gun violence. It was expected that the annual rate of homicides in Colombia, more than 80 per cent of which are caused by firearms, might decrease as a result (COLOMBIA). And so it did—by 11 per cent between 1994 and 2002 (Aguirre et al., 2005, pp. 15–16). How can these changes in people's behaviour be explained in countries whose firearms death rates have been among the highest in the world? What are the reasons for their apparent success? Can these successes be replicated elsewhere?

This chapter takes stock of the emerging debate on small arms demand[1] and points to future directions for research and policy. It focuses on demand for firearms acquisition and possession, which is distinct from demand for violence.[2] A governing assumption of the chapter is that a reduction in demand for firearms can eventually lead to a reduction in firearm-related violence and deaths.[3] The chapter shows that firearms demand reduction may be as important as, or in certain cases even more important than, the physical collection of weapons. It situates demand reduction at the intersection of security and development.[4] It assumes that efforts to reduce socio-economic inequality can contribute to sustained security.

Drawing on a growing body of empirical research, the chapter highlights a number of factors that condition demand, and considers the extent to which current interventions take these factors into consideration. It aims, above all, to generate practical insights for strengthening concrete violence reduction and arms control measures. There is little chance that measures to reduce small arms supply will succeed over the longer term if demand for these weapons remains constant. Where production constraints lead to a decrease in small arms supply, new supplies become available via other channels if demand remains constant (Muggah et al., 2005, p. 31). Moreover, if efforts to improve human security continue to be based on an expectation that the number of weapons in circulation has to be reduced, there is a clear obligation on the part of those intervening to ask why such weapons are being held in the first place, and to address these reasons.

The chapter asks the following questions:

- What are the constitutive elements of small arms demand?
- How have these elements been taken into account in specific policy and arms control interventions?
- How can a demand perspective be mainstreamed into current arms control practice?

Demand can be understood as the interplay between motivations and means. There are a number of ways of framing demand. It can be examined at the macro, or state, level; at the micro level; and from the perspective of groups and individuals. Alternatively, demand can be studied through the lens of econometrics and behavioural psychology, or from international relations, anthropological, and criminological perspectives. A truly comprehensive analysis of demand must, in other words, contend with its breathtaking scale and complexity.

This chapter draws on the preliminary findings of a project initiated by the Small Arms Survey and several partners to explore the intricacies of small arms demand. It reviews the current literature and illustrates core theoretical insights with evidence from Brazil, Colombia, Papua New Guinea, the Solomon Islands, and South Africa. The chapter's principal conclusions include the following:

- Small arms demand can be described as the 'other side' of the small arms 'coin'. Analysing small arms control from a purely supply-side perspective risks misunderstanding the issue and misdiagnosing problems.
- Demand can be understood as the interplay between motivations (deep and derived preferences) and means (prices and resources). Unless interventions to reduce demand take these factors into account, they will probably fail.
- The design, implementation, monitoring, and evaluation of small arms reduction interventions must adopt a demand-sensitive perspective. Successful arms reduction is often locally hewn, is sensitive to political and social context, draws on public-private partnerships, and takes a flexible, participatory approach.
- Sustained empirical research on demand reduction is needed. At the same time, lessons from the many innovative and dynamic interventions currently under way around the world need to be learned and disseminated.

SMALL ARMS DEMAND TO DATE

The academic literature has indirectly tackled demand issues from a variety of perspectives. Criminologists, for example, have analysed firearms crime and youth violence in the US context. One of the main conclusions of this literature is that '[g]uns *intensify* violence. And for that reason it is a worthy goal of public policy to keep guns out of violent encounters' (Cook and Ludwig, 2000, p. 29, emphasis in original). This research tends to support the notion that a reduction in demand for firearms can lead to a significant decrease in the *intensity* of violence, while it assumes that levels of violence do not necessarily change as a result of a reduction in firearms demand. It asserts that firearm-related violence is more lethal than other forms of violence, resulting in comparatively higher rates of homicide (Cook, 1991). Moreover, as the COSTING chapter makes clear, firearms violence is costly and can be understood as a 'tax' on people's standard of living (Cook and Ludwig, 2000, p. 8).

While rates of violence typically remain constant, it appears that derived preferences for firearms acquisition, possession, and use can be changed. Criminologists Wellford, Pepper, and Petrie examine a variety of intervention programmes that were applied in different cities in the United States. For example, policies to deter firearm-related crime, including tougher sentencing, can indirectly reduce demand. They increase the deterrent effect of the punishment, thereby raising the relative price of firearms acquisition and ultimately misuse by criminals. The effectiveness of these programmes is far from clear, however, and more empirical research is urgently needed (Wellford, Pepper, and Petrie, 2005, p. 230).

More recently, Fitzpatrick (2006) adopted an economic approach to appraising small arms demand. In his model, polarization and rent-seeking lead to a particular set of political and economic motivations, which in turn results in small arms demand. Polarization is defined as 'a measure of the potential for conflict between groups of opposing political and economic interests', while rent-seeking is the pursuit of income outside of labour or investment. Fitzpatrick thus explains a rise in small arms demand with an increase in polarization and rent-seeking.

By way of contrast, sociologists McIntyre and Weiss (2003) explore the motivations underpinning arms acquisition, holding, and misuse by children and youth in Southern Africa, specifically those involved in armed conflict and urban gangs. They contend that guns are preferred by such youth for three main reasons: (*a*) as a result of socio-economic exclusion, (*b*) as a 'livelihood' or 'coping strategy', and (*c*) due to exploitation by group leaders (McIntyre and Weiss, 2003, p. 2). The issue of small arms misuse by young men is discussed in depth elsewhere in this edition of the *Survey* (ANGRY YOUNG MEN).

Though the arms control and disarmament literature emphasizes that the small arms issue has both a supply and a demand side, the supply-side perspective nevertheless dominates most research, writing, and policy-making by arms control and disarmament agencies (Muggah and Brauer, 2004). But recent contributions to the literature have made clear that a demand perspective is crucial if efforts to reduce and prevent firearm-related violence are to have a durable impact (Muggah et al., 2005, p. 31). Small arms demand is not a new issue on the small arms agenda, but it brings a new perspective to the small arms issue as a whole and to specific issues already under discussion (e.g. transfers, brokering, national regulation). Increasingly, researchers, practitioners, and policy-makers have begun to include a demand perspective in their work. Many efforts to curb the misuse of small arms, though not explicitly labelled 'demand-related', are already being conducted along such lines because such an approach has come naturally to practitioners.

By and large, international policy-makers have found it easier to address the supply side of the small arms coin. Global and regional norms on arms production and transfers, for example, are much simpler to devise than comparably intangible and context-specific norms relating to demand. While the 1997 UN Panel of Experts noted the importance of demand in its report (see the first epigraph of this chapter), this observation has not yielded much in the way of international norms, although some small arms instruments do contain vague references to small arms demand and point to possibilities for demand reduction. For example, the Bamako Declaration of the Organisation of African Unity (OAU) (now the African Union, AU) recognizes that the small arms issue has both 'supply' and 'demand' dimensions (OAU, 2000, V.1.vii) and that 'comprehensive solutions to the problem of the illicit proliferation, circulation and trafficking of small arms and light weapons . . . include . . . supply and demand aspects' (sec. V.2.v).

Many efforts to curb the misuse of small arms have already integrated a demand perspective.

The UN's *Programme of Action to Prevent, Combat and Eradicate the Illicit Trade in Small Arms and Light Weapons in All Its Aspects* (UNGA, 2001) explicitly mentions 'demand' in paragraph 7 of its Preamble: States express their concern 'about the close link between terrorism, organized crime, trafficking in drugs and precious minerals and the illicit trade in small arms and light weapons' and emphasize 'the urgency of international efforts and cooperation aimed at combating this trade *simultaneously from both a supply and demand perspective*' (emphasis added). Otherwise, the *Programme of Action* indirectly refers to demand through its strong emphasis on prevention and by mentioning related issues such as development, the promotion of cultures of peace, conflict resolution, and security sector reform (SSR) (Regehr, 2004, pp. 4, 8; 2001).

But the most serious efforts to mainstream demand into arms control emerge from development and specialized UN and non-governmental agencies. Practitioners at the margins of the arms control and disarmament communities have started incorporating a demand approach in much of their work, especially the United Nations Development Programme (UNDP), the United Nations Children's Fund (UNICEF), the World Health Organization (WHO), World Vision, and Oxfam GB. These organizations recognize the clear linkages between socio-economic exclusion and armed violence, and many of their poverty-reduction strategies have been designed to reduce armed violence, or at the very least to avoid aggravating it (Muggah and Brauer, 2004).

Demand is dynamic: it fluctuates with changes in motivations and means.

A series of workshops organized by the Quaker United Nations Office (QUNO) with the collaboration of the American Friends Service Committee (AFSC) since 1999 has sought to distil the experience of small arms practitioners working in various regions of the world. They have identified the following factors as especially important in shaping small arms demand: poverty and economic inequality, lack of fundamental human rights, poor governance, malfunctioning post-conflict programmes such as disarmament, demobilization and reintegration (DDR), cultural attitudes, and identity (Jackman, 2003a; Atwood and Jackman, 2005). Drawing on their analysis of five cases—also used in this chapter—Atwood, Glatz, and Muggah (2006) suggest ways of applying a demand perspective to future small arms work.

MODELLING SMALL ARMS DEMAND[5]

There are undoubtedly many ways of conceptualizing small arms demand. For the purposes of this chapter, however, demand for small arms and light weapons is defined as the interaction of individual and group *motivations* and *means*. Motivations can be further disaggregated into deep and derived preferences, while means refers to real and relative prices, as well as resources. For demand to be present, motivations and means must coincide. Demand is dynamic; a decrease in either motivations or means can reduce demand, while an increase in either factor can stimulate it.

Motivations are preferences that are determined by an individual's beliefs and attitudes. To a certain extent, these are conditioned by the social and cultural context in which he or she lives. Individual motivations thus tend to be socially embedded. The demand model distinguishes between 'deep' and 'derived' preferences, which together constitute motivations. *Deep preferences* do not change over the lifetime of an individual. They are similar for all human beings.[6] By contrast, an individual has a choice of *derived preferences* that he or she can pursue to satisfy deep preferences—some of them including small arms acquisition and use, others excluding them. Intriguingly, arms availability can itself influence derived preferences. Cook (1998) has described this as a 'contagious' process, while Regehr (2004, p. 7) prefers the term 'local arms race': an increase in small arms supply within one group can lead to increased demand by others as they try to match the first group's firepower.

An example of a deep preference is security for oneself, one's family, or one's home. Small arms acquisition, possession, and use is only one possible response to situations in which security is threatened. An alternative derived preference is reliance on the provision of security by government or private security institutions. Which option is pursued (i.e. which derived preference is chosen) depends in each case on the availability of alternative derived preferences, as well as on resources and prices.

Means consist of resources and prices. Resources can be monetary (e.g. income, assets, and credit) as well as non-monetary (e.g. social capital, personal ingenuity, and access to small arms acquisition networks). Real and relative prices help determine an individual's or a group's purchasing power.[7] Changes in resources and prices influence whether an individual can fulfil specific derived preferences, thus conditioning small arms demand. An increase in supply almost always lowers prices, which in turn increases an individual's purchasing power. A decrease in supply has the opposite effect.

This chapter contends that the acquisition and possession of small arms is determined by an interplay of means and motivations. Yet the practical application of this framework poses a range of challenges. In order to measure demand, one could focus on the number and types of weapons acquired. But the motivations and means of individuals and groups must also be examined. Cases characterized by sudden demand reversals and behavioural shifts may be especially informative. In Brazil, for example, the 2003 Disarmament Statute raised firearm registration and renewal fees and introduced stricter penalties for the possession of illicit guns. In so doing, it increased the relative price of firearms acquisition and possession, both legal and illicit. At the same time, a public information campaign highlighted the dangers of keeping firearms at home. These factors undoubtedly help explain the high level of participation in the 2003–05 gun buy-back programme (Atwood, Glatz, and Muggah, 2006, p. 19; Lessing, 2005, pp. 203, 207).[8]

Those who acquire small arms are not always identical to those who possess and use them. In some cases, adults acquire weapons, while youth hold and use them. For example, gang leaders may acquire weapons, which are then borrowed by gang members for gang-related work. Similarly, police and military institutions acquire the weapons that officers and reservists hold and use during their shifts. In designing effective interventions, policy-makers need to be conscious of situations such as these, where the demand for firearms acquisition differs from the demand for firearms possession and (mis)use.

> Increased supply to a state, non-state armed group, or individual may influence demand by others.

ACTORS DEMANDING SMALL ARMS

Several types of actors may, at any given time, seek to acquire or possess weapons. Three categories are especially important when considering mechanisms to reduce weapons demand: states, non-state armed groups, and individuals (including groups of individuals). There are important interrelationships among these categories that must also be taken into account. Regehr has pointed out that often increased supply to one group creates new demand by other groups (Regehr, 2004, p. 5). For example, the procurement of weapons by a state for its armed forces or police services may stimulate demand for more sophisticated weapons by non-state armed groups. The resulting insecurity, whether real or perceived, can encourage individuals to arm themselves, or band together in militia, in order to protect themselves. There are many other variants of this process. The arming of rebel forces may prompt a government to increase its acquisition of weapons, as was the case in Haiti in early 2004 (Muggah, 2005, p. 13). The following subsections discuss demand by non-state armed groups and individuals. State demand is treated separately, in Box 1, because its policy implications are fundamentally different from those of the other two actors. A quantitative study of state demand for small arms and light weapons is presented elsewhere in this volume (PRODUCTION).

Non-state armed groups

Compared with state security forces, the small arms holdings of non-state groups—whether rebels, insurgents, terrorists, or organized criminal gangs—are quite modest. The Small Arms Survey estimates that such groups account for less than one per cent of total global firearm stockpiles (Small Arms Survey, 2002, p. 103). If absolute numbers are even a crude indicator of demand, their influence on global production, stockpiling, and trade is negligible. Nevertheless, the effects of arms acquisition and misuse by non-state actors reverberate at the local, national, and regional levels. Reducing demand among non-state actors is an essential component of violence prevention and improved weapons reduction efforts.

There are startling similarities between the demand expressed by politicized non-state actors in conflict situations and that by criminal gangs in ostensibly 'peaceful' societies.[9] Apart from the fluid boundaries between these two categories, particularly during 'post-conflict' transition periods,[10] both exhibit, among other things, analogous command and control structures, preference orderings for various types of weaponry, price sensitivity (both financial and deterrence-based), as well as similar resource mobilization strategies.[11] Armed gangs in Papua New Guinea, known

Paramilitary fighters of the AUC stand guard before a poster of the Virgin of Guadalupe during a ceremony to lay down their arms in Otu, north-west Colombia, in December 2005.
© Fernando Vergara/AP Photo

Box 6.1 State demand for small arms and light weapons

States are among the primary consumers of small arms and light weapons. Annual procurement by national armed forces is about one million units (PRODUCTION). In principle, the theory of demand presented here applies to all types of buyers, but where buyers represent groups the purchasing decision is not a private, individual choice but a public, collective one. State demand is thus the sum of governmental agency demands that may or may not be coordinated in any centralized fashion.

From a traditional International Relations perspective, state-level demand is driven by exogenous and endogenous factors, such as 'concerns about internal stability, regional hostility, general insurance and the desire to project power internationally' (Levine and Smith, 1997, p. 342). In Weberian terms, the deep preferences of the state include the maintenance of its monopoly over the legitimate use of violence (internal) and national security, i.e. the preservation of sovereignty and 'core values' against threats (external), while derived preferences refer to the size and distribution of armed forces and related auxiliaries (see Weber, 1919).

The means side of state demand is determined by the financial and other resources a state has available for small arms procurement. This includes the other budget items that small arms expenditures must compete with. Furthermore, weapons prices for small arms at the time a procurement decision is made are important since they will also affect the state's ability to purchase small arms.

At the domestic level, the state has the monopoly of force, and it is the malfunctioning of the state that may lead an individual to seek small arms for his or her protection and security. At the international level, each state has to provide for its own protection and defence in the absence of an overarching authority. A state's derived preferences become a problem once they threaten the security of other states or the security of its own citizens. As indicated above, collective choice vastly complicates the study of the decision-making process, especially in the case of weapons-related choices, which are often conducted in secret. While a particular decision may be (imperfectly) observed, the underlying means and motivations are difficult to pin down. Moreover, neither state budgetary resources nor weapons prices (relative to other goods or services, whether security-related or not) are transparent. Nor is it always clear exactly who constitutes the collective that takes the weapons-procurement decision at whose budgetary expense.

In order to develop an initial understanding of state demand, it would be useful to conduct a set of case studies that would compare and contrast weapon-acquiring agencies within the same state. What procurement policies and procedures exist within and across those agencies? How are the weapons requirements of specific agencies determined? What does a needs assessment ('motivations') consist of? How does a weapons budget fit within an overall agency budget, and how is that budget integrated into the state's total budget? What budgetary control or oversight mechanisms exist at the state or agency level, and how are these enforced or circumvented? In the abstract, it is easy to speak of 'state' demand for small arms and light weapons. In practice, the answers to such questions will vary greatly, depending on the state and agencies studied.[12]

In policy terms, state demand also differs fundamentally from non-state group and individual demand. Policy-makers typically seek to reduce demand for small arms by non-state actors as they expect this will reduce firearms violence. The case of state demand is far less clear-cut. On the one hand, if a state increases its weapons purchases this may yield a reduction of (firearms) violence as a result of overall improvements in internal security. On the other hand, a government may use newly acquired small arms to suppress its own citizens or pursue aggressive foreign policy goals (Regehr, 2004, p. 5).

Contributor: Jurgen Brauer

---

colloquially as *raskols*, are a case in point as they sometimes operate in ways similar to organized guerrilla groups, for example the Bougainville Revolutionary Army (BRA), which fought in Bougainville province.[13] Armed criminal gangs in Central America are another prominent example of non-state actors with sophisticated transnational connections between members, robust revenue diversification, and similar organizational structures, just like the rebel groups that preceded them. This is also true of the guerrilla and paramilitary groups in Colombia (COLOMBIA).

The factors that condition the preferences of non-state actors include relative levels of state legitimacy and the credibility of public security sector entities. The relative stability of state institutions can also influence the real and relative prices of weaponry, while access of non-state groups to international and domestic revenue (whether legal or illicit) determines the resources available for weapons acquisition.

In Iraq, armed conflict has spurred demand for small arms among non-state armed groups and civilians alike. A power vacuum in many parts of the country has enabled various insurgent groups to exert control over territory, increasing their demand for small arms in order to ensure its defence. Competition between different armed groups has sparked a further rise in weapons demand. Since armed groups have targeted civilians, in addition to government and US institutions, real and perceived rates of victimization and insecurity have fuelled an increase in small arms demand among civilians as well. The situation in Iraq is thus a prime example of a 'local arms race' among both armed groups and civilians (Regehr, 2004, p. 7; HRW, 2005; Jackman, 2003b). Researchers affiliated with the Boston Gun Project observed a surprisingly similar trend among urban gangs in the Boston area (Kennedy, 1997, p. 452; Braga and Kennedy, 2002).

In Papua New Guinea and the Solomon Islands, as in many other countries of the South Pacific, some armed groups were motivated to acquire weaponry due to widespread misgivings about ethnic biases in the armed forces and police. Grievances also arose over the imposition of national legal regimes that trumped existing customary norms, further undermining public confidence in the state and the meting out of justice locally. Moreover, as Alpers (2005) and LeBrun and Muggah (2005) have shown, the costs of acquiring a weapon declined significantly in both countries due to systemic police corruption (including the leaking of weapons to kinsmen)[14] and the tendency of political elites to foster stockpiling and personal possession among constituents. According to Alpers, '[i]n the Southern Highlands, this same elite of politicians, civil servants and conflict entrepreneurs is also responsible for the proliferation and misuse of small arms—especially at election time' (Alpers, 2004, p. 9).

Individual demand

The majority of all known stockpiled firearms—approximately 60 per cent—are in the hands of civilians. This figure includes legal as well as illicit weapons (Small Arms Survey, 2002, p. 104). Small arms demand as expressed by individuals is thus a vitally important piece of the puzzle. Yet individual demand is socially embedded; it cannot be isolated from collective preferences or collectively-determined prices and resources (Brauer and Muggah, 2006).[15] While, like supply, individual demand for small arms can be influenced by national regulation, such rules must address the factors that condition demand if they are to be effective. The reverse also appears to be true. Cukier (2001) has argued that national gun control legislation 'not only reflect[s], but also shape[s] prevailing social values'. National laws typically stipulate the conditions under which civilians are permitted to acquire and hold firearms, whether for

Civilians hold about 60 per cent of all known stockpiled firearms. self-defence, hunting, collection, or other purposes. At the same time, national legislation may reflect broad social norms defining acceptable civilian demand. In most US states, for example, rules governing firearms ownership are relatively lax, in contrast to those adopted in Australia, Canada, and the UK (see LCAV, 2004; SAFER-Net, 2002; 2003; 2004; OSI, 2000).

Individual small arms demand is also influenced by the legitimacy of the state and, above all, by the state's ability to ensure public order and security. Where firearms legislation is strict and legal penalties for firearms misuse high, the strength of demand for firearms—at least for criminal purposes—can decrease considerably (see second epigraph). This is particularly true where the perceived probability of punishment, rather than its severity, is increased (Cook and Ludwig, 2004, p. 604).

In contrast, where state security services are perceived as ineffective or predatory, preferences for self-protection (against armed criminals or even police and soldiers) are likely to increase. For example, in the Garissa district of Kenya, where state presence is minimal, nomadic and pastoral communities have filled the security gap. Clan members have acquired weapons in order to protect their livestock and property and to exert control over natural resources, including water (Mkutu, 2003; Haji Aden, 2001; see also Eavis, 2002, p. 253).

MOTIVATIONS AND MEANS: INDICATORS OF SMALL ARMS DEMAND

This section examines demand for small arms by non-state armed groups and individuals, using a series of major indicators (see Table 6.1). As described above, demand can be analysed through the prism of motivations and means. On the motivations side, the section is structured according to the following deep preferences: personal security; social and economic security; individual status; and political identity, representation, and group status. Derived preferences include a desire for small arms, which may be replaced by other choices to realize a particular deep preference. For each deep preference, examples from Small Arms Survey case study research are cited that illustrate how and why small arms acquisition and possession were chosen over other possible derived preferences.[16] On the means side, the section is divided into the following subsections: relative monetary value of firearms; individual cost of/benefit from illicit possession and misuse; social cost of/benefit from (legal and illicit) firearms acquisition and ownership; difficulty/ease of access; and income and wealth, including credit. Table 6.1 offers a provisional framework designed to sketch the relationships between motivations and means on the one hand, and interventions designed to influence demand on the other. Table 6.2 provides an overview of the case studies, the measures that were deployed, and their effects on small arms demand.

It is important to note that small arms demand is always a product of several factors. For demand to exist, both motivations and means have to be present, but this is only a necessary—not a sufficient—condition of demand. At this stage, it is not possible to formulate general rules that determine, for any given situation, whether small arms demand will be present. For example, high unemployment rates on their own do not always foster higher demand for firearms. Other factors have to be present, with the combination of factors needed to spur demand differing from case to case. It thus makes sense to speak of demand 'indicators' when rendering conclusions about an increase or decrease in demand. Indicators are considered separately in order to identify specific entry points for interventions.

Firearms demand is often related to a deep preference for personal security.

Motivations

Personal security. Firearms acquisition and possession, and thus firearms demand, are often related to a deep preference for personal security. This deep preference cannot be changed, but it can be satisfied by a broad range of derived preferences other than firearms acquisition and possession. Depending on the circumstances, these alternatives can include state police and judicial services or private security companies. If a government fails to provide security, citizens may choose to acquire and hold firearms in order to fill this gap (Bendaña, 2001, p. 65).

Strong individual demand for firearms to ensure personal security has been seen in many places, including Brazil, Colombia, Papua New Guinea (PNG), and South Africa. Since the negotiated settlement in 1994, South African citizens have increasingly reported that they feel insecure and do not trust the state to provide for their security. Negative reactions among licensed gun owners to the new gun-control legislation in 2005 have confirmed that perceived insecurity remains high (Carroll, 2005). This has been paralleled by evidence of ineffective policing in several local cases (Kirsten et al., 2006).

Similarly, middle-class citizens in Brazilian cities have expressed a lack of confidence in the country's police and judiciary. In responding to a 2002 survey, 61 per cent of Brazilians said 'the police do not do a good job controlling crime' in their neighbourhood (Lessing, 2005, p. 205). The judiciary allegedly focuses on crimes committed by relatives or acquaintances of the victims, while crimes committed by career criminals are seldom investigated (Soares, 1996, p. 239; Lessing, 2005, p. 205). This perception has helped spur a strong derived preference for firearms acquisition

Table 6.1 Small arms demand indicators and possible policy responses[17]

Motivations	Deep preferences (unchangeable)	Possible policy responses (leading to derived preferences other than acquisition of small arms and light weapons)
	Personal security	• institute or strengthen community policing • reform justice (courts, penal) and security (police, military) sectors • take firearms out of circulation • improve public infrastructure (e.g. street lighting) • promote gun-free zones in schools, workplaces, churches, markets, shopping places, and sporting facilities
	Social and economic security	• provide education and employment opportunities, particularly for youth • stigmatize corruption • support reintegration of ex-combatants and ex-criminals
	Individual status	• challenge norms of violent masculinity and offer alternatives • reverse the role of media, entertainment, and recreation in normalizing and endorsing gun possession and misuse • encourage social customs dissociating guns from power, pride, and manhood
	Political identity, representation, and group status	• improve public access to and participation in government at the municipal and national levels • acknowledge and act to redress inequalities, injustice, and human rights abuses, which can influence recourse to (armed) violence • increase capacity for non-violent conflict resolution
Means	**Prices**	**Possible policy responses**
	Monetary value of firearms (relative to other goods, particularly substitutes)	Restrict supplies and thereby effect an increase in the prices of small arms
	Individual cost of/benefit from illicit possession and misuse	Increase the penalties of illicit possession and misuse: • strengthen national gun control laws to ensure small arms ownership is subject to specific criteria (e.g. age, proof of need, safe storage, spousal approval, background checks, trigger and safety locks), renewable licensing, and universal firearms registration • improve response and efficiency rates of police to requests for assistance from citizens
	Social cost of/benefit from firearms acquisition and ownership	Make gun acquisition and ownership more costly by strengthening social and customary controls: • engage communities in development and disarmament schemes (e.g. weapons for development programmes) rather than buy-backs • initiate public education programmes to stigmatize guns and gun violence and to generate support for alternative behaviour (e.g. through the establishment of gun-free zones)
	Resources	
	Income and wealth (including credit)	• combat illicit trafficking in drugs • cautionary use of monetary incentives
	Social capital	• reinforce social controls (council of elders, women's associations) • support for non-violent conflict management • introduce community policing, neighbourhood watch

Map 6.1 **South Africa**

Legend:
- ·—··— International boundary
- —— Province boundary
- ▣ Capital city
- ● Provincial capital
- G. GAUTENG
- MPU. MPUMALANGA

Map 6.2 **Colombia**

Legend:
- ·—··— International boundary
- ▣ Capital city
- ● Department capital

and possession among certain segments of the Brazilian population.

The case of Bogotá (Colombia) provides another example of strong firearms demand driven by the perceived failure of governmental authorities to satisfy a deep preference for security and protection. Following several mayoral violence reduction and disarmament programmes during the 1990s that drew on innovative legislation described in Box 6.2, these perceptions changed remarkably; 66 per cent of respondents to a 2003 'Quality of Life Survey' stated that the mayoral interventions had improved their perception of security. According to respondents, this was partly due to a stronger police presence, along with the renovation of old police stations and the construction of new ones (Aguirre et al., 2005, p. 24). The combination of increased enforcement and changing perceptions appears to have contributed to a reduction in firearms demand (COLOMBIA).

In PNG as well, dysfunctional police and judicial systems have led to an increased civilian preference for firearms acquisition and possession for purposes of self-defence and protection *against* police or defence forces (Alpers and Twyford, 2003, p. 118). There is considerable evidence of widespread corruption within the public security forces, along with numerous cases where members of the police and of the defence forces have reportedly committed human rights violations using firearms (PAPUA NEW GUINEA).

Social and economic security. Another deep preference that is potentially important for firearms demand is social and economic security. If educational and professional opportunities are lacking, in particular for male youth, firearms acquisition and possession may be the derived preference of choice for income generation. Unemployment can thus be used as a proxy for firearms demand in certain contexts. In South Africa, firearms are demanded by male youth who do not have access to education and the formal economy (Kirsten et al., 2006). In PNG as well, high unemployment rates appear to have contributed to firearms demand (PAPUA NEW GUINEA).

In the Brazilian *favelas,* poverty and lack of educational and professional opportunities are prevalent, and the drug trade is the only way to obtain income and status and thereby fulfil the deep preference for social and economic security. Lessing argues that in the *favela,* when joining a drug faction, an individual also 'decides', albeit indirectly, to acquire a firearm. This is true even though the individual does not actually own the gun. It is the faction that collectively owns firearms, loans them to faction members for gang-related work, and controls the ammunition (Lessing, 2005, p. 214).

Individual status. Demand for firearms can also stem from a deep preference for individual status. This deep preference is closely linked to social and economic security. Income often determines individual status. If other derived preferences—such as professional opportunities—are unavailable, firearms are chosen as a means of achieving power, social status, and masculinity.

In South Africa, for example, firearms have come to symbolize masculine identity and power. In this context, the Gun-free Zones (GFZ) project has not only sought to reduce firearms demand by making communities more secure, but has also allowed male youth to contribute actively to that process. They have, in other words, been offered an alternative route to acquiring individual status (Kirsten et al., 2006). In Brazil, disarmament campaigns, such as those conducted by the NGO Viva Rio, have attempted to disassociate masculinity and firearms by promoting such messages as 'Real men don't need guns' and 'Choose gun-free: your weapon or me', or 'Mothers, disarm your sons'. Such initiatives seek to wean traditional male or family roles from their frequent association with weapons (Lessing, 2005, p. 204).

Political identity, representation, and group status. Political factors may also serve as demand indicators, in particular deep preferences for political identity, representation, and group status. If alternative options for achieving these deep preferences do not exist or are not sufficiently attractive, people may well choose to acquire and hold (and perhaps use) firearms. In PNG, for example, violent tribal conflict has a long history and dates back to the time before the introduction of firearms in the 1980s. The introduction of firearms only exacerbated the violence

Map 6.3 **Brazil**

Fighters on the front line in West Honiara, Solomon Islands, 2000. © Bryn Evans/Panos Pictures

resulting from tribal conflict (Capie, 2003, p. 92). In the Solomon Islands, an important factor driving firearms demand has been group pressure, a result of the *wantok* system[18]. This can work both ways, however. Positive group pressure has contributed to the success of the Weapons-free Village (WFV) campaign (Nelson and Muggah, 2004).

In South Africa, firearms have come to symbolize political power, citizenship, and political identity during the apartheid era and after. During apartheid, the state was highly militarized, with weapons held by white government soldiers, white civilian-military commandos, and leaders of homelands. In the 1980s, members of the liberation movements, including young members of self-defence or self-protection units, acquired weapons as well, which then served as symbols of freedom. Even in the post-apartheid period, this emotive attachment to firearms has continued to be influential (Kirsten et al., 2006).

Means

Relative monetary value of guns. An important determinant of small arms demand on the means side is the relative monetary value of firearms. This value is relative because it depends on the trade-off one has to make in purchasing a firearm against the purchase of such other goods as food or clothing. At face value, the higher the relative monetary value of a firearm, the less likely a person is to choose the firearm over other goods that are crucial for survival. However, simply raising relative prices through supply-side measures is clearly insufficient from a demand perspective. If other factors are stimulating demand for firearms, people will tend to acquire weapons that have a lower relative

Box 6.2 Background: firearms regulation in Colombia

The main pillar of the current regulatory framework in Colombia is state ownership of all firearms and limited civilian access. Only the state is authorized to produce, import, and sell firearms. Under National Government Decree 2535 of 1993, the carrying of arms must be authorized by the state; it is not an inherent right. The Decree distinguishes three types of firearms: those that can only be used by the armed forces (*uso privativo*), guns the use of which is restricted (*uso restringido*), and civilian guns (allowed for self-defence, sports shooting, or collection). What is more, the license for holding or carrying a firearm can be given, suspended, or removed whenever the competent authority deems necessary (Pardo, 1995, p. xxv). Consequently, civilians and firms must demonstrate the need for a firearm before receiving a licence. Licences are issued by the Ministry of Defence Office for Control and Trade of Arms (OCCA).

Previously, the system of record-keeping was very poor. Decree 2535 tightened controls over civilian arms possession, encouraging civilian registration through a special amnesty, which included a buy-back programme and temporary gun permits regardless of a holder's legal situation. Around 190,000 permits were issued under the amnesty. All licences issued before 1993 also had to be updated (Bulla Rodríguez, 1995, p. 239).

Decree 2535 stipulates that a gun permit authorizes its owner to hold a weapon in a declared building that may be his or her residence or place of work. A civilian can obtain only two such licences of ten years' duration. An individual must justify wanting to carry a gun. A restricted-use gun licence can be given to individuals who can show they face a risk of death. Article 4 of the Decree specifies that, although the state is the sole owner of the firearm, the licensee is fully responsible for its use.

The Colombian regulatory framework also enables political and administrative authorities, under an agreement with the military and police, to temporarily restrict or even ban the carrying of firearms in order to control urban crime (Bulla Rodríguez, 1995, p. 238). This has been one of the main forms of arms control in the city of Bogotá.

Source: adapted from Aguirre et al. (2005, p. 8)

value. Supply-side measures thus need to be complemented with demand-side ones. Weapons acquisition can also represent an economic investment where a firearm enables its owner to acquire other goods at a lower relative price. This, then, lowers the relative price of the firearm.

In PNG, prices are high for factory-made firearms because they are in short supply. As a result, many home-made weapons, which are less expensive in relative terms, have been substituted for factory-made ones. More recently, the supply of factory-made weapons has increased (PAPUA NEW GUINEA). In Colombia, the monetary prices of legal firearms are much higher than those of illicit guns. A legal firearm may cost several times the minimum wage, while the price of an illegally obtained weapon may be as little as ten per cent of its legal market price (Aguirre et al., 2005, p. 6).

Individual cost of/benefit from illicit possession and misuse. Penalties for illicit weapons possession and misuse can be an important component of demand reduction measures. In Brazil, the December 2003 Disarmament Statute increased the costs of weapons registration and renewal, and strengthened penalties for illicit possession. These measures can have unintended consequences, however. An increase in registration and

Map 6.4 Papua New Guinea

renewal costs may have led to an increase in the number of unregistered weapons among lower-income classes rather than a reduction in demand. Overall, it appears that increased registration and renewal costs account for only part of the recent reduction in firearms demand in Brazil (Lessing, 2005, pp. 207–08). In Colombia, registration and renewal costs are also rather restrictive (see Box 6.2), but not enough data is currently available to assess the impact they may have on demand.

A more targeted approach, applied in the United States, focuses on increasing penalties for the *criminal use* of firearms. The assumption is that increased legal liability for firearms misuse has a deterrent effect on criminals and leads them to substitute firearms for other weapons or to alter their behaviour. While it may not always be possible to distinguish individual criminal from non-criminal demand for firearms, the Boston Gun Project/Operation Ceasefire, which specifically targeted criminals, was associated with a clear reduction of firearms violence (Braga and Kennedy, 2002, pp. 276–77). The project increased the risk of prosecution for firearms misuse, for example through tracing or the collection of ballistic information before sale (Cook and Ludwig, 2004, pp. 603–04).[19] In the Solomon Islands, the Regional Assistance Mission to the Solomon Islands (RAMSI), deployed in 2003, has worked to enforce new arms control legislation by raising the cost of illicit firearm possession and misuse through deterrence based on strong penalties (Nelson and Muggah, 2004).

A woman walks past a mobile gun collection station in Rio de Janeiro, Brazil, in November 2004. The sign reads: 'Finish with your gun before your gun finishes you off.' © Douglas Harrison Engle/WPN

Map 6.5 **Solomon Islands**

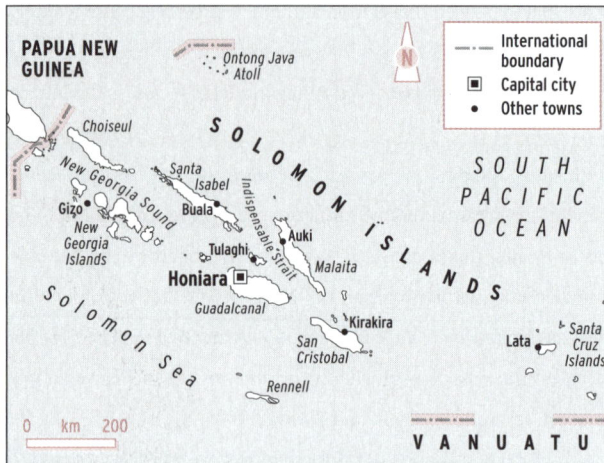

Social cost of/benefit from (legal and illicit) firearms acquisition and ownership. This indicator overlaps in large part with the motivations-side indicator 'individual status', discussed above. This complex of issues can be analysed in straightforward cost–benefit terms. Whether an individual decides to acquire (or forgo acquiring) a weapon is determined by the social cost or benefit that he or she is likely to experience as a result of such an acquisition (or non-acquisition). For example, the WFV campaign in the Solomon Islands has increased the social cost of acquiring and possessing firearms. Building on a strong tradition of customary regulation, the WFV campaign has transformed norms sanctioning firearms acquisition and possession into norms that condemn both (Nelson and Muggah, 2004). The campaign by the NGO Viva Rio in Brazil, which targeted the traditional association of guns and masculinity, as well as the GFZ project in South Africa (both cited above), have had a similar effect.

Difficulty/ease of access. This indicator is closely linked to the relative monetary value of firearms. The more difficult it is to gain access to firearms supplies, the higher the relative monetary price of firearms. As mentioned above, in PNG factory-made firearms have become increasingly available. When access to factory-made firearms was difficult, demand was satisfied through the production of home-made firearms (Muggah, 2004).

Rebel Papua New Guinea Defense Force soldiers guard the entrance to the sleeping quarters of a barracks in Port Moresby, March 2001.
© Torsten Blackwood/AFP/Getty Images

Table 6.2 Intervention programmes targeting small arms demand

Location	Intervention programme	Duration	Target group(s)	Main deep preference(s) as proxy for demand	Outcomes/effects
Brazil (Lessing, 2005)	Disarmament Statute, including national buy-back programme; referendum on ban on civilian ownership	Since December 2003; referendum on 23 October 2005	General population	Security; protection; social and economic security	Buy-back programme successful, indicating a decrease in demand (people wanted to get rid of their weapons and obtain financial compensation). Ban on civilian ownership rejected in referendum: the 'no' vote may reveal a general discontent with government rather than a preference for firearms.
Colombia (Bogotá) (Aguirre et al., 2005; see also Atwood, Glatz, and Muggah, 2006)	Various mayoral disarmament programmes based on new 1993 constitution	Since 1993	Urban population of Bogotá	Security; protection	Improved real and perceived levels of security and a reduction in firearms homicides translate into a reduction in demand.
South Africa (Kirsten et al., 2006)	Gun-free Zones (GFZ) project; Firearms Control Act (FCA), including Firearms-free Zones (FFZ)	Since 1995 2000 (in force since 2004)	General population	Political identity and representation	Reduced firearms demand in areas where GFZ are located.
PNG (Muggah, 2004)	Mendi Peace Agreement	3 May 2002	Armed non-state tribal groups; general population	Group status	Few successful demand reduction efforts.
Solomon Islands (Nelson and Muggah, 2004)	RAMSI WFV Campaign	Since July 2003 Since 2002	Armed non-state tribal groups; general population	Group status	Demand reduction successful through the combination of voluntary measures (WFV campaign) and enforcement (RAMSI).

Licensing provisions in national firearms legislation largely determine the difficulty or ease of access to firearms. An example is the 2000 South African Firearms Control Act (FCA; in force since 2004). It has increased the legal age of firearms ownership from 16 to 21, stipulates that most people may hold only one weapon for self-defence and up to four others for hunting and other purposes, and requires licence applicants to undergo a training course and police background check (Carroll, 2005). In Colombia, licensing is even more restrictive under the 1993 legislation (see Box 6.2; Aguirre et al., 2005, p. 8).

Income and wealth (including credit). While this factor certainly plays a role, it is perhaps the least promising of the demand-reduction intervention entry points. All case studies show that low earned income does not prevent arms acquisition and possession if other demand factors are strong. Those who have a strong desire for firearms will find some means of obtaining them. In PNG, for example, incomes have generally been low, but firearms demand has nevertheless remained high. As discussed above, in the Brazilian *favelas* firearms themselves represent an investment in income generation. Income results from the drug trade and membership in a drug faction is directly linked to weapons acquisition (Lessing, 2005, p. 213). Though raising a host of difficult ethical dilemmas, a decrease in people's income could, in fact, lower demand.

CONCLUSION

Demand reduction is increasingly recognized as an integral element of arms control and disarmament. Though consciousness of its importance has been slow to evolve in multilateral negotiations, practitioners on the ground have begun to incorporate, at least implicitly, a demand perspective in their small arms work. This chapter has provided a general overview of current debates on small arms demand and distilled a number of lessons from Brazil, Colombia, Papua New Guinea, the Solomon Islands, and South Africa. It shows that arms control interventions must endeavour to incorporate both supply- and demand-side approaches if they are to achieve meaningful, sustainable results.

The chapter conceives of small arms demand as a function of motivations (deep and derived preferences) and means (resources and prices). It analyses a series of specially commissioned case studies with the use of a range of demand indicators. In addition to accounting for the resources and prices associated with arms acquisition, the chapter finds that DDR, arms control, and violence-reduction programmes must also address the question of preferences. While arms buy-back programmes can have an indirect effect on means by raising the relative prices of firearms, they must be complemented by explicit measures designed to reshape derived preferences if they are to generate any dividends. Individual demand can be curbed through national gun-control legislation combined with disarmament and citizenship campaigns, as in the cases of Brazil and Colombia. Group pressure can push individual members of a society to dispense with firearms, as has occurred in Papua New Guinea, the Solomon Islands, and South Africa.

Small arms demand is a complex issue that touches upon virtually every aspect of the vast small arms agenda. This chapter has attempted to point the way to future research and policy work that is more resolutely informed by a demand perspective. With respect to research, more case studies from a variety of national and local settings are needed, including contexts in which no efforts have been made to influence demand. Such 'control studies' would facilitate the comparison of demand patterns between intervention and non-intervention settings and improve over-all understanding of the effectiveness and efficiency of intervention measures. A larger, more thematically and geographically diverse set of case studies would also help to distinguish case-specific findings from lessons that could be applied to general programme design.

A major obstacle to research on small arms demand is the issue of measurability. The indicators examined in this chapter provide a useful starting point for examining small arms demand. This framework should be refined through the inclusion of more specific indicators to facilitate an assessment of demand in quantitative terms. In particular, specific aspects of demand need to be untangled in future research, including: demand by weapon model or type (for light weapons as well as small arms), demand for small arms ammunition (as opposed to the weapons themselves), demand aspects of small arms transfers (authorized as well as illicit), and demand according to societal group (men/women, youth/adults, first-time owners/current owners, collectors/users).

On the policy side, the case studies have shown that intervention programmes work best where they are designed from the bottom up in a participatory fashion. In such cases, affected populations must feel ownership of the intervention programme in order for it to be successful. Government–civil society partnerships are critical to the effective implementation of demand reduction programmes, as exemplified by the GFZ project. A participatory approach can help identify derived preferences that satisfy deep preferences, but do not involve firearms. ✍

LIST OF ABBREVIATIONS

AFSC	American Friends Service Committee	PNG	Papua New Guinea
AU	African Union	QUNO	Quaker United Nations Office
BRA	Bougainville Revolutionary Army	RAMSI	Regional Assistance Mission to the
DDR	disarmament, demobilization and		Solomon Islands
	reintegration	SSR	security sector reform
FCA	Firearms Control Act (South Africa)	UNDP	United Nations Development
FFZ	Firearms-free Zones		Programme
GFZ	Gun-free Zones	UNICEF	United Nations Children's Fund
OAU	Organisation of African Unity	WFV	Weapons-free Village
OCCA	Office for Control and Trade of Arms	WHO	World Health Organization
	(Colombia)		

ENDNOTES

1   The term 'small arms demand' is used here interchangeably with 'firearms demand' and refers to demand for small arms as well as light weapons.

2   Firearms can be used—and demanded—for purposes other than armed violence: for example, for hunting or sports shooting.

3   To be sure, a reduction in demand for firearms *acquisition* refers only to weapon flows and thus may not necessarily lead to a reduction in firearm-related violence if individual and collective stocks remain high. After all, firearms are highly durable goods. Depending on context, however, firearm flows can be crucial in sustaining violence. For example, Cook and Ludwig point out that in the United States firearm acquisition and use are closely linked. This holds true especially for firearms used in crime, which are usually acquired through illicit channels shortly before use (Cook and Ludwig, 2004, p. 602).

4   The close link between development and security was emphasized by the UN High Level Panel on Threats, Challenges and Change (UN, 2004, p. viii).

5   This section is based on Muggah and Brauer (2004) and Brauer and Muggah (2006).

6   Here the model follows the framework proposed by Stigler and Becker (1977), who contend that 'tastes' (or human preferences) typically remain stable, even if phenomena such as addiction, habitual behaviour, advertising, and fashion suggest otherwise.

7   Standard economic theory states that supply and demand determine the price and quantity traded of the relevant good. Yet in this demand model price is understood as a factor which itself helps determine demand. The two types of demand are in fact very different. Whereas the model is used to measure the demand of an individual person or group, prices vary as a function of the *aggregated* demand of a whole society.

8   While a majority of Brazilians voted against banning most civilian gun sales in a 23 October 2005 referendum, most analysts attribute this decision to general discontent with the government (personal communication with Benjamin Lessing, formerly Viva Rio, December 2005).

9   For more on the motivations of armed groups, see the separate chapter on this issue (ARMED GROUPS).

10   See, for example, Small Arms Survey (2005, ch. 10) for a critical review of 'post-conflict' environments and the morphing of insurgent groups into organized criminal gangs.

11   This assessment is based on Demetriou, Muggah, and Biddle (2001), Dowdney (2003), Godnick (2002), Hillier and Wood (2003), and Aguirre et al. (2005), among others.

12   Paul Collier identifies five factors that determine military expenditure, and thus state demand for weapons. These factors vary across countries: '1. Active international warfare, 2. Peacetime military budget inertia, 3. Neighbourhood effects (arms races), 4. Internal rebellion or civil war, 5. Beneficiaries and vested interests' (Collier, 2006, p. 10).

13   Bougainville was granted autonomy in June 2005.

14   Large amounts of weaponry have been leaked by the Papua New Guinea Defence Forces (PNGDF) to politicians and criminals over the past decade (Alpers, 2005, p. 26).

15   A wide range of individuals may demand small arms, including hunters and sports shooters, those wanting small arms for self-defence, and criminals. As previously indicated, demand for small arms is not identical to demand for violence.

16   The cases of Brazil, Colombia, PNG, the Solomon Islands, and South Africa are presented in more detail in Atwood, Glatz, and Muggah (2006).

17   This table is based in large part on Atwood, Muggah, and Widmer (2005, pp. 100–1).

18   'The term "wantok" (one talk) in Melanesian Pidgin literally means someone who speaks the same language. In popular usage it refers to the relations of obligation binding relatives, members of the same clan or tribe, as well as looser forms of association' (Dinnen, 1997, p. 12).

19   To be sure, this kind of measure focuses first and foremost on reducing the misuse of firearms. Yet this often implies a reduction in firearms demand itself: as criminal behaviour is reduced, demand for firearms that are sought for criminal purposes decreases as well.

BIBLIOGRAPHY

AFSC (American Friends Service Committee). 2001. 'Curbing the Demand for Small Arms: A Middle East Seminar.' A workshop organized by the American Friends Service Committee and the Regional Human Security Center at the Institute of Diplomacy, Amman, Jordan, 8–9 July.

——. 2002. 'Traditional Cultural Practices and Small Arms in the Middle East: Problems and Solutions.' A workshop organized by the American Friends Service Committee and the Regional Human Security Center at the Institute of Diplomacy, Amman, Jordan, 3–4 November.

Aguirre, Katherine, Óscar Becerra, Simón Mesa, and Jorge A. Restrepo. 2005. *Assessing the Effect of Policy Interventions on Small Arms Demand in Bogotá, Colombia*. Background paper (unpublished). Centro de Recursos para el Análisis de Conflictos (CERAC), Bogotá. Geneva: Small Arms Survey. 29 October.

Alpers, Philip. 2004. *Gun Violence, Crime and Politics in the Southern Highlands: Community Interviews and a Guide to Military-style Small Arms in Papua New Guinea*. Background paper (unpublished). Geneva: Small Arms Survey.

——. 2005. *Gun-Running in Papua New Guinea: From Arrows to Assault Weapons in the Southern Highlands*. Special Report No. 5. Geneva: Small Arms Survey.

—— and Conor Twyford. 2003. *Small Arms in the Pacific*. Occasional Paper No. 8. March. Geneva: Small Arms Survey.

Archer, Dane, Rosemary Gartner, and Marc Beittel. 1983. 'Homicide and the Death Penalty: A Cross-National Test of a Deterrence Hypothesis.' *Journal of Criminal Law and Criminology,* Vol. 74, No. 3. Autumn, pp. 991–1013.

Atwood, David, Anne-Kathrin Glatz, and Robert Muggah. 2006. *Demanding Attention: Addressing the Dynamics of Small Arms Demand*. Occasional Paper No. 18. Geneva: Small Arms Survey.

—— and David Jackman. 2000. 'Working Where it Hurts: Perspectives From the Field on Small Arms Demand.' *Disarmament Forum,* No. 2. Geneva: UNIDIR. <http://www.unidir.ch/pdf/articles/pdf-art149.pdf>

—— and David Jackman. 2005. *Security Together: A Unified Supply and Demand Approach to Small Arms Control.* Working Paper. Geneva: Quaker United Nations Office. February.

——, Robert Muggah, and Mireille Widmer. 2005. 'Motivations and Means: Addressing the Demand for Small Arms.' In *Missing Pieces: Directions for Reducing Gun Violence through the UN Process on Small Arms Control.* Geneva: Centre for Humanitarian Dialogue, pp. 93–104.

Bendaña, Alejandro. 2001. 'Demand Dimensions of Small-Arms Abuse.' *Seton Hall Journal of Diplomacy and International Relations,* Vol. 2, No. 2. Summer/Fall, pp. 63–67.

Braga, Anthony A. and David M. Kennedy. 2002. 'Reducing Gang Violence in Boston.' In Winifred L. Reed and Scott H. Decker, eds. *Responding to Gangs: Evaluation and Research.* Washington, DC: National Institute of Justice, pp. 265–88. July. <http://www.ncjrs.org/pdffiles1/nij/190351.pdf>

Brauer, Jurgen and Robert Muggah. 2006 (forthcoming). 'Completing the Circle: Building a Theory of Small Arms Demand.' *Contemporary Security Policy,* June.

Bulla Rodríguez, Patricia. 1995. 'Control de armas y seguridad ciudadana.' In Juan Gabriel Tokatlían and José Luis Ramírez, eds. *La violencia de las armas en Colombia.* Bogotá: Fundación Alejandro Ángel Escobar, pp. 230–52.

Capie, David. 2003. *Under the Gun: The Small Arms Challenge in the Pacific.* Wellington: Victoria University Press.

Carroll, Rory. 2005. 'Gun Curbs Enrage South Africans.' *Guardian,* 6 January. <http://www.guardian.co.uk/southafrica/story/0,13262,1384059,00.html>

Collier, Paul. 2006. 'War and Military Expenditure in Developing Countries and Their Consequences for Development.' *Economics of Peace and Security Journal,* Vol. 1, No. 1, pp. 10–13.

Cook, Philip J. 1991. 'The Technology of Personal Violence.' *Crime and Justice,* Vol. 14, pp. 1–71.

——. 1998. 'The Epidemic of Youth Gun Violence.' Lecture delivered at the Perspectives on Crime and Justice Lecture Series. Washington, D.C. 5 May. <http://www.ncjrs.org/txtfiles/172851.txt>

—— and Jens Ludwig. 1997. 'Guns in America: National Survey on Private Ownership and Use of Firearms.' NIJ Research in Brief. Rockville, MD: National Criminal Justice Reference Service (NCJRS). May. <http://www.ncjrs.org/txtfiles/165476.txt>

——. 2000. *Gun Violence: The Real Costs.* Oxford: Oxford University Press.

——. 2004. 'Principles for Effective Gun Policy.' *Fordham Law Review,* Vol. 73, No. 2. November, pp. 589–613.

Cukier, Wendy. 2001. *National Firearms Controls.* Background paper (unpublished). Geneva: Small Arms Survey.

Demetriou, Spyros, Robert Muggah, and Ian Biddle. 2001. *Small Arms Availability and Trade in the Republic of Congo.* A Study Prepared by the Small Arms Survey for the International Organisation for Migration and United Nations Development Programme. September. <http://www.undp.org/bcpr/smallarms/docs/roc_small_arms_study.pdf>

Dinnen, Sinclair. 1997. *Law, Order and State in Papua New Guinea.* Discussion Paper No. 97/1. Canberra: State, Society and Governance in Melanesia Project, Research School of Pacific and Asian Studies, Australian National University. <http://eprints.anu.edu.au/archive/00002018/01/ssgmdinnen.pdf>

Dowdney, Luke. 2003. *Children of the Drug Trade: A Case Study of Children in Organised Armed Violence in Rio de Janeiro.* Rio de Janeiro: ISER/ Viva Rio.

Eavis, Paul. 2002. 'SALW in the Horn of Africa and the Great Lakes Region: Challenges and Ways Forward.' *Brown Journal of World Affairs,* Vol. 9, No. 1. Spring, pp. 251–60.

Eshete, Tibebe and Siobhan O'Reilly-Calthrop. 2000. *Silent Revolution: The Role of Community Development in Reducing the Demand for Small Arms.* Working Paper No. 3. World Vision International, September.

Fitzpatrick, Christopher. 2006 (forthcoming). *The Economics of Small Arms Demand: Polarization and Rent-Seeking in Haiti and Latin America.* Bonn: Bonn International Centre for Conversion.

Godnick, William, with Robert Muggah and Camilla Waszink. 2002. *Stray Bullets: The Impact of Small Arms Misuse in Central America.* Occasional Paper No. 5. Geneva: Small Arms Survey. October.

Haji Aden, Ebla. 2001. 'Small Arms Proliferation in Garissa District—Reasons Behind the Supply and Demand.' Paper presented at the BICC Conference 'Curbing the Demand Side of Small Arms in IGAD States: Potentials & Pitfalls.' Addis Ababa, 26 April.

Hemenway, David. 2004. *Private Guns, Public Health.* Ann Arbor: University of Michigan Press.

Hillier, Debbie and Brian Wood. 2003. *Shattered Lives: The Case for Tough International Arms Control.* London/Oxford: Amnesty International/Oxfam. <http://web.amnesty.org/aidoc/aidoc_pdf.nsf/index/ACT300012003ENGLISH/$File/ACT3000103.pdf>

HRW (Human Rights Watch). 2005. *A Face and a Name: Civilian Victims of Insurgent Groups in Iraq.* New York: HRW, Vol. 17, No. 9(E). October. <http://hrw.org/reports/2005/iraq1005/iraq1005.pdf>

Instituto Sou da Paz. 2005. *Disarmament News.* Year 11, No. 2. São Paulo: Instituto Sou da Paz. March.

Jackman, David. 2003a. 'Lessening the Demand for Small Arms and Light Weapons: Summary of International Workshops, 1999–2002.' Unpublished paper. Geneva: Quaker United Nations Office.

——. 2003b. 'Small Arms and Security in Iraq.' *Ploughshares Monitor,* Vol. 24, No. 3. Autumn.
   <http://www.ploughshares.ca/libraries/monitor/mons03a.htm>

Kennedy, David M. 1997. 'Pulling Levers: Chronic Offenders, High-Crime Settings, and a Theory of Prevention.' *Valparaiso University Law Review,* Vol. 31, No. 2. Spring, pp. 449–84.

Kingstone, Steve. 2005. 'Rare Drop in Gun Deaths in Brazil.' BBC News, São Paulo. 3 September. <http://news.bbc.co.uk/2/hi/americas/4210558.stm>

Kirsten, Adèle, Lephophotho Mashike, Knowledge Raji Mtshedisho, and Jacklyn Cock. 2006. *Islands of Safety in a Sea of Guns: Gun-free Zones in Fothane, Diepkloof, and Khayelitsha.* Working Paper No. 3. Geneva: Small Arms Survey.

Laming, Andrew. 2005. *The Autonomous Bougainville Government Elections: International Observer Team Individual Report.* May.
   <http://www.andrewlaming.com.au/icreate/documents/BougainvilleFINAL.pdf>

LeBrun, Emile and Robert Muggah. 2005. *Silencing Guns: Local Perspectives on Small Arms and Armed Violence in Rural Pacific Islands Communities.* Occasional Paper No. 15. Geneva: Small Arms Survey.

LCAV (Legal Community Against Violence). 2004. 'State & Local Laws.' <http://www.lcav.org/content/state_local.asp>

Lessing, Benjamin. 2005. 'The Demand for Firearms in Rio de Janeiro.' In Rubem César Fernandes, ed. *Brazil: The Guns and the Victims.* Rio de Janeiro: Viva Rio, pp. 202–20.

Levine, Paul and Ron Smith. 1997. 'The Arms Trade: Winners and Losers.' *Economic Policy,* Vol. 12, No. 25. October, pp. 337–70.

Logan, Sam. 2004. 'Brazil's Gun Buy-Back Program: Will It Work?' Americas Program, International Relations Center (IRC). Silver City, NM.
   <http://americas.irc-online.org/pdf/articles/0410brazilgun.pdf>

McIntyre, Angela and Taya Weiss. 2003. *Exploring Small Arms Demand: A Youth Perspective.* ISS Paper 67. Pretoria: Institute for Security Studies. March.
   <http://www.iss.co.za/Pubs/Papers/67/Paper67.pdf>

Mkutu, Kennedy. 2003. *Pastoral Conflicts and Small Arms: The Kenya-Uganda Border Region.* London: Saferworld. November.
   <http://www.saferworld.org.uk/publications/Pastoral%20conflict.pdf>

Mockus, Antanas. 2001. *Cultura ciudadana, programa contra la violencia en Santa Fe de Bogotá, Colombia, 1995–1997.* Washington, DC: Inter-American Development Bank. July. <http://www.iadb.org/sds/doc/Culturaciudadana.pdf>

Muggah, Robert. 2004. *Diagnosing Demand: Assessing the Motivations and Means for Firearms Acquisition in the Solomon Islands and Papua New Guinea.* Discussion Paper 2004/7. State, Society and Governance in Melanesia Project. Canberra: Australian National University/Research School of Pacific and Asian Studies.

——. 2005. *Securing Haiti's Transition: Reviewing Human Insecurity and the Prospects for Disarmament, Demobilization, and Reintegration.* Occasional Paper No. 14 (updated). Geneva: Small Arms Survey. October.

—— and Jurgen Brauer. 2004. *Diagnosing Small Arms Demand: A Multi-disciplinary Approach.* Discussion Paper No. 50. Durban: School of Economics and Management, University of KwaZulu-Natal. June.

——, Jurgen Brauer, David Atwood, and Sarah Meek. 2005. 'Means and Motivations: Rethinking Small Arms Demand.' *HFG Review: Small Arms and Light Weapons: A Call for Research.* Spring, pp. 31–38. <http://www.hfg.org/hfg_review/5/hfgsmallarms.pdf>

Nelson, Carol and Robert Muggah. 2004. *Solomon Islands: Evaluating the Weapons Free Village Campaign.* Independent report commissioned by the Small Arms Survey. Geneva: Small Arms Survey.
   <http://www.smallarmssurvey.org/AddRes/Weapons%20Free%20Villages%20Evaluation%20Nelson%20Muggah.pdf>

New Zealand Herald. 2005. 'Fiji Hired Guns Were "Advance" Platoon.' 21 November.

Observatorio de Cultura Urbana en Bogotá/Comisión de Cultura Ciudadana. 2002. *La cultura ciudadana en Bogotá: Resultados de la primera aplicación dels sistema de medición. Líneas de base y metas del objetivo de cultura ciudadana del plan de desarrollo 2001–2004 'Bogotá para vivir todos del mismo lado'.* Bogotá: Instituto Distrital de Cultura y Turismo, Alcaldía Mayor de Bogotá, D.C. Abril.
   <http://univerciudad.redbogota.com/bajar-pdf/013/investiga-culturaciudadana.pdf>

OSI (Open Society Institute). 2000. *Gun Control in the United States: A Comparative Survey of State Firearm Laws.* New York: Open Society Institute/ Center on Crime, Communities & Culture/Funders' Collaborative for Gun Violence Prevention. April.
   <http://www.soros.org/initiatives/justice/articles_publications/publications/gun_report_20000401>

OAU (Organization of African Unity). 2000. *Bamako Declaration on an African Common Position on the Illicit Proliferation, Circulation and Trafficking of Small Arms and Light Weapons* (Bamako Declaration). Bamako. 1 December.
   <http://www.amaniafrika.org/nbi_declaration_pdf/bamako.pdf>

Pardo Rueda, Rafael. 1995. 'Prólogo.' In Juan Gabriel Tokiatlán and José Luis Ramírez, eds. Violencia de las armas en Colombia. Bogotá: Fundación Alejandro Ángel Escobar, pp. xv–xxvi.

Pézard, Stéphanie. 2004. 'The Demand for Small Arms.' *Taking Stock of Small Arms Research: A Small Arms Survey Perspective.* Prepared for the Harry Frank Guggenheim Foundation Workshop on Expanding the Knowledge Base on the Problem of the Proliferation, Trade, and Misuse of Small Arms and Light Weapons, 27–29 June, New York.

QUNO (Quaker United Nations Office). 2003. 'A Summary of Lessons on Small Arms Demand and Youth.' *Small Arms Demand in the Caribbean: Special Focus on Haiti and Youth Issues.* Port-au-Prince: QUNO, 8–13 June.

Regehr, Ernie. 2001. 'The UN and a Small Arms Program of Action: Measuring Success.' *Ploughshares Monitor,* Vol. 22, No. 4. December. <http://www.ploughshares.ca/libraries/monitor/mond01e.html>

——. 2004. *Reducing the Demand for Small Arms and Light Weapons: Priorities for the International Community.* Working Paper 04-2. Waterloo, ON: Project Ploughshares. <http://www.ploughshares.ca/libraries/WorkingPapers/wp042.pdf>

Republic of Colombia. 1993. *Decreto numero 2535 de 1993 (diciembre 17) por el cual se expiden normas sobre armas, municiones y explosivos.* 17 December. <http://www.mindefensa.gov.co/nuevoweb/normatividad/DEC.%202535%20DE%201993.htm>

SAFER-Net. 2002. 'Australia.' <http://www.research.ryerson.ca/SAFER-Net/regions/Oceania/Ala_OR02.html>

——. 2003. 'England and Wales.' <http://www.research.ryerson.ca/SAFER-Net/regions/Europe/Eng_MH03.html>

——. 2004. 'Canada.' <http://www.research.ryerson.ca/SAFER-Net/regions/Americas/Can_JY04.html>

Sandler, Todd and Keith Hartley. 1995. *The Economics of Defense.* Cambridge: Cambridge University Press.

Small Arms Survey. 2001. *Small Arms Survey 2001: Profiling the Problem.* Oxford: Oxford University Press.

——. 2002. *Small Arms Survey 2002: Counting the Human Cost.* Oxford: Oxford University Press.

——. 2005. *Small Arms Survey 2005: Weapons at War.* Oxford: Oxford University Press.

Soares, Luiz Eduardo. 1996. *Violence and Politics in Rio de Janeiro.* Rio de Janeiro: ISER.

Stigler, George J. and Gary S. Becker. 1977. 'De gustibus non est disputandum.' *American Economic Review,* Vol. 67, No. 2. March, pp. 76–90.

UN (United Nations). 1999. 'United Nations International Study on Firearm Regulation.' United Nations Crime and Justice Information Network. <http://www.uncjin.org/Statistics/firearms/>

——. 2004. *A More Secure World: Our Shared Responsibility. Report of the Secretary-General's High-level Panel on Threats, Challenges and Change.* New York: UN. <http://www.un.org/secureworld/report2.pdf>

UNGA (United Nations General Assembly). 1997. *Report of the Panel of Governmental Experts on Small Arms.* A/52/298 of 27 August. <http://www.smallarmssurvey.org/source_documents/UN%20Documents/Other%20UN%20Documents/A_52_298.pdf>

——. 2001. *Programme of Action to Prevent, Combat and Eradicate the Illicit Trade in Small Arms and Light Weapons in All Its Aspects ('UN Programme of Action').* 20 July. Reproduced in UN document A/CONF.192/15.

Weber, Max. 1919. *Politik als Beruf* [Politics as a Vocation]. Munich: Duncker & Humblot.

Wellford, Charles F., John V. Pepper, and Carol V. Petrie. 2005. *Firearms and Violence: A Critical Review.* Washington, DC: National Academies Press.

ACKNOWLEDGEMENTS

Principal authors

Anne-Kathrin Glatz and Robert Muggah

Contributors

Katherine Aguirre, David Atwood, Jurgen Brauer, Jackie Cock, Philip J. Cook, Nicole Haley, David Hemenway, Adèle Kirsten, Benjamin Lessing, Carol Nelson, and Jorge A. Restrepo

A father, craft firearm in hand, stands by his children near the goldmine where he works.
© Chris Sattlberger/Panos Pictures

Jumping the Gun
ARMED VIOLENCE IN PAPUA NEW GUINEA

7

INTRODUCTION

In response to chronic armed violence in the country's urban centres and Highlands provinces,[1] the Papua New Guinea (PNG) government initiated a far-reaching review of its firearms legislation and controls in 2005. This review included the establishment of a National Guns Control Committee, a road-show that brought the issue of firearms control to the top of the national agenda and culminated in a National Gun Summit, and a lengthy report to Parliament—which has yet to draw a response. Firearms, while few in real numbers relative to the population, were determined to be a major contributor to real and perceived insecurity, and an obstacle to good governance, improved state security, and sustainable development. This chapter traces out the findings of an armed violence assessment administered in the country's National Capital District (NCD) and Southern Highlands Province (SHP) by the Small Arms Survey, with support from the United Nations Development Programme.[2] The assessment contributes to and furthers the public debate and policy interest as manifest in the recent National Gun Summit, builds upon the growing body of research on small arms in the Pacific,[3] and offers a comprehensive mapping of armed violence in two of PNG's most affected areas. It also advances a forward-looking agenda by raising an array of findings that call into question received wisdom concerning armed violence in PNG, and comes at a time when there is a rare opportunity to take some decisive steps towards addressing the problems identified. Specifically, it challenges donors, practitioners, researchers, and advocates alike to reconsider conventional arms control and disarmament approaches in contexts such as PNG.

PNG is a young nation that is chronically affected by armed violence. Its citizens are currently experiencing victimization rates that are among the highest in the world. Firearms (both craft and factory-made) have played a significant role in exacerbating the country's law and order problems and have given rise to invidious forms of violence, including maiming, abduction, and kidnapping for ransom.[4] The costs in both human and material terms have been substantial. Though the prospects for improving security in PNG might at first appear rather bleak, armed violence there is heterogeneous and diverse, comprising many different law and order environments (Dinnen, 2001). It is through recognizing this that interventions might be better developed and targeted. To this end, this chapter yields a number of critical and counter-intuitive findings that should also usefully inform and influence prospective interventions designed to mitigate insecurity and reduce arms availability and demand. The chapter's key findings include the following:

- Victimization rates among households in NCD and SHP are higher than previously reported. More than half of all households surveyed reported being victimized in the previous six months.

- More than 80 per cent of all victimization events in NCD and SHP involved the use of a weapon, though not necessarily a firearm.

Map 7.1 **Papua New Guinea**

- Domestic violence is the primary contributor to insecurity, though other factors such as social conflict and armed criminality are also important.

- Inter-group fighting and criminal violence are key sources of insecurity in both NCD and SHP. As such, the received wisdom that associates tribal fighting with the Highlands provinces and criminal violence with urban areas must be reconsidered.

- The calls to turn the Hela region into a province represent a potential source of conflict in SHP. People there are arming themselves with the intention of causing widespread civil unrest should their calls continue to fall on deaf ears.

- The demand for firearms is robust in both NCD and SHP. Coupled with this, ammunition prices are lower than previously reported, and appear to be declining.

- Weapons reduction programmes do not, as yet, have widespread support in SHP or NCD. Without improvements in law and order, people in NCD and SHP are generally unwilling to disarm.

- Successful interventions will need to be targeted and based on a sound, evidence-based diagnosis of PNG's law and order problems region by region.

METHODOLOGY

The aim in undertaking this particular violence assessment, as with others, was to generate quantitative and qualitative data that would better inform geographic and demographic understandings of armed violence in PNG. The overall methodology employed was adapted from methodologies previously employed by the Small Arms Survey in more than 25 countries around the world. The primary tools included two survey instruments: a large-scale household survey on armed violence, and participatory focus groups supplemented by individual interviews. Specifically, the household survey comprised structured questions about victimization, weapons misuse and armed violence, personal views on guns, law and order, and the effects of violence, while the focus groups explored five key thematic areas: types of violence, causes of violence, costs of violence, ways to reduce violence, and the circumstances under which communities would willingly disarm.

Because of the need to generate preliminary findings in time for the National Gun Summit, it was determined that the assessment should be purposive and non-random, consisting of two area-based assessments——one urban, the other rural—that would assess the causes, scale, and distribution of armed violence in two of the most conflict- and crime-riven parts of the country. NCD and SHP were chosen because what little objective crime data there is available makes evident that the urban centres of Port Moresby and Lae, and the Highlands provinces, are acutely affected by armed violence.[5]

In all, 292 household surveys were completed at 12 survey locations in NCD, with surveys being undertaken in six suburbs—Gerehu, Ensisi, Waigani, Gordons, Tokorara, and Hohola—and six settlements: Vadavada, Kaugere,

Map 7.2 Southern Highlands Province survey area with households reporting an armed robbery or armed assault, December 2004–May 2005

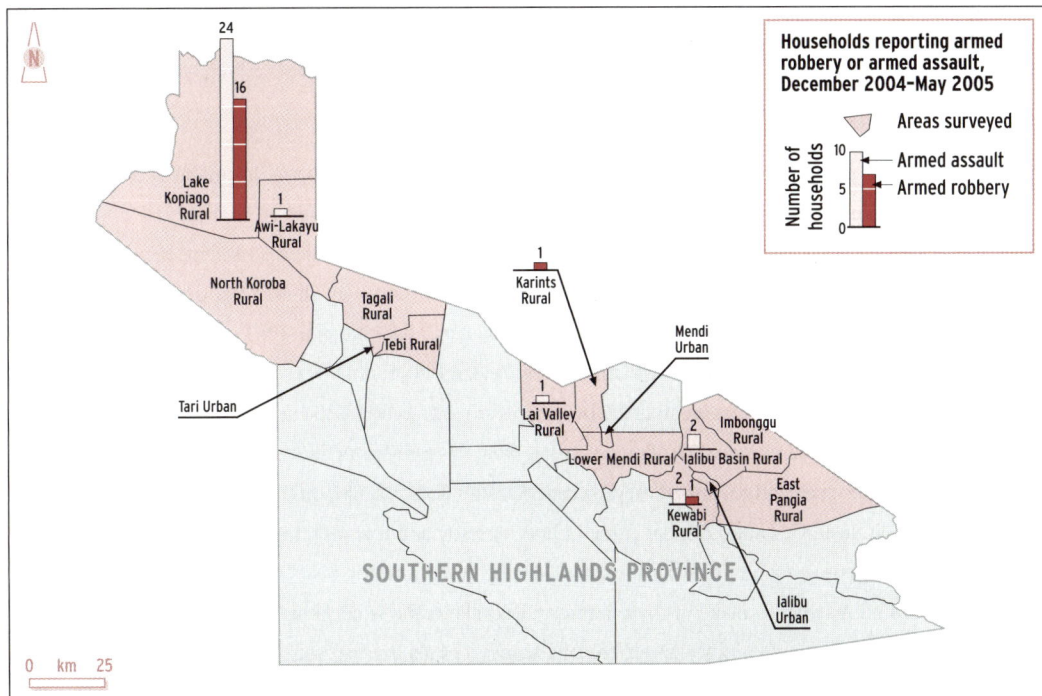

Figure 7.1 **Reported crime incidents, 2000–04**

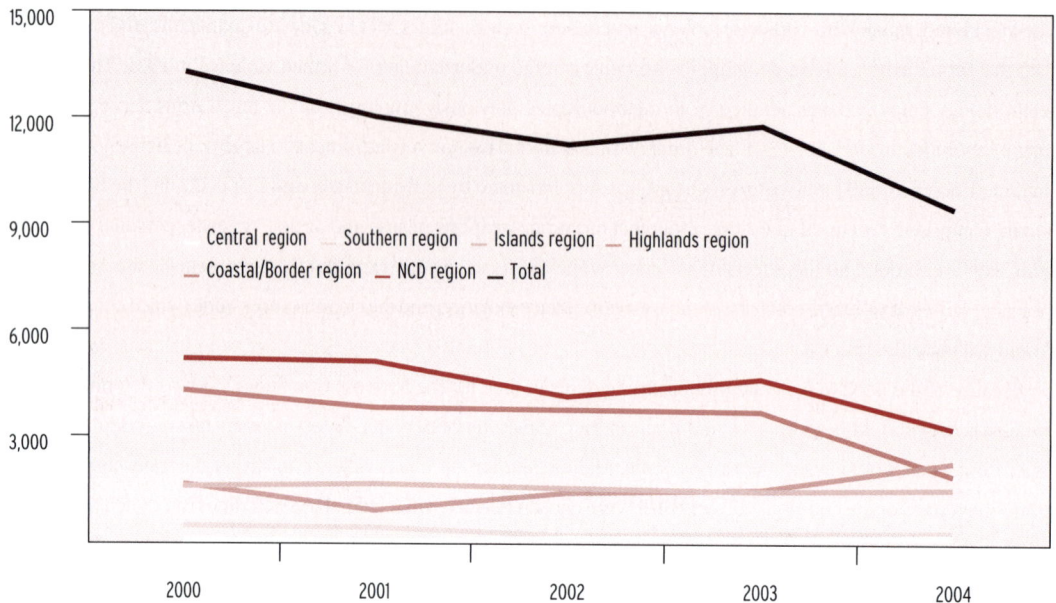

Central region   Southern region   Islands region   Highlands region
Coastal/Border region   NCD region   Total

Sources: Gomez (2005); UNDSS (2005)

Oro, Two-Mile, Morata, and Nine-Mile.[6] In SHP, some 235 household surveys were completed. They were adminis-tered and focus groups held in 5 of the province's 8 districts (Ialibu/Pangia, Imbonggu, Koroba/Kopiago, Mendi, and Tari) and in 15 of its 30 Local Level Government (LLG) areas (see Map 7.2).

VIOLENCE TRENDS

PNG's law and order situation has attracted much national and international attention. Invariably it is described as deteriorating, with many commentators asserting that crime rates have steadily increased over the last 20 to 30 years (Levantis, 1998; Sikani, 2000; Windybank and Manning, 2003). Certainly, local and international media reports suggest that criminal violence is rampant, small arms proliferation rife, and security at an all-time low due to rising levels of armed violence. Travel advisories from around the world recommend that visitors to PNG exercise a high degree of 'caution' (Australia, 2005, p. 1) or 'vigilance' at all times (New Zealand, 2005, p. 1). These advisories tend also to reflect the received wisdom that urban centres are troubled by criminal violence, whereas the mostly rural Highlands provinces are troubled by ethnic or tribal violence. The findings arising from this assessment directly challenge such assumptions.

Available surveillance data in the form of official crime statistics does not support the widespread view that crime is at an all-time high. Instead, it suggests that crime in Port Moresby and regional crime levels are both falling, and have been since 2000.

It is difficult to know what to make of these figures. Officially, crime is declining, but data gathering, where it exists, is limited, incomplete, notoriously unreliable, and shows a marked urban bias.[7] Massive under-reporting and under-coverage means that less than half of all crime is now recorded.[8] That said, this survey and the 2004 Port

Box 7.1 Previous victimization surveys

Prior to the assessment discussed in this chapter, three previous studies had considered victimization and insecurity in PNG. All three had an urban focus. They offer opportunities for comparison with this survey. For instance, PNG was one of 13 countries covered in the 1992 International Crime Victimization Survey (Zvekic and Alvazzi del Frate, 1995) conducted by the United Nations Interregional Crime and Justice Research Institute (UNICRI). That survey, which involved 1,583 PNG respondents–597 of whom were resident in Port Moresby–revealed that PNG's victimization prevalence rates with respect to assaults and threats (10.3 per cent), robbery (9.8 per cent), and sexual incidents (11.8 per cent) were higher than those reported in the other 12 cities surveyed–Kampala, Dar es Salaam, Johannesburg, Cairo, Tunis, Buenos Aires, Rio de Janeiro, Costa Rica, Beijing, Bombay, Jakarta, and Manila.

Unlike the UNICRI assessment, which focused specifically on armed violence, two other studies considered both violent and non-violent crime. In doing so, the Safer Port Moresby Initiative's Youth and Crime Survey found that 38 per cent of crimes in Port Moresby were perpetrated with violence and that 61 per cent of offenders used a weapon (UNDP, 2004, p. 52). Similarly, the 2004 Port Moresby Community Crime Survey found that crimes of violence accounted for 46 per cent of victimization incidents (NRI, 2005, p. 3).[9] It also found that two-thirds of NCD households had been victims of crime in the previous 12 months; 57 per cent of households had been the victims of multiple crimes; 33 per cent of repeat crime; and 1 household in 8 had been a victim of crime ten or more times.

Moresby Community Crime Survey (NRI, 2005) indicate that reporting rates in Port Moresby are much the same as they were a decade or so ago (Zvekic and Alvazzi del Frate, 1995), suggesting that crime in NCD might well be falling or at least stabilizing. Participants in both surveys agreed.

The situation in the Highlands is much less ambiguous. In contrast to the national crime trends reviewed above, reported crime in the Highlands is at an all-time high. This is in spite of the fact that only a small fraction of actual crimes occurring in the Highlands are reported. To let one example stand for many, SHP's western regional administrator has revealed that there were over 200 conflict-related deaths in Tari District alone in 2003–04 (Phillip Moya, cited in Lewis, forthcoming), yet the National Crime Incidents Summary for 2004 (Annexe 1) records only 38 murders for the entire province.

Despite the inherent limitations of PNG's crime data, it is possible to render crude cross-country comparisons. For example, there were 97 reported murders (33 per 100,000) in NCD in 2004 (see Annexe 1). By contrast, the Australian Capital Territory (ACT) with a population of 324,000—comparable to that of NCD—recorded no murders in 2004 (ABS, 2005).[10] There were 862 reported armed robberies, or 294 per 100,000 persons, in NCD, whereas in the ACT there were only 106 armed robberies, or 33 per 100,000 (ABS, 2005).

VIOLENCE AND VICTIMIZATION

Far from being generalized and prolific throughout the country, armed violence in PNG is geographically and demographically specific. The present assessment, for instance, revealed that NCD's settlements and SHP's western Hela region (which is represented by the Kopiago and Tari Districts) are disproportionately affected by armed violence. In NCD, the assessment found that 60–65 per cent of the most prevalent types of violence—domestic violence, intergroup fighting, armed robbery, and armed assault—occur in the settlements.[11] Sikani (2000, p. 41) reports that offenders too are disproportionately represented in the settlements—accounting for 90 per cent of police arrests between 1996 and 1998. That the settlements are disproportionately affected by violent crime was also borne out by the Port Moresby Community Crime Survey (NRI, 2005).

Table 7.1 Violence types and reported victimization rates in SHP and NCD, December 2004–May 2005		
Violence type	**% of SHP households victimized**	**% of NCD households victimized**
Domestic violence	26	18
Intergroup fighting	17	18
Armed assault	13	12
Armed robbery	8	15
Threats/Intimidation/Extortion	6	9
Attempted murder	9	3
Sexual assault/Rape	8	3
Murder	5	1
Abduction	0.5	1

Note: For purposes of this study, a victimized household was one in which one or more household members were reported as having experienced the relevant violent crime.

Though armed violence is concentrated in specific regions and among certain groups, the present assessment also found that overall victimization rates among households in NCD and SHP are higher than previously reported. The reasons for this are two-fold: official data gathering in PNG, as in other underdeveloped countries, is extremely weak, and few assessments to date have used the multisource diagnostic tools employed by the Small Arms Survey for this study. These findings have implications for researchers and policy-makers alike, and alert them to the fact that official data should always be analysed with caution, as it will not always accurately reflect violence trends or victimization rates.

Table 7.1 provides a typology of violence and comparative victimization rates as reported in the two survey areas. It shows that domestic violence is the principle type of victimization and the primary contributor to insecurity in both areas, but also indicates the role that social conflict and armed criminality play as catalysts for insecurity.

It is important to review these victimization trends in detail. Household surveys revealed that 50 per cent of NCD households and 51 per cent of SHP households were the victims of violent crime in the six months to May 2005. What is more, over a quarter of all households (26 per cent in NCD and 28 per cent in SHP) had been victimized more than once.

The survey also revealed that four out of five (80 per cent) instances of victimization involved the use of a weapon, though not necessarily a firearm. Of those victimization incidents where weapons were involved, bladed weapons and firearms were the weapons of choice. Twenty-seven per cent of all SHP households participating in the study and 28 per cent of all NCD households reported some form of victimization involving a bush knife or axe in the preceding six months, while 23 per cent of SHP households and 19 per cent of NCD households reported some form of victimization involving a firearm. These findings suggest that violence prevention initiatives that focus on firearms alone (e.g. pure collection or buy-back programmes) are missing much of the broader picture in PNG.

The household survey data concerning the typology of victimization is confirmed by hitherto unpublished hospital and epidemiological data sourced from various hospitals and clinics. This data specifically enables researchers to

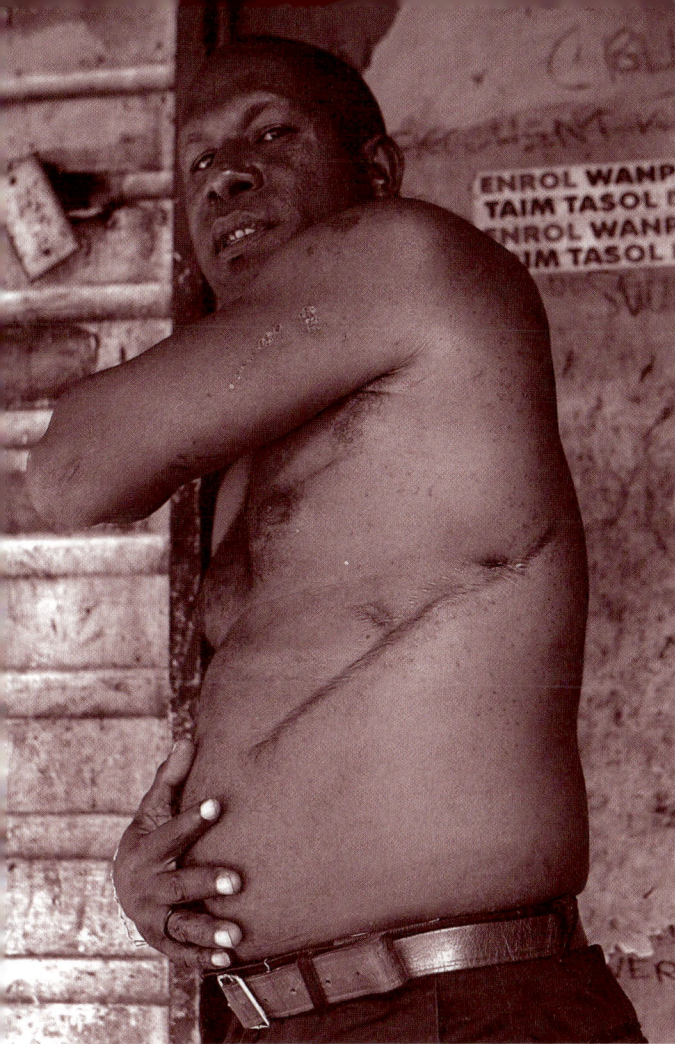

The leader of a feared raskol gang displays his bullet wound and battle scars in a Port Moresby shanty town, July 2004. © Torsten Blackwood/AFP/Getty Images

distil the role of firearms in victimization as compared to other weapons. For instance, in-patient records and reports from Mendi and Tari Hospitals in the SHP confirm that victimization involving bush knives and axes results in significantly more trauma admissions than that involving firearms: at Mendi Hospital, gunshot and bladed injuries resulted in 12 and 41 per cent of all in-patient admissions, respectively (Embiap, 2005), while at Tari Hospital, they accounted for 12 and 25 per cent of serious trauma cases, respectively.[12]

Despite the fact that bush knives and axes result in far more trauma admissions, firearm-related trauma is more likely to result in a lethal outcome. Indeed, while 82 per cent of the gunshot injuries treated at Tari Hospital between 2003 and 2005 proved fatal, only 30 per cent of the bush knife and axe injuries did so. The Tari Hospital data also revealed that men are disproportionately affected by gunshot injuries—accounting for more than 80 per cent of all case of external injuries. There was no such bias in the case of bush knife and axe trauma. The Mendi Hospital data revealed similar patterns, although women accounted for a greater proportion of violence-related trauma admissions. Indeed, while women accounted for 47 per cent of all trauma admissions at Mendi Hospital between May 2004 and May 2005, they accounted for some 54 per cent all violence-related admissions (Embiap, 2005, p. 4).

Domestic and family violence

As noted above, domestic and family violence, including that between co-wives, emerged as the chief cause of victimization and related insecurity across the survey of both areas. Specifically, domestic and family violence were reported to have affected 18 and 26 per cent, respectively, of all households in NCD and SHP in the six months prior to the survey. Although survey and focus group participants appeared willing to discuss frankly the issue of domestic violence, it is submitted that victimization rates are still most likely under-reported, as further qualitative investigations indicated that only the most serious cases—those that resulted in injury or trauma of some kind—were being reported. This is not altogether surprising. Research undertaken by the PNG Law Reform Commission (PNG LRC) in 1983–84 revealed that a third to two-thirds of men in PNG, 95 per cent of men in the Highlands, and 57 per cent of rural women considered it acceptable for husbands to beat their wives (Ranck and Toft, 1986, p. 24; Bradley,

Figure 7.2 **Summary of trauma reports at Tari Hospital, 2003–05**

NUMBER OF REPORTS

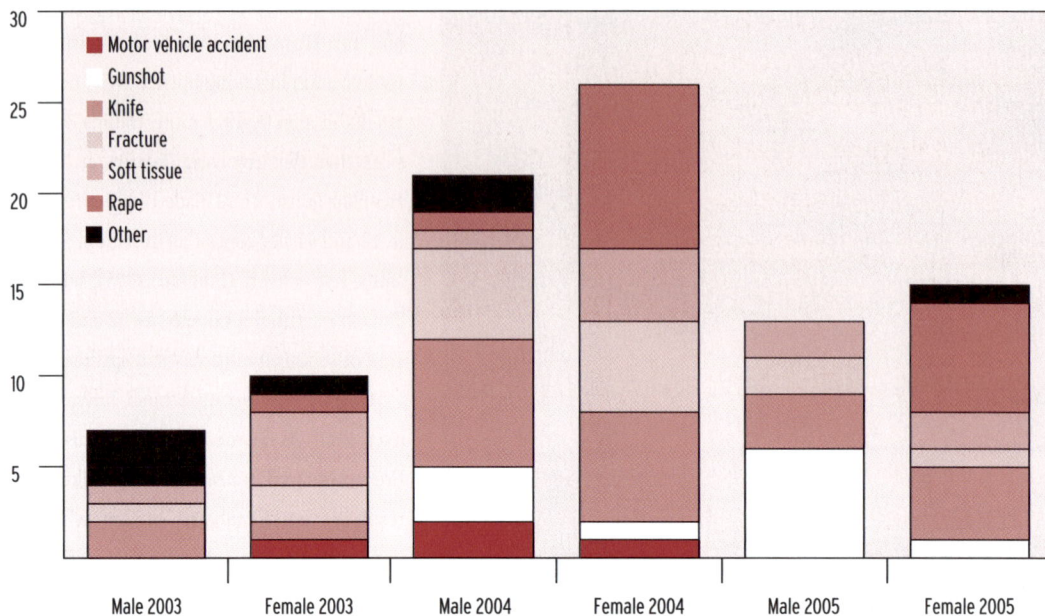

Legend:
- Motor vehicle accident
- Gunshot
- Knife
- Fracture
- Soft tissue
- Rape
- Other

X-axis: Male 2003, Female 2003, Male 2004, Female 2004, Male 2005, Female 2005

Note: The records for 2005 cover January to mid-May only.

1988). The present assessment noted too that in many parts of PNG, cases of domestic violence only ever go before the village court if there is obvious trauma, such as broken bones.

The assessment also revealed that domestic violence is likely to involve the use of weapons. Of those households reporting domestic violence, 63 per cent in SHP and 77 per cent in NCD reported the use of a weapon such as a firearm, bush knife, blunt instrument, fire, or 'red hot' metal. These findings not only confirm those of the PNG LRC (1992) survey on domestic violence, which evinced significant levels of victimization involving the use of weapons,[13] but also reveal increasingly high levels of serious domestic/family violence—something policy, advocacy, and research communities will need to address more purposively in future.

Domestic violence is not, it seems, distributed evenly across all regions and households. In NCD, for example, a greater proportion of affected households were located in the settlements. The assessment also found a correlation between province or region of origin of NCD households and the likelihood that they had been affected by domestic violence. Specifically, it found that 24 per cent of households originally from the Highlands or Gulf Provinces and 23 per cent of Central Province households were affected by domestic violence. By contrast, no Momase or Islands-origin households reported domestic violence.

Intergroup fighting

Intergroup fighting also emerged as a key source of insecurity in both NCD and SHP. In SHP, this finding was not particularly surprising, given the attention tribal fighting in the province has attracted. The prevalence of intergroup fighting in Port Moresby is, however, a worrying new trend. The assessment found that in NCD in the six months

A victim of domestic violence shows her head wound patched up with tape in a women's shelter in Port Moresby, October 2005. © David Gray/Reuters

prior to the survey, 18 per cent of respondent households had been victims of intergroup fighting, 27 per cent included someone who had been involved in intergroup fighting as a combatant, and 15 per cent included someone who had changed residence due to intergroup fighting. The assessment further revealed that 65 per cent of victimized households and 75 per cent of households that included a combatant or someone who had changed residence due to intergroup fighting were located in the settlements.

These findings warrant further investigation, especially as 'tribal' or intergroup fighting in PNG has tended to be viewed as a Highlands problem. Was it predominantly households that were originally from the Highlands that were affected by intergroup fighting? The answer, surprisingly, is 'no', though it remains the case that a significant proportion of households originally from the Highlands were affected. Importantly, the assessment also found that almost one-third of victimized households, one-third of households with combatants, and over half of all the households that included someone who had changed residence due to intergroup fighting were originally from Central Province. It also found that households originally from Central, Gulf, and the Highlands Provinces were all disproportionately affected by, and involved in, intergroup fighting, such that approximately one-third of all households from these provinces reported being affected by intergroup fighting (see Table 7.2). By contrast, only 13 per cent of households from the remaining 12 provinces were similarly affected. Tribal or intergroup fighting is not, in other words, a Highlands-specific problem.

Table 7.2 NCD households affected by intergroup fighting by province of origin, December 2004–May 2005

Province(s) of origin	% of total households in the NCD survey	% of NCD households from specific provinces with combatants*	% of all victimized NCD households	% of victimized NCD households from specific provinces with combatants	% of all NCD households that had changed residence due to intergroup fighting
Central Province	27	34	29	34	53
Gulf Province	22	33	29	26	14
Highlands Provinces	26	29	29	28	28
All other provinces	25	13	13	12	5

* The percentages shown here are the percentage of households from a particular province or group of provinces that had a household member who had been a combatant in the six months prior to the survey, e.g. the percentage of households originally from Gulf Province that included a combatant. In this case, the survey revealed 63 households originally from Gulf Province. Of these 63 households, 21 households (33 per cent) reported that a household member had been involved in intergroup fighting as a combatant.

Intriguingly, intergroup fighting in SHP now appears to be less prolific than it is in NCD. Specifically, in the preceding six months, 17 per cent of SHP households reported having been the victim of intergroup fighting, 24 per cent included someone who had been involved in intergroup fighting as a combatant, and 11 per cent included someone who had changed residence due to intergroup fighting. In the case of SHP, the assessment found that the majority of households affected by and involved in intergroup fighting were clustered in the west of the province.

Armed assault and armed robbery

Armed assault and armed robbery were the next most commonly experienced forms of victimization in SHP and NCD. Surveyed SHP households reported that in the six months to May 2005, 13 per cent had been affected by armed assault and 8 per cent by armed robbery. By contrast, 12 per cent of NCD households had been affected by armed assault and 15 per cent by armed robbery. In both cases, victimization events were clustered in specific geographic areas. In SHP, four out of five reported events occurred in the Hela region, while three out of five events in NCD occurred in the settlements.

Murder and attempted murder

Households experiencing a murder or attempted murder were also concentrated in certain areas. In SHP, at least four out of five victimization events occurred in the west of the province. Alarmingly, 13 per cent of households in the Hela region reported an attempted murder in the six months prior to the survey, while 7 per cent reported an actual murder. This compared with just 3 per cent and 1 per cent of households respectively in the east. Overall victimization rates in NCD were similar to those found in eastern SHP. Incidents were also clustered in particular areas, such that all households reporting a murder and almost three-quarters of those reporting an attempted murder were located in NCD's settlements.

Sexual assault and rape

Rapes and sexual assaults in SHP and NCD typically involve the use of weapons, including firearms.

Sexual assault and rape remain a major concern in PNG. More than 8 per cent of responding households in SHP and 3 per cent in NCD reported that someone in their household had been the victim of sexual assault or rape in the six months to May 2005. In both cases, reported victimization events were geographically clustered—89 per cent of reported SHP incidents occurred in the Hela region, while 60 per cent of NCD's victimized households were located in the settlements. Alarmingly, the survey also revealed that rapes and sexual assaults in SHP and NCD typically involve the use of weapons, particularly firearms and bush knives. This was so in all but one case reported in the survey, which also revealed that the vast majority of reported rapes occur in the village, and often in the home, and that in most cases the assailant(s) are known.

Although the survey did not obtain data concerning the age or sex of the victims or details about the contexts in which the assaults took place, researchers were able to obtain some supplementary data from Tari and Kainantu Hospitals that suggests that young girls are disproportionately affected. When considering possible intervention strategies, victim profiles are critical. Taken collectively, the Tari and Kainantu data suggests that three-quarters of sexual assault and rape victims are under 20 years of age. In the much smaller Tari sample, four out of five victims were actually 15 or younger, and 35 per cent were under 5. In the Kainantu sample, 30 per cent of rape victims were 10 years or under, while 45 per cent were aged between 11 and 20. By way of comparison, a survey undertaken by Riley, Wohlfahrt, and Carrad (1985), cited in the PNG LRC report (1992, p. 15), found that girls aged between 11 and

15, 8 and 11, and under 8 years accounted for 22, 13, and 12 per cent, respectively—i.e. just under half—of all of the rapes and sexual assaults reported to Port Moresby General Hospital in the first quarter of 1985. Seemingly, then, young and very young girls make up a somewhat greater percentage of rape victims than they did two decades ago.

Where violence occurs

The home emerged as the principal location for victimization in this violence assessment, just as it had in the 2004 Port Moresby Community Crime Survey (NRI, 2005, p. 24). Indeed, 26 per cent of all victimization events reported to the NCD assessment team occurred in the home and another 18 per cent in the immediate vicinity of the home. A further 30 per cent of victimization events were reported to have occurred within the home community. In the SHP, 40 per cent of reported victimization events occurred in the home, 7 per cent in its immediate vicinity, and 29 per cent in the home community. In both cases, three-quarters of all victimization events occurred in the home community.

FIREARMS-RELATED VIOLENCE

In order to isolate the role of firearms in armed violence and victimization, the survey also explored community views on guns and the prevalence of firearms (DEMAND). It found that the demand for firearms is robust in both NCD and SHP—more than 40 per cent of all respondents indicated that guns make them feel safer, while 41 and 34 per cent of those surveyed in NCD and SHP, respectively, claimed that they would acquire such a weapon if they could. Many already own guns, claiming to hold them, among other reasons, for self-protection, enforcing compensation arrangements, and status.

Table 7.3 Firearms types and frequency with which they have been seen, December 2004–May 2005

Firearm type	% of SHP households having seen	% of NCD households having seen
Home-made shotgun	66	66
Factory-made shotgun	29	28
Handgun	13	28
Pump-action shotgun	9	17
M16/AR15	14	16
Self-loading rifle (SLR)	8	8
.303	8	4
.22	7	8
Hand grenade	3	8
Machine gun (M-60/.50)	4	2

Note: While the frequency with which particular firearms types are seen is suggestive of how prevalent those particular weapons are, it should be noted that several people might in fact see the same weapon. As such, it is not possible to estimate total gun numbers based on reported sightings alone. For instance, although ten SHP survey respondents reported seeing a machine gun, on further discussion it became evident that only four separate machine guns had actually been seen.

Perceptions concerning the prevalence of firearms, however, varied between regions, and even within them. For example, in SHP, half of respondents felt that gun ownership was on the rise, while 45 per cent felt that gun numbers were static. For the most part, it was people in the west of the province who felt that gun ownership was increasing. By contrast, only 21 per cent of those surveyed in NCD felt this. Almost half felt that gun numbers were static, and close to a third actually felt they were decreasing. Table 7.3 highlights the firearms types seen and the proportion of respondents claiming to have seen a particular firearm in the six months prior to the survey.

As Table 7.3 suggests, there are variations in the types of weapons and the frequency with which they are seen in various parts of PNG. For example, although the reported presence of home-made shotguns, factory-made shotguns, M16s and AR15s, SLRs, and .22 rifles was remarkably similar in SHP and NCD, there were differences with respect to pump-action shotguns, handguns, and hand grenades—all of which were twice as likely to be seen in Port Moresby. There were differences, too, as regards the types of weapons being seen in different parts of SHP (see Map 7.3). For instance, respondents in the Hela region mostly reported seeing home-made shotguns, single-shot factory-made shotguns, and to lesser extent pump-action shotguns, and reported seeing them at least five times as often as an M16 or AR15. In the east, however, where people were claiming to see firearms relatively infrequently, high-powered assault rifles, such as M16s and AR15s or SLRs, accounted for a greater proportion of all weapons seen. Although firearms sightings differed from region to region in SHP, no such variation was found in NCD.

Another indicator of firearms prevalence and/or desirability is market price. Predictably, the more sought after the firearm, the higher its price. Table 7.4 lists reported firearms prices by type in mid-2005. Variations in price for particular firearms types were evident across districts in SHP. Prices also varied within NCD's settlements and suburbs,

Map 7.3 **Frequency of firearm displays in SHP communities, December 2004–May 2005**

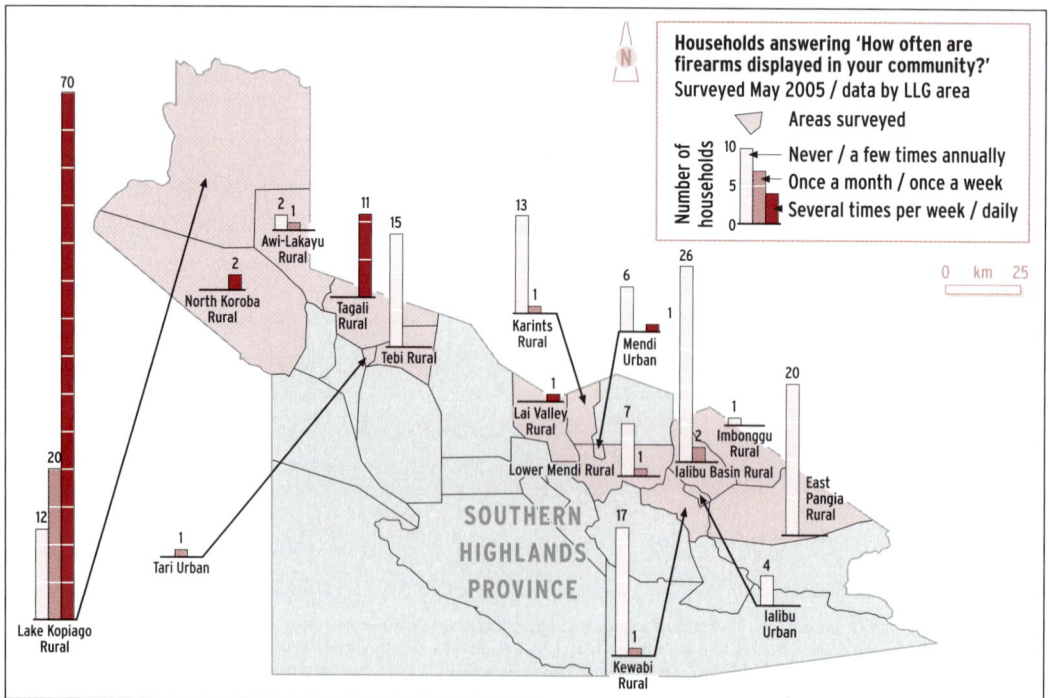

Port Moresby police hold home-made weapons that were used in armed robberies, October 2005. © David Gray/Reuters

although there were no significant differences between settlements and suburbs.

It was evident too that other forms of 'currency'—particularly women and locally grown marijuana—are also being used to acquire firearms and store-bought goods of various kinds in SHP's remote areas, where there is little cash and few income earning opportunities.[14]

Worryingly, the survey also revealed that supplies of ammunition, previously thought to be tight (Alpers, 2005), are in fact

Table 7.4   Firearms types and their market prices (PGK*) in SHP and NCD by district, December 2004–May 2005

Firearm type	SHP range	SHP average	Kopiago District (SHP) average	Tari District (SHP) average	Mendi District (SHP) average	Ialibu/ Pangia & Imbonggu Districts (SHP) average	NCD range	NCD average
Home-made shotgun	100–500	157	145	157	234	178	100–800	309
Home-made shotgun #2**	700–1,500	983	983	–	–	–	–	–
Factory-made shotgun	400–5,000	1,043	–	558	2,667	1,329	200–6,000	2,105
Pump-action shotgun	1,500–10,000	4,344	–	5,200	4,250	3,785	1,500–14,000	6,393
M16/AR15	4,000–15,000	8,045	–	6,750	10,750	8,715	4,500–20,000	8,365
SLR	5,000–14,000	9,438	–	8,125	11,000	10,667	8,000–20,000	13,295
Handgun	200–3,000	1,025	–	1,120	500	1,033	1,000–7,000	2,720
Hand grenade	1,000–1,800	1,267	–	–	1,000	1,400	400–2,000	800
Machine gun	20,000–40,000	27,500	–	–	40,000	25,000	12,000–30,000	23,625

* PGK 100 are worth about USD 33.

** These are a recent innovation: bolt-action home-made shotguns specifically designed to take 5.56 x 45 mm ammunition, which is the cheapest and most commonly available ammunition in the far western end of SHP.

Table 7.5 Ammunition types and their typical market prices (PGK*) by district, December 2004–May 2005

Ammunition type	Kopiago District (SHP)	Tari District (SHP)	Mendi District (SHP)	Ialibu, Pangia & Imbonggu Districts (SHP)	NCD
5.56 x 45 mm	5–12 (7)	5–12 (8)	15–20 (18)	30 (30)	10–50 (25)
7.62 x 51 mm	–	8–14 (11)	20–25 (23)	30 (30)	10–50 (25)
12 gauge shotgun cartridge	10–12 (11)	10–20 (12)	12–25 (19)	5–10 (6)	3–20 (9)

Note: Average price given in brackets.

* PGK 100 are worth about USD 33.

comparatively abundant, and that prices are rapidly declining, especially in the Hela region, where new supply routes have opened up. Table 7.5 lists the three main ammunition types and their black market prices as at mid-2005.

Ironically, as new trade routes encompassing Hela have opened up in the west of the region, allowing North Atlantic Treaty Organisation (NATO) standard ammunition (5.56 x 45 mm and 7.62 x 51 mm) to flow more readily onto the black market, shotgun cartridges have become increasingly hard to come by relative to other ammunition types. This price differential has given rise to some quite innovative practices, including the production of more sophisticated home-made firearms—weapons that take 5.56 x 45 mm ammunition and versions that operate like traditional bolt-action rifles—as well as the production of modified ammunition, whereby NATO standard ammunition is fitted into spent shotgun cartridges for use in home-made shotguns and pistols. Collectively, these innovations represent potential catalysts for increased violence in the near future.

CAUSES OF AND RESPONSES TO ARMED VIOLENCE

This and past research has shown that there is no single cause of armed violence in PNG. Microeconomic analyses of urban crime in PNG have by and large taken the view that rapid population growth generates crime, concluding that marginal youth are propelled into crime as a consequence of limited economic and employment opportunities (Clifford, Morauta, and Stuart, 1984; UNDP/ILO, 1993; Dinnen, 2000, p. 70). PNG criminologist Richard Sikani (1997; 2000) concurs, pointing out that urban drift, the growth of large squatter settlements,[15] the breakdown of traditional values, limited employment opportunities for high school graduates, structural inequalities between the 'haves' and 'have nots', and contestation over resource ownership are all contributing to contemporary criminality in PNG. Anthropological accounts, however, have shown that other factors such as prestige (Goddard, 1992; 1995; Dinnen, 1995; 2000)[16] or the desire to escape the expectations or demands of kin (Monsell-Davis, 1993) often come into play as well. They have also shown that 'fighting is both a recurrent and legitimate means of prosecuting claims and seeking restitution for many PNG societies' (Goldman, 2003, p. 2).

Just as there were regional differences with respect to armed violence, victimization, illegal gun ownership, and ammunition prices, the assessment revealed that the attributed causes of armed violence differ from area to area. It is instructive, then, to compare the eastern and western areas of SHP, and the suburbs and settlements of NCD, in order to highlight the differentiated causes and responses to armed violence.

Southern Highlands boys brandish their home-made weapons, May 2005. © Nicole Haley

In eastern SHP, men and women alike asserted that there were few guns in their respective communities and that they were not particularly troubled by armed violence, except during national elections and when travelling. They also reported that tribal fighting seldom occurs, because large compensation payments are made within the community to aggrieved parties in a speedy fashion so as to 'cool off' any motivations for fighting. Instead, they cited domestic and family violence as the most common and troublesome forms of violence in their communities. Women attributed domestic violence to polygamy and/or promiscuity, and to drug and alcohol abuse, while men considered jealousy, gambling, alcohol, marijuana, and adultery as causal factors. In the east, jealousy and sorcery accusations also emerged as key causes of violence, with sorcery being invoked to explain motor vehicle fatalities and other accidental deaths.

There was also general consensus that local leaders in eastern SHP are strong and that they continue to command respect. Specifically, it was asserted that they are quick to attend to disputes and that they work tirelessly to see conflicts resolved quickly, so as to diffuse the likelihood of further conflict. It was also asserted that they take family conflicts seriously, intervening quickly before the extended family becomes involved. Because of this, the researchers found considerable respect for customary law and practices, despite deteriorating confidence in the police and judicial systems. Communities in the east, for instance, were prepared to have disputes dealt with by the village courts rather than take matters into their own hands.

Despite their willingness and desire to solve disputes peacefully without recourse to violence, people in eastern SHP felt that it is increasingly difficult to do so due to inflating compensation demands. Many lamented the contradictory nature of compensation, pointing out that while it contributes to the management of disputes, it is also crippling local economies. Men and women alike felt something should be done to limit compensation payments, before they become completely unmanageable. They noted that compensation demands in the event of a death are now routinely in the order of PGK 40,000–50,000 (USD 13,000–16,000) plus pigs and up to PGK 100,000 (USD 33,000) in the case of motor vehicle accidents. To put this in perspective, average incomes in the Ialibu/Pangia and Imbonggu districts amount to less than PGK 100 (USD 33) per person per year (Hanson et al., 2001, p. 93). Local public servants at Ialibu also saw correlation between the weakness of the state and these high compensation payments: 'The law is weak so we must pay more in order to maintain peace.'

Compensation was also found to be causing problems in the Hela region, where the rule of law is comparatively weak and local leadership undermined by young thugs with guns (ANGRY YOUNG MEN). There compensation had become so inflated that people are having difficulty paying the amounts demanded, thus ratcheting up the potential for renewed violence. Kopiago men were particularly vocal in asserting that threats, intimidation, and extortion with firearms are the most virulent form of violence in their community. They further explained that local leaders had become powerless due to the proliferation of weapons, and that with guns, people are able to extract increasingly large compensation payments. Clearly, then, there are variations in the respective capacities of local communities to manage conflict and maintain peace in their own areas.

Respondents link polygamy, gambling, and marijuana to armed violence in their communities.

Armed violence was of particular concern to people in western SHP, though men and women tended to envision the root causes differently.[17] Women, for instance, saw escalating bride prices as a huge problem, pointing out that large payments keep them in bad marriages (because they can't possibly repay the bride price in the event of a divorce), causing them and their children to suffer. Women were also very much concerned about the growing incidence of domestic and family violence. They saw three key factors as contributing to the problems they were experiencing— polygamy, gambling, and marijuana. They were most concerned about polygamy and promiscuity, and linked with it the risk of exposure to HIV/AIDS.[18]

Men also identified these same three factors, but ranked them in the reverse order. They felt that marijuana was the key cause of armed violence in their communities. Indeed, village leaders, local mediators, and peace and good order committees complained over and over again that their attempts to maintain law and order were being undermined by 'marijuana boys' who turn up at village court cases and support one or other of the involved parties on a 'no win, no fee' basis. Because of this, local leaders now feel compelled to arm themselves when hearing disputes. Indeed, throughout the province, but particularly in the west, researchers encountered the view that the law and order situation had deteriorated to the point where 'good people' are arming themselves 'for their own protection'. That demand appears to be growing more robust has real implications for potential disarmament (DEMAND).

In the SHP, conflict over resources has also emerged as a key contributor to insecurity and as a subsequent reason for acquiring firearms, particularly in the west. It was generally agreed that the upsurge in fighting in SHP coincided with the advent of resource development. Certainly, many of the province's ongoing conflicts concern the distribution of oil and gas royalties and access to the benefits of resource development. Much of the current conflict can also be attributed to state failure: the failure to provide basic services, such as health and education; the failure to mediate land and royalty disputes; the failure to address existing law and order problems; and a lack of accountability in relation to the expenditure of funds generated by resource development (Haley and May, forthcoming). Throughout

SHP, but most particularly in the Hela region, researchers regularly encountered the sentiment that the money being generated through resource development is not being channelled back into rural areas, and this is an ongoing source of discontent.

The state's failure to maximize and equitably distribute the benefits of resource development or to provide essential services in the Hela region has led to calls for the establishment of a separate Hela province, with many suggesting that if the Hela people continue to be denied essential services and a more equitable share of the benefits of resource development, then SHP will go the way of Bougainville.[19] Certainly, this was the mantra the assessment team encountered in the Tari area.

There are no services here . . . . We are ready for a fight. If they don't give us a Hela province we will fight . . . . If we don't get our province, forget about the gas pipeline project. It won't go . . . . We men and boys are gathering guns. If we don't get our own province by 2007 we will fight. It will be bigger than Bougainville. They had only a few guns; we have plenty—not just home-made guns, but high-powered weapons. They had plenty of languages—we have only one. We will fight. It will be bigger than Bougainville. Bougainville was a small fight. We are ready to fight . . . . We number 300,000. The government must hear us.[20]

PNG police brandish M-16s as they patrol the streets for raskol gangs in August 2004. © Torsten Blackwood/AFP/Getty Images

There can be no doubt that the Hela province issue presents a potential source of conflict in SHP. People in the Hela region are arming themselves with the express purpose of causing widespread civil unrest should the national government fail to respond to their calls for the establishment of a separate province prior to the 2007 elections (Agiru, 2005).

The problems in the Hela region also highlight the fact that service delivery failure contributes to insecurity in that it gives rise, among other things, to contestation over non-state resources and access to non-government service providers (Haley and May, forthcoming). In SHP, service delivery failure has also seen public services become privatized, creating yet another source of conflict. Increasingly, those without cash resources are finding it difficult to utilize the judicial system and to access non-violent solutions to their problems. As such, there can be no doubt that state failure is contributing to armed violence in SHP.

As in SHP, the NCD focus group participants were asked to identify the types and causes of violence in their communities. Some eight key causal factors were identified, including unemployment, school drop-outs, alcohol abuse, drug trafficking and abuse, financial difficulties, family problems, lack of funds to pay school fees, and peer pressure. Across NCD, unemployment, particularly youth unemployment, was held to be the single most important factor contributing to violence. Financial difficulties and alcohol and/or drug abuse were also widely seen to be the next most significant factors contributing to armed violence. NCD focus group participants expressed particular concern about the availability and consumption of locally brewed alcohols such as 'steam'.

Most respondents said they would not give up their guns unless the law and order situation improved.

Despite these grim findings, affected communities agreed that there are a number of pragmatic and immediate interventions that might alleviate these problems. These included employment generation, programmes to reduce alcohol consumption, restrictions on carrying weapons, increased police presence and law enforcement, and compulsory education. As in SHP, views concerning the causes of violence tended to differ by gender, with men and women also diverging in their views about the most effective ways of combating violence. For instance, women in NCD advocated a total ban on the sale of alcohol, citing alcohol as a key cause of violence in their communities. Women, too, were most vocal about wanting to see the Vagrancy Act reactivated and about stopping migration from rural to urban areas. Men, however, saw unemployment as the key contributor to violence. More so than the women, they felt that the creation of employment and income generation opportunities would see violence reduced.

NCD residents also pointed out that their communities are actively involved in trying to improve the security situation in their own local areas. Many had established community groups, neighbourhood watch initiatives, and peace and good order committees to mediate minor conflicts. The peace committees were particularly active in the settlements. Although people generally felt that they had strengthened law and order, it was also felt that compensation was on the rise, with some communities meeting up to five times a week at local parks to solve disputes, to discuss how compensation monies could be raised, or to agree upon demands to be imposed on other groups. As in SHP, it was felt that compensation demands were steadily inflating, such that current levels of compensation were held to be unsustainable. It was also felt that rising compensation payments were contributing to greater cycles of retribution and violence.

Clearly, any effort to enact civilian disarmament will be a challenging enterprise in PNG. Despite the havoc caused by guns, concern about rising compensation payments, and recognition that they are crippling livelihoods and contributing to the escalation of violence, the vast majority of respondents claimed that they would be unwilling to give up their guns unless the law and order situation improved. Specifically, they rejected the idea of gun buy-backs or 'weapons for development' programmes, arguing that such schemes would not work unless everyone disarmed simultaneously. Instead of being seen as part of the problem, guns were seen by many to be offering a degree of protection, with respondents in both SHP and NCD asserting that their guns were their 'protection' and 'bodyguard'.

That said, respondents in NCD felt that firearms reduction might prove more successful if people received payment for their firearms and were provided with employment or other income earning opportunities. Significantly, it was repeatedly noted that one of the many motivating factors for firearms acquisition is the need to make a living: many NCD focus group participants reported that people in their communities had obtained guns with the express purpose of leasing them out or using them to eke out a living through crime. This suggests that economic factors are contributing to the proliferation of firearms in NCD in a way that they are not in SHP, and potential interventions will need to take account of this.[21]

CONCLUSION: LOOKING FORWARD

As Dinnen (2004, p. 71) rightly notes, external interventions need 'to be grounded in a thorough understanding of the socio-economic and political complexities of the recipient countries.' Given the sheer size of mainland PNG, large-scale external interventions of the kind undertaken in the Solomon Islands and on Bougainville are unlikely to succeed or be financially viable. Further, there is as yet little local support for gun surrenders or weapons reduction initiatives. Certainly, past initiatives have either failed or have tended to yield mainly home-made weapons. Indeed, despite popularity among donors and policy-makers, weapons reduction initiatives often target the wrong people, tend to enjoy only limited success, and can even do more harm than good if they do not take proper account of the local context (Muggah, 2005, p. 289). For instance, reducing firearms numbers may well have little impact on violence against women—for as this survey has shown, SHP women are five times more likely to be admitted to hospital with a bladed injury than a gunshot injury. This is not to say that weapons reduction initiatives are doomed or that they have no merit, merely that they should be considered alongside a broader range of alternatives. In PNG, small-scale local interventions focused on violence reduction and support to customary law—as compared to straightforward gun collection—will most likely prove productive. This was particularly evident in NCD, where many communities had established neighbourhood watch patrols and peace and good order committees to mediate minor conflicts, as a means of improving and taking back responsibility for maintaining security in their communities.

> Some community residents obtain guns with the express purpose of leasing them out for profit.

In the absence of a strong state, civil society groups are seeking to combat the growing law and order problems in their own areas. Women's, church, and other community-based organizations, for instance, have taken on a particularly active role in the area of peace-making, mediation, and conflict resolution (Dinnen and Thompson, 2004, p. 12). While undertaking this assessment, researchers encountered several small-scale local initiatives that have not only proved successful, but also self-sustaining. Researchers are of the view that such initiatives might well be replicated elsewhere, and that properly targeted donor support would not just see them grow, but even increase in effectiveness. Successful projects share several key characteristics: they were designed and developed locally, such that they value and draw upon local knowledge and cultural resources; they engage and gainfully occupy the local youth; they have empowered local leaders; and they have enhanced customary governance. This last feature is particularly critical, because when local leadership is strong, law and order can be maintained, even in the absence of effective state police and judicial systems. The different prevailing security situations in SHP demonstrate this.

In sum, this chapter finds that weapons reduction initiatives, such as the national gun amnesty that has been recommended by the National Guns Control Committee (NGCC, 2005), are likely to yield comparatively limited results in PNG's most conflict-riven areas, unless some degree of law and order is first re-established and locally specific factors fuelling demand are addressed (DEMAND). In much of SHP and NCD, the two are intrinsically linked. Many responsible citizens, including local mediators, are arming themselves for protection and expressing a general unwillingness to disarm unless the law and order situation improves. In seeking to reduce violence in these regions, it will therefore be important to 'think small'. Modest community-level interventions that build upon local capacity and seek to enhance customary governance will likely prove most productive. ◼

ANNEXE 1. NATIONAL CRIME INCIDENTS SUMMARY, 2004, BY PROVINCE

Incident type	NCD	Lae	Western Highlands Province	SHP	Eastern Highlands Province	Simbu Province	Enga Province	Other 15 provinces	Sub-total
Murder	97	34	57	38	38	11	29	303	607
Serious sex offence	253	130	153	79	69	14	17	428	1,143
Grievous bodily harm	418	78	134	120	99	27	49	327	1,252
Armed robbery	862	523	123	48	87	20	3	339	2,005
Breaking and entering	500	300	135	65	77	21	16	467	1,581
Vehicle hijacking	221	8	41	20	11	1	3	51	356
Stealing (PGK 1,000+)	166	31	54	5	19	8	2	121	406
Fraud	52	7	4	0	7	3	0	12	85
Escape from prison	17	8	0	0	1	0	0	33	59
Drug offence	339	138	22	8	57	20	4	231	819
Firearms offence	190	55	6	7	14	4	3	51	330
Escape from police	76	23	14	10	42	7	10	111	293
Abduction	7	3	10	12	12	1	1	22	68
Armed ambush	0	0	7	3	10	0	0	2	22
Exchange of gunfire	25	0	25	30	31	0	0	14	125
Shooting incident	25	25	25	33	31	0	0	28	167
Factional fighting	3	0	43	21	20	3	6	3	99
Other incidents	14	5	5	0	3	0	0	11	38
Total	**3,265**	**1,368**	**858**	**499**	**628**	**140**	**143**	**2,554**	**9,455**

Source: UNDSS (2005)

LIST OF ABBREVIATIONS

ACT	Australian Capital Territory	PNG LRC	Papua New Guinea Law Reform
LLG	Local Level Government		Commission
NCD	National Capital District	SHP	Southern Highlands Province
PGK	Papua New Guinea kina	USD	United States dollar
PNG	Papua New Guinea		

ENDNOTES

1   There are five Highlands provinces: Western Highlands Province, Eastern Highlands Province, Southern Highlands Province, Enga Province, and Simbu Province.

2   The Small Arms Survey team included Nicole Haley and Robert Muggah, with a team of international and local researchers affiliated with the Australian National University's State, Society and Governance in Melanesia Project and the University of Papua New Guinea. Preliminary findings were presented at the PNG Gun Summit held in Goroka in July 2005.

3   See, for example, Capie (2003), Alpers and Twyford (2003), Alpers (2004), Dinnen and Thompson (2004), Muggah (2004), Twyford and Muggah (2004), LeBrun and Muggah (2005), and Alpers (2005).

4   See Goldman (2003), Haley (2004), Alpers (2004), Muggah (2004), and LeBrun and Muggah (2005).

5   Between January 2004 and March 2005, approximately half of all crimes reported nationally occurred in the urban centres of Port Moresby and Lae, while close to a quarter occurred in the Highland provinces (UNDSS, 2005). Annexe 1, which provides the National Crime Incidents Summary for 2004, reveals that Port Moresby—the country's largest and fastest-growing urban centre, with a projected population now exceeding 300,000 (NSO, 2002)—currently accounts for some 34 per cent of all nationally reported crimes, despite being home to only 5 per cent of PNG's 5.5 million people.

6   Settlements are unsanctioned and unplanned residential areas lacking basic amenities and services that were established by migrants and squatters.

7   See Clifford, Morauta, and Stuart (1984), Dinnen (2000, p. 67), Walker (1985), and Muggah (2004).

8   The present assessment found that 48 per cent of the victimization events reported to the NCD team and only 16 per cent of those reported to the SHP team had been reported to the police. It also found that reporting rates are geographically and demographically differentiated, such that crimes occurring in areas where there is a police presence were more likely to be reported than those occurring in remote rural areas where there are no longer any police.

9   This is roughly consistent with the National Crime Incidents Summary for 2004 (Annexe 1), in which the violent crimes of murder, serious sexual offence, grievous bodily harm, and armed robbery account for 53 per cent of reported criminal events. It is worth noting, though, that violent crime tends to account for less than a quarter of all offences in other countries. In 2004 it accounted for 21 per cent of reported crime in the United States (US Bureau of Justice, 2005) and 11 per cent of reported crime in Australia (ABS, 2005).

10 An international comparison of criminal justice statistics, undertaken by Barclay and Tavares et al. (2003), reveals that the murder rate in Port Moresby/NCD is significantly higher than in most other cities around the world. They cite the following homicide rates per 100,000 people: Sydney (1.6), London (2.6), New York (8.9), Moscow (18.4), and Washington, DC (42.8).

11 The NCD survey included slightly more settlement households (156) than households in the suburbs (136), meaning that settlement households accounted for 53 per cent of the sample.

12 Data provided by hospital staff, 2005.

13 That survey found that 57 per cent of rural wives, 37 per cent of low-income urban wives, and 23 per cent of urban elite wives who had been hit by their husbands reported being injured with a weapon or implement (PNG LRC, 1992, pp. 16–17).

14 While undertaking this assessment, researchers received several first-hand reports from people who had traded marijuana in order to acquire an assault rifle. Typically, they had traded 40–60 kg of marijuana for a single M16 or SLR. Although this might seem high, people in the same areas reported trading 10 kg of marijuana for a pair of steel-capped work boots.

15 UNDP (2004, p. 16) estimates that 50 per cent of NCD residents now live in squatter settlements.

16 Embiap (2005, p. 1) concurs, suggesting that violence is used 'as a way of earning respect, pride and wealth'. It is worth noting, too, that prestige can also motivate people to abandon criminal activities. Dinnen, for example, has shown that public weapons surrenders are seen by some as a strategy for accessing resources, 'project funds and employment opportunities' (Dinnen, 1995, p. 109), and a means of building prestige (p. 116).

17 Previous participatory assessments administered by the Small Arms Survey in South Asia, South-East Asia, and the South Pacific have found similar gender and demographic differences in how problems are conceived and ultimately acted upon. See, for example, Banerjee and Muggah (2001), Moser-Puangsuwan and Muggah (2003), and LeBrun and Muggah (2005).

18 The burgeoning HIV/AIDS epidemic is a factor fuelling violence in SHP and elsewhere in PNG. Anecdotal and qualitative evidence indicates that the incidence of HIV/AIDS is rising dramatically in SHP—recent figures suggest a five-fold increase since 2003. Already people in rural areas are dying of AIDS-related illnesses, yet sadly national awareness and prevention campaigns are yet to reach these areas due to service delivery

failure. In some parts of PNG, AIDS deaths are now being attributed to witchcraft. At Kopiago, for instance, accusations of witchcraft have resulted in a spate of brutal witch trials over the past two years. In the first of these trials, six women were held captive and tortured over the space of a fortnight. They were beaten, stabbed, cut with bush knives, and burnt with heated reinforcing iron (Haley, forthcoming).

19   Violent conflict, sparked in part by secessionist sentiments and discontent about the distribution of benefits from the highly successful Bougainville Copper Mine, raged in PNG's Bougainville Province from 1988 until the late 1990s. During that time, thousands of people were killed. In mid-2005, after a lengthy peace process, an Autonomous Bougainville Government was elected and sworn in. In 2020 there is to be a constitutionally guaranteed referendum on the issue of independence. There is extensive literature on the origins and unfolding of the conflict; see, for instance, May and Spriggs (1990) and Regan (1998).

20   Interviewee statement, Tari, May 2005.

21   Although people in SHP did not report acquiring weapons to lease out, both weapons and mercenary gunmen are available for hire and are serious and unpredictable variables in intergroup conflict (Alpers, 2005).

BIBLIOGRAPHY

ABS (Australian Bureau of Statistics). 2005. *Recorded Crime: Victims, Australia 2004*. Canberra: ABS.

Agiru, Anderson. 2005. 'Implications of Autonomy on PNG.' *The National* (Port Moresby). 3 June.

Alpers, Philip. 2004. *Gun Violence, Crime and Politics in the Southern Highlands: Community Interviews and a Guide to Military-style Small Arms in Papua New Guinea*. Unpublished background paper. Geneva: Small Arms Survey.

——. 2005. *Gun-running in Papua New Guinea: From Arrows to Assault Weapons in the Southern Highlands*. Special Report No. 5. Geneva: Small Arms Survey.

—— and Conor Twyford. 2003. *Small Arms in the Pacific*. Occasional Paper No. 8. Geneva: Small Arms Survey.

Australia. Department of Foreign Affairs and Trade. 2005. *Travel Advice Papua New Guinea*. Updated 15 September.
<http://www.smartraveller.gov.au/zw-cgi/view/Advice/Papua_New_Guinea>

Banerjee, Dipankar and Robert Muggah. 2002. *Small Arms and Human Insecurity: Reviewing Participatory Research in South Asia*. Geneva/Colombo: Small Arms Survey and Regional Centre for Strategic Studies.

Barclay, G. and Cynthia Tavares et al. 2003. *International Comparisons of Criminal Justice Statistics 2001*. Issue 12/03. London.

Bradley, Christine. 1988. 'Wife-beating in Papua New Guinea: Is It a Problem?' *Papua New Guinea Medical Journal*. Vol. 31, No. 4, pp. 257–68.

Capie, David. 2003. *Under the Gun: The Small Arms Challenge in the Pacific*. Wellington: Victoria University Press.

Clifford, William, Louise Morauta, and Barry Stuart. 1984. *Law and Order in Papua New Guinea*. Discussion Paper No. 16, Vols. 1 & 2. Port Moresby: Institute of National Affairs and Institute of Applied Social and Economic Research.

Dinnen, Sinclair. 1995. 'Praise the Lord and Pass the Ammunition: Criminal Group Surrender in Papua New Guinea.' *Oceania*. Vol. 66, No. 2, pp. 103–18.

——. 2000. 'Threatening the State: Crime and Public Order in Papua New Guinea.' In Beno Boeha and John McFarlane, eds. *Australia and Papua New Guinea: Crime and the Bilateral Relationship*. Canberra: Australian Defence Studies Centre, pp. 65–74.

——. 2001. *Law and Order in a Weak State: Crime and Politics in Papua New Guinea*. Adelaide: Crawford House Publishing.

——. 2004. 'Australia's New Interventionism in the Southwest Pacific.' In Nancy Sullivan, ed. *Governance Challenges for PNG and the Pacific Islands*. Madang: Divine Word University Press.

—— and Edwina Thompson. 2004. *Gender and Small Arms Violence in Papua New Guinea*. State, Society and Governance in Melanesia Discussion Paper No. 2004/8. Canberra: State, Society and Governance in Melanesia Project, Australian National University.

Embiap, James. 2005. *Violence and Guns in Mendi: A Report on Violence-related Incidents and Deaths in Mendi District, PNG*. Unpublished report. Mendi: Mendi District Health Service.

Goddard, Michael. 1992. 'Big-man, Thief: The Social Organisation of Gangs in Port Moresby.' *Canberra Anthropology*. Vol. 15, No. 1, pp. 20–34.

——. 1995. 'The Rascal Road: Crime, Prestige and Development in Papua New Guinea.' *Contemporary Pacific: A Journal of Island Affairs*. Vol. 7, No. 1, pp. 55–80.

Goldman, Laurence. 2003. *'Hoo-Ha in Huli': Considerations on Commotion and Community in the Southern Highlands Province of Papua New Guinea*. State, Society and Governance in Melanesia Discussion Paper No. 2003/8. Canberra: State, Society and Governance in Melanesia Project, Australian National University.

Gomez, Brian, ed. 2005. *Papua New Guinea Yearbook 2005*. Port Moresby: The National and Cassowary Books.

Haley, Nicole. 2004. 'A Failed Election: The Case of the Koroba-Lake Kopiago Open Electorate.' In Philip Gibbs, Nicole Haley, and Abbey McLeod. *Politicking and Voting in the Highlands: The 2002 Papua New Guinea National Elections*. State, Society and Governance in Melanesia Discussion Paper No. 2004/1. Canberra: State, Society and Governance in Melanesia Project, Australian National University, pp. 16–26.

——. Forthcoming. 'Witchcraft, Torture and HIV/AIDS.' In Vicki Lucker, Sinclair Dinnen, and Alan Patience, eds. *Law, Order and HIV/AIDS in PNG*. Canberra: Pandanus.

—— and Ron May. Forthcoming. 'Introduction: Roots of Conflict in the Southern Highlands.' In Nicole Haley and Ron May, eds. *Conflict and Resource Development in the Southern Highlands of Papua New Guinea*. Canberra: State, Society and Governance in Melanesia Project, Australian National University.

Hanson, Luke, Bryant Allen, Michael Bourke, and Tess McCarthy. 2001. *Papua New Guinea Rural Development Handbook*. Canberra: Australian National University.

LeBrun, Emile and Robert Muggah, eds. 2005. *Silencing Guns: Local Perspectives on Small Arms and Armed Violence in Rural Pacific Island Communities*. Occasional Paper No. 15. Geneva: Small Arms Survey.

Levantis, Theodore. 1998. 'Tourism in Papua New Guinea: A Comparative Perspective.' *Pacific Economic Bulletin*. Vol. 13, No. 1, pp. 98–105.

Lewis, Neryl. Forthcoming. 'Conflict Vulnerability Assessment, Southern Highlands Province, Papua New Guinea.' In Nicole Haley and Ron May, eds. *Conflict and Resource Development in the Southern Highlands of Papua New Guinea*. Canberra: State, Society and Governance in Melanesia Project, Australian National University.

May, Ron and Matthew Spriggs, eds. 1990. *The Bougainville Crisis*. Bathurst: Crawford House Press.

Monsell-Davis, Michael. 1993. 'Urban Exchange: Safety-net or Disincentive? Wantoks and Relatives in the Urban Pacific.' *Canberra Anthropology*. Vol. 16, No. 2, pp. 45–66.

Moser-Puangsuwan, Yeshua and Robert Muggah, eds. 2003. *Whose Security Counts? Participatory Research on Armed Violence and Human Insecurity in Southeast Asia*. Geneva: Small Arms Survey and Nonviolence International.

Muggah, Robert. 2004. *Diagnosing Demand: Assessing the Motivation and Means for Firearms Acquisition in the Solomon Islands and Papua New Guinea*. Discussion Paper No. 2004/7. Canberra: State, Society and Governance in Melanesia Project, Australian National University.

——. 2005. 'Managing "Post-conflict" Zones: DDR and Weapons Reduction.' In Small Arms Survey, *Small Arms Survey 2005: Weapons at War*. Oxford: Oxford University Press, pp. 267–301.

NGCC (National Guns Control Committee). 2005. Unpublished background report. PNG Guns Summit, Goroka, 4–8 July.

NRI (National Research Institute). 2005. *Port Moresby Community Crime Survey, 2004*. Waigani: NRI.

NSO (National Statistical Office). 2002. *National Thematic Map Tables PNG 2000 National Census on CD-ROM*. Waigani: NSO.

New Zealand. Ministry of Foreign Affairs and Trade. 2005. *Travel Advice Papua New Guinea*. Updated 21 June. <http://www.mfat.govt.nz/travel/countries/png.html>

PNG LRC (Papua New Guinea Law Reform Commission). 1992. *Final Report on Domestic Violence*. Report No. 14. Boroko: PNG LRC.

Ranck, S. and Susan Toft. 1986. 'Domestic Violence in an Urban Context with Rural Comparisons.' In Susan Toft, ed. *Domestic Violence in Papua New Guinea*. Monograph No. 3. Port Moresby: PNG LRC.

Regan, Anthony. 1998. 'Current Developments in the Pacific: Causes and Course of the Bougainville Conflict.' *Journal of Pacific History*. Vol. 33, No. 3, pp. 269–85.

Riley, I., D. Wohlfahrt, and E. Carrad, eds. 1985. *The Management of Rape and Other Sexual Offences in Port Moresby*. Report to a workshop held at Department of Community Medicine, University of Papua New Guinea, Port Moresby General Hospital, 4 June.

Sikani, Richard. 1997. *Live to Steal, Steal to Live: Juvenile and Economic Crime*. NRI Preliminary Paper No. 3. Waigani: National Research Institute.

——. 2000. 'The Criminal Threat in Papua New Guinea.' In Beno Boeha and John McFarlane, eds. *Australia and Papua New Guinea: Crime and the Bilateral Relationship*. Canberra: Australian Defence Studies Centre, pp. 35–56.

Twford, Conor and Robert Muggah. 2004. 'Trouble in Paradise: Small Arms in the Pacific.' In Small Arms Survey, *Small Arms Survey 2004: Rights at Risk*. Oxford: Oxford University Press, pp. 277–307.

UNDP (United Nations Development Programme). 2004. *Port Moresby Diagnosis of Insecurity Report*. Port Moresby: UNDP.

——/ILO (International Labour Organization). 1993. *Papua New Guinea: Challenges for Employment and Human Resource Development*. Port Moresby: UNDP/ILO.

UNDSS (United Nations Department of Safety and Security). 2005. *National Crime Summary Report: 01 Jan 04–31 Dec 04*. Port Moresby: UNDSS.

US Bureau of Justice. 2005. *Criminal Victimization 2004*. Washington, DC: US Bureau of Justice.

Walker, John. 1985. *Crime and Justice Statistics in Papua New Guinea*. Canberra: Australian Institute of Criminology.

Windybank, Susan and Mike Manning. 2003. *Papua New Guinea on the Brink*. Issue Analysis No. 30. Centre for Independent Studies. 12 March.

Zvekic, Ugljesa and Anna Alvazzi del Frate, eds. 1995. *Criminal Victimisation in the Developing World*. UNICRI Publication No. 55. Rome: UN Inter-regional Crime and Justice Research Institute.

ACKNOWLEDGEMENTS

Principal authors

Nicole Haley and Robert Muggah

Contributors

Sarah Garap, David Kavanamur, and Elly Kinkin

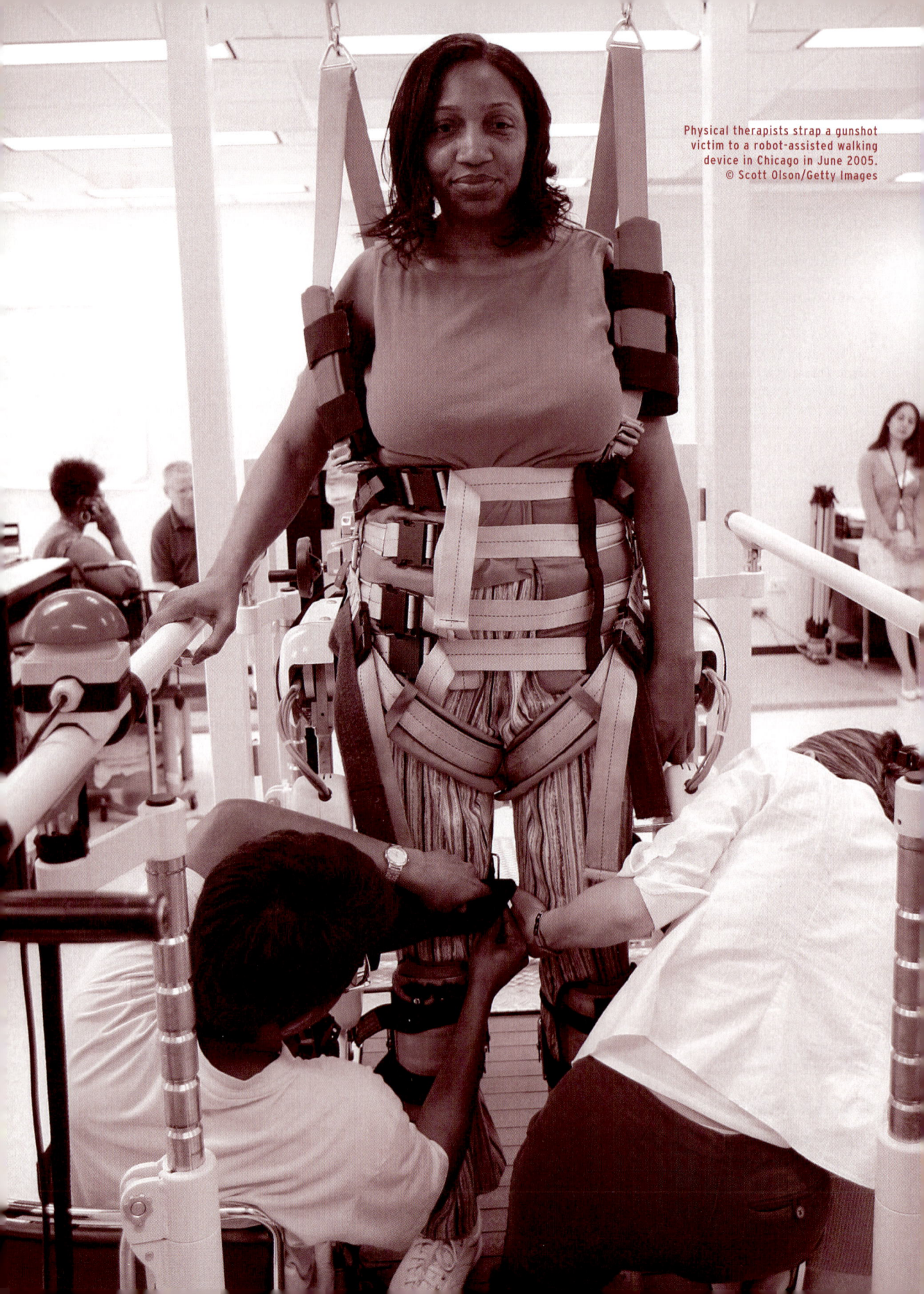

The Instrument Matters
ASSESSING THE COSTS OF SMALL ARMS VIOLENCE

8

INTRODUCTION

My life has not been easy the last few years. I've gone through three robberies and the death of my wife, who left me six children, including a baby. I was having a hard time paying for food and school. One day [. . .] a friend gave me FBU 15,000 [USD 14] to start a business making ropes. People thought I had a lot of money. [One evening], armed robbers attacked my house. [. . .] They came in and asked for money. I gave them what I had, but they still shot me. I had a wound and a broken thighbone.

The following morning, people from the church took me to the Gitega hospital, where I spent several months. After a while, the nurses asked me for money even though I had nothing. After that, the nurses refused to treat me normally. My wound and broken bone got infected. No one wanted to change my bandages. The nurses put me away from the other patients in an isolated room as my wound was festering. [. . .] I was waiting for death.

—Testimony by Déo, aged 47, Burundi[1]

There is more to armed violence than instant life or death for the individuals involved. Déo's experience illustrates how a weak public health system can eventually cost the wounded their lives. An entire family's well-being may also be deeply affected by the incapacitation of its main income provider.

This chapter examines the impacts of armed violence from an economic perspective by reviewing the medical costs of injuries; the productivity lost due to death, inactivity, and disability; and reductions in quality of life. It considers the following questions:

- How do the costs of gun violence compare to those of violence committed with other means, and what explains the difference?
- Which societies carry the highest costs?
- How can methodology be improved to increase our understanding of the global costs of gun violence?

Estimates of the costs of violence not only highlight the multiple effects of armed violence; they can also serve as key reference points for resource allocation and priority setting. In developed countries, high costs are frequently used to justify more spending on violence prevention. As Déo's case illustrates, however, developing countries that cannot afford to care for victims will probably spend less than they should. Comparing costs with levels of armed violence can thus help identify in which societies victims are the most vulnerable. The chapter's main conclusions are:

- Relatively low violence-related public health expenditures in developing countries do not necessarily mean that gun violence is less of a burden. Limited spending may actually point to unresponsive medical systems, which mean that gun injuries are less likely to be treated and more likely to be lethal.

- Small arms misuse accounts for an excessive proportion of the costs of violence. In Brazil and Colombia, the medical treatment of a firearm injury costs between 1.7 and 3 times more than that of a stabbing. Firearms injuries also tend to affect young, potentially productive segments of the population.

- The intent of small arms violence influences its lethality and cost. Accidental shootings and gun assaults are generally less fatal than suicide attempts, and therefore necessitate more medical expenses. On the other hand, premeditated killings and the high lethality of suicide attempts have important indirect effects, resulting, for instance, in significant losses of earnings.

- Misconceptions about the costs of violence abound, and methodologies need to be refined to gain a better under-standing of the global costs of gun violence.

The chapter begins with an overview of the different types of costs and methodological approaches considered in the literature on the burden of violence. The second section focuses on the contribution of firearms to the overall costs of violence. It argues that a number of factors—including the nature of gun violence and the responsiveness of medical systems—are likely to influence the scope of the problem. The third section presents the results of three pilot studies that compare the costs of violence perpetrated with firearms and sharp instruments in Bogotá and Cali (Colombia), and Rio de Janeiro (Brazil). These studies followed draft methodological guidelines developed by the Small Arms Survey for the World Health Organization (WHO) and the US Centers for Disease Control and Prevention (CDC).

Small arms misuse accounts for an excessive proportion of the costs of violence.

All monetary results in this chapter have been converted to 2003 purchasing power parity (PPP) USD (hereafter PPP USD).[2] PPP values account for price differences across countries, allowing international comparisons of real output and incomes. PPP USD 1 has the same purchasing power in the domestic economy as USD 1 has in the United States.[3]

THE COSTS OF VIOLENCE: TYPOLOGIES AND COMPARABILITY ISSUES

Defining and ordering the various types of costs of violence following a coherent approach is essential for generating meaningful comparisons. This section introduces a framework to conceptualize the various costs associated with violence, drawing from previous work conducted by the Small Arms Survey and others.[4] It further highlights difficul-ties in comparing existing estimates due to the lack of a standardized methodological approach. When reviewing knowledge to date, the section also introduces a distinction between collective violence (i.e. conflict) and societal violence. The latter includes interpersonal violence—or violence that is directed against another person, such as assaults—and self-directed violence, including suicide attempts and self-mutilation.[5]

What types of costs?

Any attempt to highlight the impacts of violence must recognize that violence affects societies at all levels, as opposed to only victims and perpetrators. Accordingly, studies documenting the economic effects of violence have covered a broad range of costs (Table 8.1).

Analysts commonly make a distinction between direct and indirect costs.[6] Direct costs are those that arise directly from acts of violence and require actual payments by individuals or institutions. They can be further divided

Table 8.1 A typology of the costs of violence

Cost category	Type of cost	Components
Direct costs	Medical	In-patient costs (hospitalization, surgery, physician fees, drugs, laboratory tests) Out-patient costs Rehabilitation Ambulance fees
	Non-medical	Costs of policing and incarceration Costs of legal services Direct perpetrator control costs Costs of foster care Private security contracts Post-conflict reconstruction costs Care provided to displaced people
Indirect costs	Tangible	Productivity losses (earnings and time) Lost investments in social capital Life insurance costs Indirect protection costs Macroeconomic costs (reduced production, property values, tourist streams, and foreign investment)
	Intangible	Health-related quality of life (pain and suffering, psychological costs) Other quality of life (reduced job opportunities, access to schools, public services, and participation in community life)

Sources: Adapted from WHO (2004a, p. 6); Lindgren (2005, p. 5)

into medical and non-medical costs, given the importance the literature has given to documenting the costs of medical treatment (WHO, 2004a; Waters et al., 2005). Direct medical costs generally comprise in-patient costs, including costs of hospitalization and surgery, physician fees, drugs, and laboratory tests; and the costs of out-patient visits, rehabilitation, and ambulance fees. Direct non-medical costs include those incurred by the criminal justice system, such as costs of policing and incarceration, legal services, direct perpetrator control, foster care, and private security contracts. In the case of conflict, the costs of rebuilding destroyed infrastructure and providing care to displaced people can be included (Lindgren, 2005, p. 5).

Indirect costs refer to lost resources and opportunities resulting from violence. Studies tend to focus on tangible costs such as reduced productivity or output. Other tangible costs include lost investments in social capital (e.g. the cost of education of the victim and perpetrator), life insurance costs, reduced productivity or output by the perpetrator, and macroeconomic costs (e.g. reduction in production, property values, tourist streams, and foreign investment due to violence and conflict). Also included are intangible costs such as reductions in quality of life. Generally speaking, quality of life includes many components such as job opportunities, access to schools, public services, and participation in community life. In the context of violence, it is usually associated with health-related quality of life, which includes the pain and suffering, both physical and psychological, that arise from violent incidents.

In sum, direct costs represent the actual economic burden imposed on society and indirect costs represent the potential loss in resources. Both direct and indirect costs are of concern, because they represent forgone monetary value to society that could have been invested in positive projects.

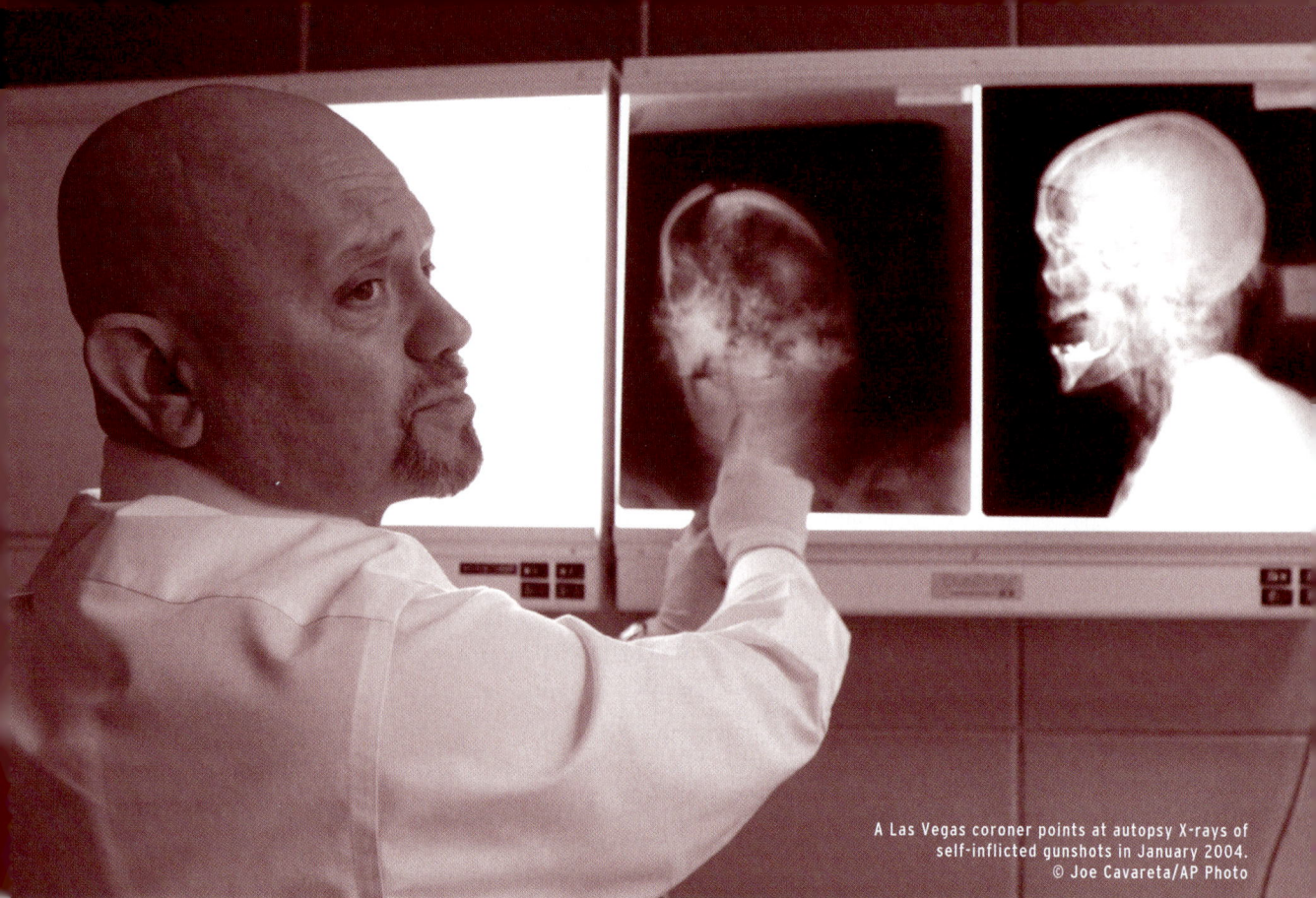

A Las Vegas coroner points at autopsy X-rays of self-inflicted gunshots in January 2004.
© Joe Cavareta/AP Photo

Methodological approaches

Methodological approaches to quantify the costs of violence include *modelling, willingness to pay,* and *accounting.* Analysts seeking to document the economic impact of conflict have developed modelling techniques to determine how the economy could have developed in the absence of war (Lindgren, 2005, p. 4). This is done by comparing a conflict-affected country's gross domestic product (GDP) both to its pre- and post-war economic trends, and to the GDPs of similar countries—such as its neighbours—not affected by conflict. The differences will generate an estimate of the costs of conflict, which are usually measured in reductions of annual growth or investment. Modelling techniques can be useful for the study of conflict, but are more difficult to apply to the study of societal violence. Levels of interpersonal and self-directed violence are relatively stable when compared to conflict, which hinders the analysis of their impact before, during, and after violence.

Willingness to pay assumes that the cost of a violent incident is the total sum of what individuals are willing to pay for reducing the risk of becoming a victim. This approach, when properly designed, can capture direct treatment costs, indirect costs, and costs associated with pain and suffering. There are three ways to estimate willingness-to-pay values. One is through surveys of individuals' willingness to pay to avoid a given problem in hypothetical situations. The second involves observing 'averting behaviour'; i.e. actual cases where individuals undertake preventive measures to avoid exposure to or mitigate the effects of violence. Investments made in preventive measures are then used as a proxy for individual willingness to pay to avoid violence. The third way involves examining court decisions on damage payments. While willingness to pay has the potential of generating a more comprehensive picture of the indirect costs of violence, it remains to be tested in developing countries and conflict situations.

The accounting approach involves counting and adding up a selection of the costs identified above. This can be done by multiplying the number of violent incidents by the estimated average cost per incident, or by focusing on macro-level expenditures—both public and private—that may be linked to violence. Productivity losses are usually calculated by multiplying the time lost due to violence by the income that victims would be generating if they had not been injured. In the costing of conflict-related violence, indirect cost calculations will also include lost production and investment, and impacts on capital flows (Lindgren, 2005, p. 5).

While accounting produces relatively conservative estimates when compared with willingness-to-pay techniques, it remains the most common approach to date. Accounting estimates are also seen as more credible among non-specialists unfamiliar with economic models or survey methods (Lindgren, 2005, p. 14), and are therefore likely to have a greater impact at the policy level.

Comparability issues

There are presently no standardized approaches to costing violence, resulting in wide-ranging and competing estimates among and even within countries. Comprehensive reviews of studies examining the costs of conflict and societal violence point to the lack of estimates that both use comparable methodological approaches and focus on the same set of costs.[7]

Reviewing 11 studies that provide 36 country-level estimates, Lindgren finds that civil war can account for anywhere between 0.3 per cent and 90 per cent of annual GDP (2005, p. 13). Different assumptions can result in dramatically different estimates, even for a single country. The estimated costs of civil war in Sri Lanka and Nicaragua, for instance, vary greatly. Depending on the person doing the costing, conflict in Sri Lanka cost 2.2–15.8 per cent of GDP per year. In Nicaragua, variations were even more pronounced: estimates range from 0.8 to 90 per cent of annual GDP (Table 8.2).

Table 8.2 Differing estimates of the cost of civil war for Sri Lanka and Nicaragua			
Country	Author	Conflict years	Cost per year as % of GDP
Sri Lanka	Richardson and Samarasinghe (1991)	1983–88	11.3
	Grobar and Gnanaselvam (1993)	1983–91	2.2
	Harris (1997; 1999)	1983–92	8.8
	Kelegama (1999)	1983–94	10.9
	Arunatilake, Jayasuriya, and Kelegama (2001)	1984–96	10.8–15.8
Nicaragua	Fitzgerald (1987)	1980–84	15.4
	Stewart and Humphreys (1997)	1965–90	4.4
	DiAddario (1997)	1980–87	17.3–25.7
	Stewart, Huang, and Wang (2000)	1977–93	0.8
	Lopez (2000)	1978–79, 1981–88	90.0

Source: Adapted from Lindgren (2005, p. 13)

Analyses of the costs of societal violence can result in equally wide-ranging findings. In the United States, for example, figures for the direct medical costs of child abuse range from PPP USD 1,965 per child per year in Washington State to PPP USD 44,173 per child per year in West Virginia (Waters et al., 2005, p. 306). Estimates will also vary greatly depending on whether indirect costs are included.

Variations in estimates are accentuated across countries, as medical costs and wages are comparatively lower in low- and middle-income countries than in high-income countries. The average cost per homicide was estimated at PPP USD 55,000 in Cape Town, sharply lower than in Australia (PPP USD 910,000) or New Zealand (PPP USD 1,426,000).[8] Analysts also face the challenge of measuring impacts on the comparatively large—yet unrecorded—informal economy.[9] These lower costs, however, do not mean that violence is relatively less of a problem. On the contrary, rates of violent injuries are particularly high among developing countries (WHO, 2002, p. 11).

Estimates of the costs of societal violence also have limited geographical coverage, with studies being undertaken primarily in developed countries and Latin American states. This makes the global economic burden of violence more difficult to assess than that of road traffic accidents, for example, for which methodological guidelines have long been established (TRL, 1995) and a large enough sample of comparable studies exists, allowing for worldwide and regional extrapolations (WHO, 2004b).

What do we know about the global costs of violence?

While methodological variations make it difficult to compare results across studies and settings, there is general consensus that violence imposes a significant economic burden on societies affected by it. Economic models using conflict data sets make it possible to measure the average impact of conflict on a country's GDP. Drawing on data for 92 countries, 19 of which faced civil war, Collier concludes that countries affected by internal conflict experience an annual decline in their GDP per capita of 2.2 per cent relative to their counterfactual (Collier, 1999, p. 181). A subsequent study of 211 countries found that wars of this type caused an average 2.4 per cent reduction in annual growth (Hoeffler and Reynal-Querol, 2003, p. 19). Other studies that account for various costs of conflict at the country level have usually found that civil wars have an even greater economic impact, averaging 10 per cent of annual GDP (Lindgren, 2005, p. 13).

Societal violence can impose an equally alarming burden. Costs for low-income countries may be underestimated, due to the small number of cases reviewed and the difficulty of comparing lost wages and income with those of high-income settings (WHO, 2004a, p. 14). The available evidence does suggest, however, that developing countries suffer more from violence than the industrialized world. Given the continent's high exposure to violence, the most

A child cries on the coffin of her father, a policeman who died of gunshot wounds in Colombia in August 2003. © Efrain Patino/ AFP/Getty Images

revealing comparative estimates originate from Latin America. A 1999 study, based on six country case studies, estimated that 20.9 per cent of Latin America's GDP was being consumed by violence in terms of destruction, diversion of resources, and loss of human and financial capital (Londoño and Guerrero, 1999, p. 3). By comparison, in the United States, despite relatively high rates of violence for an industrialized state, violence is estimated to cost between 3.3 and 6.5 per cent of GDP, even when including indirect costs such as lost earnings and psychological costs.[10]

Most studies demonstrate that direct medical costs represent only a small fraction of the total burden of violence, despite being the focus of the majority of studies. A study comparing six Latin American countries found that the direct non-medical costs of interpersonal and collective violence (including expenditures on police, security systems, and judicial services) exceeded medical costs across all case studies (Buvinic, Morrison, and Shifter, 1999, p. 20), with ratios of medical to non-medical costs ranging from 1:1.2 in El Salvador to 1:30 in Venezuela. Although it is difficult to compare indirect costs across settings, there is general agreement that indirect costs are much higher than direct costs (Waters et al., 2005, p. 305). This suggests that a comprehensive assessment of the impacts of violence should not focus only on direct expenditures to support victims and deal with perpetrators, but must also consider lost opportunities and the destruction of resources that would have otherwise been available in the future.

Studies have also sought to measure the cost of violence prevention initiatives when compared with their benefits—whether real or potential. A number of preventive interventions targeting child abuse, child crime, sexual and domestic violence, and crime in general were found highly cost-effective (WHO, 2004a, pp. 28–29). Collier and Hoeffler conducted economic evaluations of five different instruments to prevent or reduce conflict by comparing their costs to their potential benefits (2004, pp. 21–22). They conclude that external military intervention under Chapter VII of the UN Charter was the most effective, and that aid—as part of conflict prevention, but not of post-conflict recovery—was the least.

Civil wars cause a 2.2 per cent reduction in annual growth in affected countries.

COSTING GUN VIOLENCE: AN OVERVIEW OF THE ISSUES

Small arms are a common instrument in both conflict and societal violence. Globally, they are estimated to be involved in 60–90 per cent of conflict deaths, 40 per cent of homicides, and 6 per cent of suicides (Small Arms Survey, 2005, p. 230; 2004, p. 175). Firearms are also a major vector in fatal injuries following conflicts and in countries affected by acute urban violence (Small Arms Survey, 2005, p. 270; CERAC, 2005, p. 74). This section discusses whether and how the use of such weapons has implications for the costs of violence, and outlines challenges and opportunities for advancing current knowledge of the global costs of gun violence.

The costs of gun violence

The literature on the contribution of small arms violence to conflict is nascent and provides only limited insights about the associated economic burden (Small Arms Survey, 2005, pp. 228–65; ICRC, 1999). As small arms are involved in the overwhelming majority of conflict deaths, however, their contribution to the costs of conflict can only be significant. Studies examining the costs of gun violence generally adopt a public health approach and focus on the direct medical costs, productivity losses, and in a few cases the reductions in quality of life that can be attributed to societal gun violence. Geographical coverage is also extremely limited, with the majority of studies focusing on the United States. Rare exceptions include Canada, El Salvador, and South Africa.

Table 8.3 **Average direct medical costs per firearm injury by severity, selected studies (2003 PPP USD)**					
Location and source	**Sample**	**Year**	**Fatal**	**Serious (admitted)**	**Slight (emergency department only)**
Canada (Miller, 1995, table 1)	National	1991	8,828 (CAD 8,591)	30,037 (CAD 29,228)	5,224 (CAD 5,083)
El Salvador (Paniagua et al., 2005, p. 191)	San Salvador (one hospital)	2003	n/a	5,500 + 370 per bed day	n/a
South Africa (Allard and Burch, 2005, p. 592)	Cape Town (one hospital), abdominal gunshot wounds only	2005	n/a	3,427 (ZAR 10,269)	n/a
US (Miller and Cohen, 1997, p. 335)	National	1993	25,038	35,202	5,987

Most studies outside the United States focus exclusively on the direct medical costs of hospitalized gun injuries (Tables 8.3 and 8.4). Even so, the estimated cost of treating one firearm injury between developed countries such as Canada and the United States can vary significantly. Unsurprisingly, unit costs appear to be significantly lower in developing countries such as El Salvador and South Africa.

Gun violence makes a significant contribution to the overall cost of violence. In the United States, the total costs of gun violence, including productivity losses and reduced quality of life, are estimated at PPP USD 115–144 billion per year (Table 8.4). In El Salvador, treating hospitalized firearms injuries consumes more than 7 per cent of the country's entire health budget (Paniagua et al., 2005, p. 191). Treating an abdominal gunshot wound in South Africa costs 13 times the government's per capita health spending (Allard and Burch, 2005, p. 591).

As with violence in general, the indirect costs of gun violence are significantly higher than direct medical costs (see Table 8.4). Cook and Ludwig (2000), using a willingness-to-pay survey, found that the total costs of gun violence in the United States amount to PPP USD 115 billion, which is much more significant than estimates taking into account only productivity losses and direct medical costs (less than PPP USD 35 billion; see Table 8.4). Direct to indirect costs ratios can be highly inconsistent, however, even between similar countries. In Canada in 1991, productivity losses and reductions in quality of life were, respectively, 25 and 78 times more costly than direct medical costs (Miller, 1995, table 3). A US study found that productivity losses and reductions in quality of life were, respectively, 12 and 28 times higher than direct medical costs (Miller and Cohen, 1997, table 8). Overall, these findings suggest that the greatest costs of firearms violence are intangible and have to do with issues of reduced quality of life, pain and suffering, and psychological impacts that affect society as a whole.

Relatively few studies have tried to justify gun violence prevention strategies through cost-effectiveness analyses. Based on a willingness-to-pay survey, Ludwig and Cook (1999) estimate that the American public believed initiatives that would successfully reduce the number of gun injuries by 30 per cent would be worth spending PPP USD 27.3 billion, or PPP USD 860,000 per injury. In Jamaica, analysts calculated that if gunshot injuries admitted at Kingston Public Hospital were blunt instrument injuries instead, the hospital would be saving JMD 2.13–3.56 million (PPP USD 60,749–101,533) per year in medical costs (Zohoori et al., 2002, p. 260). In other words, an intervention that suc-

Table 8.4 **Total annual direct and indirect costs of gun violence, selected studies (2003 PPP USD)**				
Location and source	**Sample**	**Direct medical costs**	**Productivity losses**	**Quality of life**
US (Cook and Ludwig, 2000, pp. 70, 79, 115)	National sample, 1997	0.5–2.1 billion	22.3–30.5 billion	Total direct and indirect costs of 114.6 billion*
US (Max and Rice, 1993, p. 171)	National sample, 1990	2 billion	2.3 billion for non-fatal injuries and 24.5 billion for fatal injuries	n/a
US (Miller and Cohen, 1997, p. 337)	National sample, 1992	3.4 billion	42.4 billion	98.5 billion
Canada (Miller, 1995, table 3)	National sample, 1991	65.2 million (CAD 63.4 million)	1.6 billion (CAD 1.6 billion)	5.1 billion (CAD 5 billion)
El Salvador (Paniagua et al., 2005, p. 191)	National, extrapolated from one hospital in San Salvador (admitted only)	13.2 million	n/a	n/a
South Africa (Allard and Burch, 2005, p. 593)	National, extrapolated from one hospital in Cape Town (admitted abdominal gunshot wounds only)	66.8 million (ZAR 200 million)	n/a	n/a

* Using willingness-to-pay method; figure therefore includes direct and indirect costs.

ceeded in limiting the use of guns in violence—even if the overall number of injuries remained steady—would trigger net savings for the medical system. In Canada, while the costs of the new gun registration system largely exceeded forecasts, available estimates of the costs of gun violence for the country make the investment look much more cost-effective (see Box 8.1).

Box 8.1 **Putting numbers in perspective: the costs of controlling firearms in Canada**

Comprehensive schemes to regulate firearms are costly and the subject of considerably more debate than other interventions aimed at reducing injury and death. Canada has had relatively strict controls on handguns and required restricted weapons permits and registration to possess them since the 1930s.[11] In 1995 new legislation known as Bill C-68, supported by police and public health groups, introduced licensing to possess any firearm and registration of all firearms, and prohibited a wider range of semi-automatic firearms, along with short-barrelled handguns. The bill was hugely controversial as it was expensive to implement, but even so, it appears to be considerably more cost-effective than previously believed when considering possible savings in terms of firearms violence reduction.

Since Canada passed Bill C-68 in 1995, the costs associated with licensing two million firearms owners and seven million firearms have dramatically exceeded initial estimates. Estimates in 1995 forecast a total additional cost for Bill C-68 of CAD 119 million (PPP USD 122 million), and the project anticipated collecting fees of CAD 117 million (PPP USD 120 million), a total additional net cost of CAD 2 million (PPP USD 2 million) (Canada, 2002).[12] In December 2002 the auditor general revealed that the firearms regulation programme would cost more than CAD 1 billion (PPP USD 0.85 billion) by 2004-05 (an average of CAD

100 million, or PPP USD 85 million per year) and collect only about CAD 140 million (PPP USD 119 million) in fees for the entire period since 1995 (Canada, 2002, para. 10.3).[13] However, the auditor general did not comment on the appropriateness of the expenditure or the effectiveness of the legislation.

One of the challenges in evaluating the impact of firearms legislation is the gap between the passage of the law and its implementation. In the case of Bill C-68, for example, the law was passed in 1995, but the deadline for licensing all firearms owners was 2001 and that for registering all firearms was 2003. Consequently, an evaluation of the bill's final impact must wait several years.

Although it is too early to attribute this trend to the passage of the legislation, firearms deaths have declined dramatically from 1,125 (3.8 per 100,000) in 1995 to 816 (2.2 per 100,000) in 2002, the last year for which there is data (Wilkins, 2005, p. 42). The most pronounced changes are in youth firearms suicide rates (Wilkins, 2005, p. 38). Firearms suicide and homicide rates decreased more rapidly than non-gun suicide and homicide rates, suggesting that the legislation may be a contributing factor, and that the substitution effect was only partial. Firearms injury information is not available beyond Fiscal Year (FY) 1997–98. At that time, there was a significant decline in injuries requiring hospitalization during the period as well: from 1,125 (3.9 per 100,000) in FY 1993–94 to 767 (2.6 per 100,000) in FY 1997–98, a reduction of 32 per cent (Hung, 2005). While other factors besides legislation contribute to changes in firearms death rates, mortality and morbidity figures suggest stronger controls do contribute.[14]

Discussions to date generally focus on the costs of the legislation rather than its impact on the costs of firearms death and injury, even though these costs dwarf the investment of CAD 100 million (PPP USD 85 million) per year.[15] A 1995 study, for example, finds that gunshot wounds occurring in 1991 amounted to CAD 55.3 million (PPP USD 56.8) in direct medical costs, CAD 8.1 million (PPP USD 8.3 million) in

Figure 8.1 Estimated total annual costs of fatal and non-fatal gun violence in Canada (millions of 1993 CAD*)

CAD (MILLIONS)

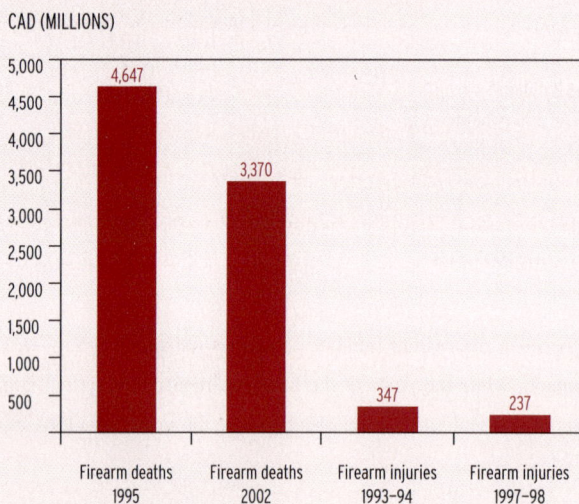

Note: * 1993 CAD 1 = 2003 PPP USD 1.03

Sources: Firearms injury and death data: Cukier and Sidel (2005); Hung (2005); Wilkins (2005); costing figures: Miller (1995, Table 1)

mental health care, CAD 1.55 billion (PPP USD 1.59 billion) in lost productivity, and CAD 4.97 billion (PPP USD 5.1 billion) in lost quality of life, for a total cost of CAD 6.6 billion (USD PPP 6.8 billion) in 1993 CAD (Miller, 1995, table 3).

So how much did Canada gain for its CAD 100 million (PPP USD 85 million) annual investment in comprehensive firearms regulation? It is too early to say. On the basis of Miller's costing study (1995), however, the savings due to the decline in firearms injuries since 1995 appear to be significant. Applying Miller's cost estimates to available firearm mortality and morbidity data highlights the amplitude of potential savings. In 2002 the annual costs of fatal gun violence were potentially reduced by 1993 CAD 1.3 billion (PPP USD 1.3 billion) when compared with 1995. The costs of non-fatal gun violence also decreased dramatically, saving Canada some 1993 CAD 110 million (PPP USD 113 million) in FY 1997–98 when compared with FY 1993–94. In sum, the potential benefits of new legislation in terms of violence prevention and reduction far outweigh its implementation costs.

Source: Cukier (2005)

Do small arms contribute disproportionately to the costs of violence?

Violence committed with firearms generates higher costs than violence committed with other instruments. In Cape Town, South Africa, for instance, injuries due to sharp objects, such as stabbings, accounted for 43 per cent of homicides, while firearms accounted for 39 per cent. Stabbings and firearms were each responsible for 43 per cent of the total economic costs of violence (including direct medical costs and productivity losses), however, suggesting that firearms homicides were more costly than homicides committed with knives (Phillips, 1998, table 11). As Max and Rice sum up, 'firearm injuries are relatively more costly compared with both other injuries and other illnesses in general' (Max and Rice, 1993, p. 183).

The seriousness and lethality[16] of firearms violence result in significant indirect costs. As Table 8.5 illustrates, the average productivity losses and quality of life reductions attributed to an injury are much higher for deaths than for non-fatal injuries. This is because non-fatal injury survivors, although deeply affected, will be able to return to a productive activity and to their communities and families after recovery in a majority of cases. The proportion of gun injuries that are lethal is much higher for firearms than for cut/stab wounds, which increases their overall costs. In the United States, more than one in every five hospitalized firearms injuries results in death, while the ratio for cut/stab wounds is of one death for every 759 injuries (Miller and Cohen, 1997, table 7). Consequently, the average gunshot injury in the United States will cost PPP USD 937,000 compared to just PPP USD 19,000 for a cut/stab wound, a ratio of almost 50 to 1. When considering only non-fatal injuries, gunshots cost PPP USD 196,000 per victim versus PPP USD 14,000 for cuts and stabbings, a ratio of 14 to 1 (Miller and Cohen, 1997, p. 335).

The relative cost of gun injuries depends greatly on intent. Average direct medical costs per injury are generally higher for unintentional shootings (PPP USD 25,670) and interpersonal injuries (PPP USD 21,086) than for self-

Table 8.5 Average costs per gunshot and cut/stab wound in the United States, by severity and category of cost (2003 PPP USD)

Cost category	Gunshot wound			Cut/stab wound		
	Fatal	Hospitalized	Emergency department only	Fatal	Hospitalized	Emergency department only (intended)
Direct medical*	25,038	35,202	5,974	25,038	21,059	3,723
Direct non-medical**	3,631	3,446	1,169	3,631	1,946	661
Productivity lost	1,166,767	56,255	2,617	1,249,487	41,478	2,351
Lost quality of life	2,370,841	222,823	82,763	2,444,691	193,538	34,378
Total	**3,566,277**	**317,726**	**92,523**	**3,722,847**	**258,021**	**41,113**

* Includes medical care, mental health care, and emergency transport.

** Includes police services and insurance administration.

Source: Adapted from Miller and Cohen (1997, table 4)

Figure 8.2 **Distribution of firearms deaths and hospitalized injuries by intent in New Zealand (1992-96) and the United States (1996-98)**

PERCENTAGE

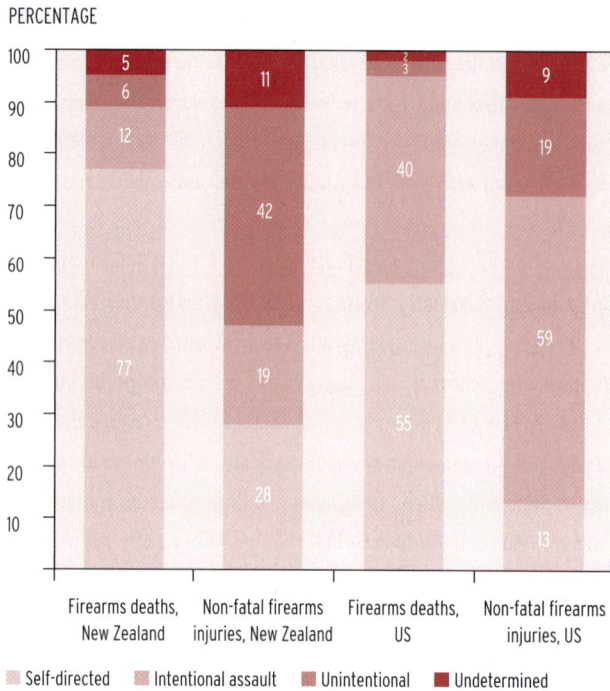

| | Self-directed | Intentional assault | Unintentional | Undetermined |

Source: Adapted from Spicer et al. (2005, p. 72)

inflicted injuries (PPP USD 6,200) (Cook and Ludwig, 2000, p. 65). As victims of firearms suicides die almost instantly, settings where a high proportion of firearms deaths are due to suicide will experience fewer non-fatal firearms injuries than countries where the majority of deaths are due to gun homicides or accidents. Even where firearms suicides represent the majority of firearms deaths, injuries due to assaults and accidents involving firearms account for the vast majority of hospitalized injuries (see Figure 8.2). High levels of unintentional and assault-related gun violence should thus result in significant medical costs. When taking into account indirect costs (productivity losses and reduced quality of life), however, self-inflicted injuries generate the highest average costs, as they most often result in the death of the victim (Miller and Cohen, 1997, table 5).

Implications for the global costs of gun violence

Given the disparate nature of existing data, it is exceedingly difficult to render a global estimate of the economic burden of violence, much less gun violence. Generating a comprehensive estimate of the global costs of societal gun violence will require expanding the existing sample of country-level estimates, which remains too limited. The available evidence does make it possible, however, to generate a broad picture of which regions may suffer the greatest burden.

Costing usually involves multiplying the number of violent incidents by average unit costs. The number of firearms deaths and injuries is therefore seen as important information for producing a global cost estimate.

Although considerable gaps remain, knowledge about the global and regional distribution of fatal gun violence by intent is increasing.[17] The global incidence of non-fatal firearm injuries is less well documented. As discussed above, however, the intent of firearms violence—which is better documented worldwide—influences the lethality of gun injuries and thus the ratio of deaths to survivors of armed violence.

Contexts where violence is meant to be lethal experience relatively fewer gun injuries, which has implications for the costs incurred by such violence. In Bogotá, Colombia, the high lethality of firearms wounds has been attributed to the 'professionalizing' of violence—reflecting the large numbers of targeted, premeditated assassinations—and there is a high proportion of particularly lethal head and abdomen wounds (Beltrán et al., 2003, p. 12). Similarly, in conflict situations, there are few survivors when firearms are used against people who are immobilized, in a confined space, or unable to defend themselves (Coupland and Meddings, 1999). Settings where a high proportion of gun violence is self-directed—such as North America—will also suffer relatively fewer non-fatal firearms injuries. In such contexts, the low numbers of injuries will translate into relatively low medical costs, while the indirect costs associated with mortality will be high.

Countries experiencing high levels of interpersonal gun violence—such as those in Latin America or Africa—should, theoretically speaking, be caring for large numbers of non-fatal injuries in their hospitals. Additional factors come into play, however. As Table 8.6 illustrates, gun injuries in Brazil and Colombia appear to be more lethal than in developed countries with high gun suicide rates such as Canada and the United States.[18] This may be the result of a strong intent to kill, as discussed above. It may also point to relatively unresponsive emergency medical systems in developing countries, which make firearms wounds less likely to be treated and more likely to be lethal. In such settings, victims of firearms violence may also refrain from going to hospitals, as they will be unable to pay for treatment. Medical expenditures therefore risk being lower than they should be among developing countries.

Table 8.6 **Incidence of firearms injuries, by severity**								
Location and source	**Sample**	**Year**	**Total fatal**	**Serious (admitted but survived)**	**Slight (emergency department only)**	**Ratio total fatal: serious**	**Ratio serious: slight**	**Ratio total fatal: non-fatal**
Brazil*	National	2002	38,088	17,793	n/a	1:0.5	n/a	n/a
Canada (Miller, 1995)	National	1991	1,450	1,244	3,031	1:0.9	1:2.4	1:2.9
Colombia**	National	2005	14,762	5,546	n/a	1:0.4	n/a	n/a
El Salvador (Paniagua et al., 2005)	National	2003	1,697	2,580	n/a	1:1.5	n/a	n/a
US (Miller and Cohen, 1997)	National	1992	37,776	61,300	72,700	1:1.6	1:1.2	1:3.5
El Salvador (Paniagua et al., 2005)	One hospital (San Salvador)	2003	n/a	623	789	n/a	1:1.3	n/a
Nigeria (Solagberu, 2003)	One hospital (Ilorin)	September 1999– October 2001	n/a	27	39	n/a	1:1.4	n/a

*Brazilian Ministry of Health data processed by Instituto de Estudos da Religião (ISER), provided in written correspondence by Luciana Phebo of ISER, 8 December 2005.

** Colombian National Police data processed by Centro de Recursos para el Analisis de Conflictos (CERAC), provided in written correspondence by Katherine Aguirre of CERAC, 8 February 2006.

Estimating the costs of gun violence in different settings will, therefore, not only increase our understanding of the global scope and nature of gun violence, but will also highlight important policy deficiencies. Comparing rates of violence with public health expenditures, in particular, can help identify where victims are the most vulnerable. While most costing literature has been used to justify increased spending on prevention in developed countries, it also offers the opportunity for low- and middle-income countries to determine whether the victims of gun violence are appropriately taken care of.

COSTING FIREARM VIOLENCE IN BRAZIL AND COLOMBIA

Working with the WHO and CDC, the Small Arms Survey has prepared a unique set of standardized guidelines to estimate the direct and indirect economic costs of interpersonal and self-directed violence (see Annexe 1 for an overview). The guidelines aim to enable low- and middle-income countries to generate robust estimates despite sometimes incomplete data. In order to test the applicability of the guidelines, the Small Arms Survey independently commissioned pilot case studies in Brazil (Rio de Janeiro) and Colombia (Bogotá and Cali) to measure the specific contribution of firearms to costs, as opposed to other means of violence.

The case studies reached conclusions that are consistent with those of similar costing work carried out in the United States and Canada. Extrapolated nationally, firearms injuries cost Brazil and Colombia's respective medical systems PPP USD 88 million (BRL 100 million) and USD 38 million (COP 29 billion) a year. The medical treatment for the average gunshot wound was between 1.7 and 3 times more expensive than that required for treating cuts or stabs, ranging

A gunshot victim lies on a stretcher in Bonsucesso Hospital in Rio de Janeiro in January 2005.
© Douglas Engle/WPN

from PPP USD 4,500 to PPP USD 11,500 per injury. These average medical costs appear relatively consistent with those of other developing countries—i.e. El Salvador and South Africa—as reported in Table 8.3.

Interestingly, however, average medical costs were more expensive in both Colombian hospitals than in Rio de Janeiro, which is counterintuitive, as the standard of living—as measured by GDP per capita—is higher in Brazil than in Colombia (UNDP, 2005). This may be partly due to the different price scales used to measure the costs of treatment. In Brazil, a national scale was used, while in Colombia, individual bills submitted to the public health system formed the basis of the estimate. Another explanation may be that the Brazilian public health system may not be as well equipped to deal with violent injuries as its Colombian counterpart. This explanation is supported by the fact that a greater proportion of victims died from their wounds in the Brazilian hospital than in the Colombian facilities.

Victims of firearms violence also lost more productive time than victims of violent cuts and stabs. Survivors of gun violence spent more days in hospital and were expected to remain inactive while convalescing longer than patients injured by sharp instruments. Consistent with other research on the victims of small arms violence (ANGRY YOUNG MEN), a particularly high proportion of patients treated for gun injuries were young men. This translates into considerable lost earnings, particularly as the average income earned in Brazil and Colombia is higher among men than women. When extrapolating results using national mortality and morbidity data, gun violence is threatening PPP USD 10 billion of future earnings (BRL 11.3 billion) in Brazil per year, and PPP USD 4 billion (COP 3,100 billion) in Colombia.

Methods and sampling

The pilot case studies involved surveying victims of violent injuries in key hospitals in both countries. These included the Hospital da Geral in Nova Iguaçu (HGNI), Rio de Janeiro, Hospital Santa Clara (HSC), Bogotá, and Hospital Universitario del Valle (HUV), Cali (see Table 8.7). Two local research institutes, ISER in Brazil and CERAC in Colombia, administered the field research. Drawing explicitly from the Small Arms Survey guidelines, the research teams elaborated standardized questionnaires to prospectively gather information from each victim of a violent act seeking care at the selected hospitals during a one-month period (12 November–12 December 2005).[19]

The research teams collected a wide variety of data, including, among others, information on the patient's demographic and socio-economic profile, the characteristics and severity of the injury, the type of care provided, and associated medical costs. In Brazil, costs for each type of treatment were calculated based on the Brazilian Medical

Table 8.7 Distribution of violent injuries by instrument, 12 November–12 December 2005

Instrument	HGNI, Rio de Janeiro	HSC, Bogotá	HUV, Cali
Firearm	25 (23%)	28 (19%)	71 (61%)
Sharp instrument	13 (12%)	83 (57%)	45 (39%)
Blunt instrument	68 (61%)	15 (10%)	0
Poisoning	2 (2%)	14 (10%)	0
Unspecified	3 (3%)	5 (3%)	0
Total	**111 (100%)**	**145 (100%)**	**116 (100%)**

Source: Small Arms Survey calculations based on ISER (2006b); CERAC (2006c)

Association payment scale and interviews with HGNI and Rio Fire Department personnel (ISER, 2006a; 2006b). In Colombia, costs were calculated based on the final medical expenses that hospitals sent to the public medical system for each patient (CERAC, 2006a).[20] Values are presented in 2003 PPP USD.[21]

Preliminary findings

All three pilot studies confirmed that firearms injuries trigger higher medical costs than injuries inflicted by bladed weapons. As Table 8.8 shows, the average gun injury cost the surveyed hospitals between 1.7 and 3 times more than a cut/stab wound. Treating the average firearm injury in Rio costs seven times Brazil's per capita public health spending; in Bogotá and Cali the costs of treating a single firearm injury reach 13 and 21 times Colombia's per capita medical expenditures.[22]

Predictably, the substantially higher medical costs of firearms injuries are primarily the result of their relative severity. Among the 12 patients that died on their way to or in the hospital as a result of their wounds, 10 were shot, while only 2 had been cut or stabbed. As Table 8.9 illustrates, in all three hospitals, the average length of stay (days spent in hospital) and the percentage of patients requiring blood transfusions were higher for victims of firearms violence than for those injured by bladed weapons.

A greater proportion of victims with firearms wounds in Rio required surgery than those wounded by sharp instruments, and operations lasted on average 1.5 hours longer. While fewer victims of firearms violence underwent surgery in Bogotá, operations were more complex for firearms wounds and cost the hospital on average 1.2 times more. Intriguingly, in both Colombian hospitals, a greater proportion of patients wounded by a bladed weapon used an ambulance than firearms violence victims.[23]

Table 8.8 Average medical costs per injury by instrument (2003 PPP USD)

	HGNI, Rio de Janeiro		HSC, Bogotá		HUV, Cali	
	Firearm	Sharp instrument	Firearm	Sharp instrument	Firearm	Sharp instrument
Ambulance	219	119	111	129	176	229
Bed*	2,044	702	0	0	2,470	1,355
Consultations	82	58	108	79	362	222
Examinations	195	161	681	337	1,229	384
Surgery	845	372	1,932	1,602	3,323	2,427
Medication	1,074	85	1,739	563	3,839	1,004
Transfusions**	37	8	0	0	0	0
Other*	24	24	2,233	1,291	4	7
Total	**4,521**	**1,529**	**6,804**	**4,001**	**11,403**	**5,628**

* In Bogotá, bed costs are included under 'Other'.

** In Bogotá and Cali, the costs of transfusions are included in other costs, such as those of surgery.

Source: Small Arms Survey calculations based on ISER (2006b); CERAC (2006c)

Table 8.9 Where do firearms make a difference?

	Instrument	HGNI, Rio de Janeiro	HSC, Bogotá	HUV, Cali
% deaths in hospital	Firearm	28	11	0
	Sharp instrument	0	2	0
% patients who used an ambulance	Firearm	44	32	51
	Sharp instrument	23	37	67
Average length of stay	Firearm	6.7	6	11
	Sharp instrument	2.3	4	5.2
% patients requiring consultations with specialists	Firearm	100	75	94
	Sharp instrument	96	57	96
% patients requiring examinations or tests	Firearm	80	93	99
	Sharp instrument	85	87	89
% patients requiring surgery	Firearm	52	75	100
	Sharp instrument	15	85	100
% patients requiring medication	Firearm	80	96	100
	Sharp instrument	92	95	100
% patients requiring blood transfusions	Firearm	32	25	37
	Sharp instrument	8	19	33
Estimated number of inactive days	Firearm	23	30	32.5
	Sharp instrument	7.8	21.5	22.1
Percentage permanently disabled	Firearm	8	4	10
	Sharp instrument	0	0	7
Average age of patient at time of injury (average number of productive life years lost)*	Firearm	28 (38)	29 (37)	28 (38)
	Sharp instrument	24 (42)	30 (36)	30 (36)
Percentage men	Firearm	90	96	96
	Sharp instrument	75	85	93

Note: Figures in red indicate instrument with worst impact.

* Assuming people can be 'productive' until the age of 65 in both countries.

Source: Small Arms Survey calculations based on ISER (2006b); CERAC (2006c)

Several key indicators were identified in the three pilot studies to evaluate the productivity losses of victims of violence. These include the number of days spent in hospital, the number of days they cannot work as they recover, and the number of productive life years lost due to death or disability. The surveys allowed for a comparison of the number of days spent in hospital, as well as the average ages of victims. Given the relatively short time frame of the pilot studies, however, the number of inactive days and the proportion of disabled patients are based primarily on

Table 8.10   Average productivity losses per injury, by severity and instrument (2003 PPP USD)*

	HGNI, Rio de Janeiro		HSC, Bogotá		HUV, Cali	
	Firearm	**Sharp instrument**	**Firearm**	**Sharp instrument**	**Firearm**	**Sharp instrument**
Non-fatal						
Time spent in hospital	265	83	196	124	360	168
Inactivity after discharge	910	281	982	664	1,063	712
Total non-fatal	**1,175**	**364**	**1,178**	**788**	**1,423**	**880**
Fatal or permanently disabled	325,045	311,406	268,835	250,291	272,779	260,823

* Productivity losses were calculated using the average PPP USD income for women and men in Brazil and Colombia, as reported in UNDP (2005). Figures also take into account informal economy income, which is estimated at 39.8 per cent of gross national product (GNP) in Brazil and 39.1 per cent in Colombia (Schneider, 2002, p. 11). Average income figures were then applied according to the gender distribution of victims treated at each selected hospital. For comparative purposes, researchers assumed that people are productive until age 65 in both Brazil and Colombia when calculating years of productive life lost due to death or disability. For productivity losses due to death or disability, a 3 per cent discount rate was applied (see Annexe 1).

Sources: Small Arms Survey calculations based on ISER (2006b); CERAC (2006c); UNDP (2005); Schneider (2002)

Table 8.11   Extrapolated total costs of/losses from gun violence for Brazil and Colombia (2003 PPP USD)

Type of injury	Number of cases and type of cost	Brazil	Colombia
Non-fatal (admissions only)	*Number of cases (year)*	*19,534 (2002)*	*5,546 (2005)*
	Average medical costs	4,521	6,804
	Total medical costs	88,309,893	37,735,821
	Average productivity losses	1,175	1,178
	Total productivity losses	22,953,094	6,532,010
	Total non-fatal costs	**111,262,987**	**44,267,831**
Fatal	*Number of cases (year)*	*30,855 (2002)*	*14,762 (2005)*
	Average productivity losses	325,045	268,835
	Total fatal non-hospitalized costs	**10,029,249,309**	**3,968,538,684**
Total	**Total medical costs**	**88,309,893**	**37,735,821**
	Total productivity losses	**10,052,202,404**	**3,975,070,694**
	Total	**10,140,512,297**	**4,012,806,515**
	Total as % of national income*	**0.5**	**1.0**

* Percentage of national income calculated from PPP USD GDP figures (from UNDP, 2005) and the estimated ratio of informal income to GNP as reported in Schneider (2002, p. 11).

Sources: Costing data: Tables 8.6, 8.8, and 8.10 (for Colombia, average costs are those of HSC, Bogotá)

doctors' estimates at the time the victim was still at the hospital. The figures presented in Table 8.10 are therefore exploratory and would benefit from the further monitoring and surveillance of patients after discharge.

It appears that firearm injuries nevertheless generate higher productivity losses than cut/stab wounds in all pilot case studies. This is primarily because in most cases, victims of firearms violence lost more productive time (Table 8.10). In addition, a greater proportion of gun violence victims were men, for whom average earned income was also higher in both countries (UNDP, 2005). This raises the importance of including the value of unpaid productive activities such as housework in such estimates, for which data was unfortunately not available for these pilot studies.

Implications for the costs of gun violence in Brazil and Colombia

Findings on the direct medical costs and productivity losses at the selected hospitals make it possible to produce a rough estimate of the costs of gun violence at the national level in Brazil and Colombia. Multiplying the number of fatal and non-fatal injuries by the average cost per injury illustrates the significance of the problem. Non-fatal injury data for Brazil and Colombia is incomplete, however, and only covers admitted patients. Actual medical costs would therefore be higher if they included information on patients receiving care at emergency departments.

Based on the available data, firearms injuries cost Brazil and Colombia's medical systems a combined PPP USD 125 million a year. Productivity losses are much more significant: they amount to PPP USD 10 billion a year in Brazil and PPP USD 4 billion in Colombia. When these two cost categories are combined, the costs of gun violence in Brazil and Colombia amount to 0.5 and 1 per cent of their respective annual national income.

> Annual productivity losses amount to PPP USD 10 billion in Brazil and PPP USD 4 billion in Colombia.

CONCLUSION

Examining the impacts of gun violence from an economic perspective can serve as an essential component in the design, monitoring, and evaluation of violence prevention and reduction initiatives. It highlights how every gunshot wound has implications that go far beyond victim and perpetrator, and thus helps justify investment in gun violence prevention and reduction. Small arms violence affects society as a whole, inflicting material costs to survivors, family, and institutions; jeopardizing future output and productivity; and affecting mindsets and well-being.

Unfortunately, very few estimates of the costs of gun violence exist. Existing studies also have different purposes, do not focus on the same costs, rely on methods that have not yet been standardized, and result in findings that are difficult to compare. Systematic data gathering on the costs of gun violence, particularly in developing countries, would represent a significant step forward in our understanding of the impacts of small arms violence.

Despite these limitations, there is ample evidence that small arms make violence worse for societies by increasing the average cost of violent injuries. Small Arms Survey pilot studies in Brazil and Colombia confirm that this is not only the case in developed countries, but applies in other regions. Medical costs are significantly higher for gunshot wounds than for other violent injuries, and victims of gun violence are younger than the average victim of violence, resulting in many lost opportunities.

Countries and regions pay very different price tags, however. Indirect costs such as lost earnings are particularly high among countries affected by highly lethal forms of gun violence, such as assassinations, mass killings, and suicides. The total medical costs of gun violence in low- and middle-income countries tend to be lower than their high levels of small arms violence might suggest. In such settings, costing studies can help identify insufficiencies in poorly resourced medical and rehabilitation systems. Improving the responsiveness of public health systems to gun violence is crucial, as it will both decrease the victims' suffering and increase their probability of surviving their wounds. ☒

ANNEXE 1. MEASURING THE COSTS OF FIREARMS VIOLENCE: A MODEL

The Small Arms Survey, together with the WHO and CDC, is currently developing guidelines to estimate the economic costs of violence. The model is meant to enable researchers to generate estimates in developing countries. The basic formula for arriving at an estimate is the following:

Total costs = number (incidence) of violent incidents × average unit cost per incident

The costs considered include direct medical costs and loss of productivity due to injury and death. While this approach does not take into account all the costs incurred by small arms violence, it is particularly valuable to the study of firearms violence. Indeed, these cost categories help highlight the disproportionate costs of gun violence when compared with other forms of violence, as firearms injuries require more intensive medical treatment and generally affect young, potentially productive segments of the population.

Obtaining incidence data

For fatal violence, the absolute minimum data required to produce an economic cost estimate should include the total number of violence-related deaths available for the study area and the average age at death of the victims. A more meaningful estimate will be produced when the data is disaggregated by intent, age, and sex of the deceased and instrument of injury (for our purposes, firearms vs. other weapons).

Data on the incidence of non-fatal violence will generally be much more difficult to obtain than data on violence-related deaths. Hospitals are likely to be the most readily accessible source of data, but will only reflect incidents leading to injuries that result in hospital treatment. Rapid assessment procedures for estimating the total number and incidence of violence-related injuries, irrespective of severity, seen in hospitals, may be required (Matzopoulos et al., forthcoming).

Obtaining costing data

In a majority of countries, even the minimum required costing data will probably not be available from an existing source. It will therefore be necessary to generate the costing data by studying costs for violence-related injuries treated in a small sample of facilities.

The first step involves selecting one or several medical facilities that are believed to be generally representative of the area under study. Costing data may then be obtained from registers or patient surveys to be administered by hospital staff in addition to their usual paper records, or by a trained team of researchers. The sampling strategy used in this chapter's three pilot studies involves capturing data on approximately 100 patients as they present themselves at the hospitals.

The following minimum information should be collected for each violent injury treated at the facility:

- sex and age of the patient;

- injury intent (interpersonal, self-directed, collective, unintentional, undetermined);

- injury severity (emergency treatment only, hospitalization only, death);

- instrument used to inflict injury (firearm, bladed weapon, other);

- length of hospital stay in days;

- use of ambulance;

- operations carried out on the patient;

- drugs given to the patient during and after the stay;

- number of examinations (e.g. X-rays) carried out on the patient;

- number of blood transfusions given to the patient;

- number and type of physicians consulted during the stay;

- estimated number of days the patient will be convalescing (i.e. not be able to work) after leaving the facility; and

- estimated number of out-patient visits the patient will undertake after leaving the facility.

The last two items may be requested from the personnel treating the patient. If time allows, however, conducting follow-up interviews with patients after they leave the facility would be preferable and more accurate.

In addition, a certain number of unit costs need to be obtained from the facility personnel. These include:

- average 'hotel cost' per bed day (excluding drugs, operations, and physicians);

- average ambulance costs;

- costs of the various drugs identified;

- average cost per type of operation, examination, and blood transfusion;

- average cost per physician consultation; and

- average cost per out-patient visit.

Extrapolating results

Based on the sample of injuries surveyed, it is possible to calculate average medical costs per violent injury. These average medical costs are then multiplied by the total number of injuries treated in the area under study to generate a rough estimate of total medical costs. Average costs may also be disaggregated by severity of injury or, as is done in this chapter, by instrument used. The latter option will produce the most revealing insights regarding the cost of firearms injuries when compared with those of wounds caused by other weapons.

Lost productivity is calculated by multiplying the amount of productive time lost due to injury or death by average earnings. Time lost is calculated from the average age at death from a violent injury, as well as the average number of convalescing days among survivors. In this chapter, average earned income figures were obtained from UNDP (2005). Values were adjusted according to the gender distribution of the sample of patients. Figures also included estimated income generated through the informal economy, based on country ratios included in Schneider (2002). Ideally, however, researchers will seek to determine the average income of victims of violence, as they may originate from lower social classes. Future loss of productivity—i.e. that due to death or permanent disability—must also be discounted to give its present value (Corso and Haddix, 2003). This chapter used a discount rate of 3 per cent per year.

Productivity losses can therefore be calculated using the following formulas.

For non-fatal injuries: *Losses = number of non-fatal injuries × average number of days lost × average daily earnings*

For fatal injuries: *Losses = number of fatal injuries × average annual earnings × discount factor*

Where: *discount factor = 1/0.03 – 1/[0.03 × (1.03)ᵃ]* (where: *a = years lost = 65 – average age at death + 1*)

LIST OF ABBREVIATIONS

BRL	Brazilian real	HSC	Hospital Santa Clara (Colombia)
CAD	Canadian dollar	HUV	Hospital Universitario del Valle
CDC	Centers for Disease Control and		(Colombia)
	Prevention (US)	ISER	Instituto de Estudos da Religião (Brazil)
CERAC	Centro de Recursos para el Analisis de	JMD	Jamaican dollar
	Conflictos (Colombia)	PPP	purchasing power parity
COP	Colombian peso	USD	United States dollar
GDP	gross domestic product	WHO	World Health Organization
GNP	gross national product	ZAR	South African rand
HGNI	Hospital da Geral in Nova Iguaçu (Brazil)		

ENDNOTES

1   Translated by the present author and adapted from MSF (2004, p. 3). Déo was eventually transferred to a free Médecins sans Frontières facility and survived. His case is not an isolated one, however. Many Burundians unable to pay their medical bills have been 'imprisoned' in hospitals (FIACAT, 2005). This situation prompted Françoise Ngendahayo, the Burundian minister in charge of national solidarity, to order the release of all such prisoners on 23 December 2005 (Netpress, 2005).

2   Unless stated otherwise, PPP USD values were computed using PPP conversion rates and the US consumer price index as reported in IMF (2005), with 2003 as the base year.

3   See UNDP (2005).

4   See Small Arms Survey (2002, p. 159; 2003, p. 131), WHO (2004a, p. 6), and Lindgren (2005, p. 5).

5   This typology is based on the WHO definition, which considers violence as '[t]he intentional use of physical force or power, threatened or actual, against oneself, another person, or against a group or community, that either results in or has a high likelihood of resulting in injury, death, psychological harm, maldevelopment or deprivation' (WHO, 2002, p. 5).

6   See Waters et al. (2005), WHO (2004a), Rice (2000), and Fleurence (2003).

7   For a review of the limitations of existing studies, see Lindgren (2005), WHO (2004a), and Waters et al. (2005).

8   Phillips (1998), Walker (1997), and Fanslow et al. (1997), as reported in Waters et al. (2005, p. 305).

9   According to one study, the informal economy is equivalent to 42 per cent of GDP in Africa, 41 per cent in Central and South America, 29 per cent in Asia, 35 per cent in transition countries, 18 per cent in Western European Organization for Economic Co-operation and Development (OECD) countries, and 13.5 per cent in North American and Pacific OECD countries (Schneider, 2002, p. 45).

10 Miller, Cohen, and Wiersema (1996); Miller, Fisher, and Cohen (2001); as quoted in WHO, 2004a, pp. 13–14.

11 Over the past 30 years, a series of measures have progressively tightened controls over rifles and shotguns. In 1977 new legislation was introduced that required a Firearms Acquisition Certificate before purchasing rifles or shotguns. At the same time, fully automatic weapons were prohibited. In 1991 the screening processes associated with getting a Firearms Acquisition Certificate were tightened to include a range of risk factors associated with suicide and domestic violence. The 1991 bill also strengthened safe storage provisions and prohibited a wider range of military weapons, including semi-automatic variants.

12 These were the estimated additional costs over and above those paid by the federal government for the operation of the existing system, including restricted weapons registry and transfer payments to provinces for the administration of the Firearms Acquisition Certificates, but did not consider other costs at the local level.

13   The audit notes that this estimate does not include all financial impacts on the government, and these were, according to the Department of Justice, due to 'major delays in making regulations, provinces opting out of the Program, the need for additional initiatives, incorrect assumptions about the rate at which it would receive applications for licences and registrations, and an excessive focus on regulation and enforcing controls' (Canada, 2002, para. 10.4).

14   Researchers have undertaken studies to assess the impact of previous changes to firearms laws in Canada. Leenaars and Lester (1997; 2001) examined trends in firearms deaths following the passage of the 1977 law. According to their studies, firearms homicide rates for victims aged 15–34 and 45–74 decreased significantly after the introduction of Bill C-51 in 1977. The studies find that the overall homicide rates decreased significantly after the introduction of Bill C-51, even when taking into account a series of social factors that include birth, marriage, divorce rate, unemployment rate, family income, and the percentage of males aged 15–21 in the population. In a subsequent study of suicide in Canada, Leenaars et al. (2003) find that firearms suicide rates decreased significantly after Bill C-51. In addition, the percentage of suicides by firearms also decreased significantly, although the impact on specific age groups and genders differed.

15   It is also important to contextualize the expenditures. The Province of Quebec spent CAD 125 million (PPP USD 106 million) to inoculate citizens against meningitis in 2002 after 85 cases were reported. New Brunswick invested CAD 485 million (PPP USD 413 million) in a segment of highway referred to as 'Suicide Alley' where 43 people had died over five years (unpublished letter, Antoine Chapdelaine et al., 10 January 2003). No one knows how much is invested to keep highways safe—be it through licensing drivers, registering vehicles, and operating safety programmes—but it is well into the billions of dollars annually.

16   Serious injuries in this chapter are those that require hospital admission, as opposed to receiving care only in emergency departments. Lethality refers to the proportion of injuries that result in deaths.

17   See Richmond, Cheney, and Schwab (2005) and Small Arms Survey (2004, pp. 172–211; 2005, pp. 228–65).

18   For evidence of these high gun suicide rates in Canada and the United States, see Small Arms Survey (2004, pp. 199–200).

19   Table 8.7 reveals the distribution of a sample of patients treated at HGNI (111 cases), HSC (145), and HUV (116). HGNI and HSC are both city hospitals that treat primarily patients coming from their surrounding area, which in many cases belong to poor social classes. HUV is the region's largest referral hospital and treats primarily serious injuries. Thus it received a majority of patients injured by small arms (61 per cent), while the other two hospitals treated injuries caused by a wider range of instruments and included data on patients receiving emergency room treatment only. For comparative purposes, however, the analysis focuses on the costs of gunshot wounds versus those of sharp instrument injuries. In all cases where intent could be defined, firearms injuries were interpersonal. In all three hospitals combined, only 23 injuries were the result of suicide attempts and these involved sharp instruments (10 cases) or poisoning (13 cases), but no firearms.

20   Luciana Phebo of ISER, and Katherine Aguirre and Jorge Restrepo of CERAC also responded to numerous follow up queries in written correspondence, January–February 2006.

21   The rates used here were computed from UNDP (2005).

22   Using per capita public health expenditures as reported in UNDP (2005).

23   One plausible explanation is linked with the rapidity of reaction of Colombian ambulances. In Bogotá, for instance, it takes on average 11 minutes for an ambulance to reach an incident scene. Taxis, on the other hand, are easily available at every street corner. While a victim of a stabbing may be able to wait for an ambulance without endangering her/his life, a gunshot victim may be more likely to be rushed to hospital by taxi (written correspondence with Jorge Restrepo, CERAC, 13 February 2006).

BIBLIOGRAPHY

Adam, Taghreed, David B. Evans, and Christopher J. L. Murray. 2003. 'Economic Estimation of Country-specific Hospital Costs.' *Cost Effectiveness and Resource Allocation*. Vol. 1, No. 3. 26 February. <http://www.pubmedcentral.nih.gov/articlerender.fcgi?artid=156022&tools=bot>

Allard, Denis and V. C. Burch. 2005. 'The Cost of Treating Serious Abdominal Firearm-related Injuries in South Africa.' *South African Medical Journal*. Vol. 95, No. 8, August, pp. 591–94.

Arunatilake, Nisha, Sisira Jayasuriya, and Saman Kelegama. 2001. 'The Economic Cost of the War in Sri Lanka.' *World Development*. Vol. 29, No. 9, pp. 1483–500.

Beltrán, Isaac de Leon, Ana Maria Fernandez, Maria Victoria Llorente, and Eduardo Salcedo Albaran. 2003. *Homicidio e intencion letal: un studio exploratoria de heridas mortals a partir de los protocolos de necropsia en Bogotá*. Borradores de MÉTODO: área de crimen y conflicto. Bogotá: Grupo Transdisciplinario de Investigación en Ciencias Sociales, MÉTODO. <http://www.dotec-colombia.org/articulo.php?id=1126>

Brand, Sam and Richard Price. 2000. *The Economic and Social Costs of Crime*. Home Office Research Study No. 217. <http://www.homeoffice.gov.uk/rds/pdfs/hors217.pdf>

Buvinic, Mayra, Andrew Morrison, and Michael Shifter. 1999. 'Violence in Latin America and the Caribbean: A Framework for Action.' In Andrew Morrison and Loreto Biehl, eds. *Too Close to Home: Domestic Violence in the Americas*. New York: Inter-American Development Bank, pp. 6–20.

Canada. Office of the Auditor General. 2002. 'Chapter 10: Department of Justice: Costs of Implementing the Canadian Firearms Program.' In *Report of the Auditor General of Canada*. <http://canadaonline.about.com/gi/dynamic/offsite.htm?site=http://www.oag%2Dbvg.gc.ca/domino/reports.nsf/html/20021210ce.html>

CERAC (Centro de Recursos para el Analisis de Conflictos). 2005. *Aproximacion a la situacionde violencia e inseguridad en Bogotá D.C.* Bogotá: CERAC. July.

——. 2006a. *Proyecto piloto 'Estimacion del costo de las lesiones violentas en Colombia.'* Background paper. Geneva: Small Arms Survey. 6 January.

——. 2006b. *Violent Injuries at HUV and HSC*. Database. Geneva: Small Arms Survey.

Collier, Paul. 1999. *On the Economic Consequences of Civil War*. Oxford Economic Papers No. 51. Oxford: Oxford University Press, pp. 168–83.

—— and Anke Hoeffler. 2004. *The Challenge of Reducing the Global Incidence of Civil War.* Copenhagen Consensus Challenge Paper. Revised. 26 March.

Cook, Philip and Jens Ludwig. 2000. *Gun Violence: The Real Costs.* Oxford: Oxford University Press.

Corso, Phaedra S. and Anne C. Haddix. 2003. 'Chapter 6: Time Preference.' In Anne C. Haddix, Steven M. Teutsch, and Phaedra S. Corso, eds. *Prevention Effectiveness: A Guide to Decision Analysis and Economic Evaluation,* 2nd edition. New York: Oxford University Press.

Coupland, Robin and David R. Meddings. 1999. 'Mortality Associated with the Use of Weapons in Armed Conflicts, Wartime Atrocities, and Civilian Shootings: Literature Review.' *British Medical Journal.* Vol. 319, August, pp. 407–10.

Cukier, Wendy. 2005. *Putting Numbers in Perspective: The Costs of Controlling Firearms in Canada.* Background paper. Geneva: Small Arms Survey.

—— and Vic Sidel. 2005. *The Global Gun Epidemic: From Saturday Night Specials to AK 47s.* New York: Praegar.

DiAddario, Sabrina. 1997. 'Estimating the Economic Costs of Conflict: An Examination of the Two-gap Estimation Model for the Case of Nicaragua.' *Oxford Development Studies.* Vol. 25, No. 1, pp. 123–41.

Fanslow, Janet, Carolyn Cognan, Brent Miller, and Robyn Norton. 1997. 'The Economic Cost of Homicide in New Zealand.' *Social Science and Medicine.* Vol. 45, No. 7, pp. 973–77.

FIACAT (Fédération internationale de l'action des chrétiens pour l'abolition de la torture). 2005. *BURUNDI: A Bujumbura, des hôpitaux transformés en prison.* November. <http://www2.fiacat.org/fr/article.php3?id_article=207>

Fitzerald, Edmund Valpy Knox. 1987. 'An Evaluation of the Economic Costs to Nicaragua of US Aggression: 1980–1984.' In Rose Spalding, ed. *The Political Economy of Revolutionary Nicaragua.* Boston: Allen and Unwin, pp. 195–213.

Fleurence, Rachel. 2003. 'An Introduction to Health Economics.' *Pharmaceutical Journal.* Vol. 271, pp. 679–81.

Grobar, Lisa Morris and Shiranti Gnanaselvam. 1993. 'The Economic Effects of the Sri Lankan Civil War.' *Economic Development and Cultural Change.* Vol. 41, No. 2., pp. 395–405.

Harris, Geoff. 1997. 'Estimates of the Economic Cost of Armed Conflict: The Iran–Iraq War and the Sri Lankan Civil War.' In Jurgen Brauer and William G. Gissy, eds. *Economics of Conflict and Peace.* Aldershot: Avebury, pp. 269–91.

——. 1999. 'The Costs of Armed Conflict in Developing Countries.' In Geoff Harris. *Recovery from Armed Conflict in Developing Countries: An Economic and Political Analysis.* London: Routledge, pp. 12–28.

Hoeffler, Anke and Marta Reynal-Querol. 2003. *Measuring the Costs of Conflict.* Oxford: Centre for the Studies of African Economies.

Hung, Kwing. 2005. *Firearm Statistics, Updated Tables.* Ottawa: Research and Statistics Division, Department of Justice, Canada. January.

ICRC (International Committee of the Red Cross). 1999. *Arms Availability and the Situation of Civilians in Armed Conflict.* Geneva: ICRC.

IMF (International Monetary Fund). 2005. *World Economic Outlook Database.* September. <http://www.imf.org/external/pubs/ft/weo/2005/02/data/index.htm>

ISER (Instituto de Estudos da Religião). 2006a. *Report on Emergency Service and Hospital Admission in Rio de Janeiro.* Background paper. Geneva: Small Arms Survey.

——. 2006b. *Report on the Costs of Ambulance Service in Rio.* Background paper. Geneva: Small Arms Survey.

——. 2006c. *Violent Injuries at HGNI.* Database. Geneva: Small Arms Survey.

Kelegma, Saman. 1999. 'Economic Costs of Conflict in Sri Lanka.' In Robert Rotberg, ed. *Creating Peace in Sri Lanka: Civil War and Reconciliation.* Cambridge, MA: Brookings Institution Press.

Leenaars, Antoon and David Lester. 1997. 'The Effects of Gun Control on the Accidental Death Rate from Firearms in Canada.' *Journal of Safety Research.* Vol. 28, No. 3, pp. 119–22.

——. 2001. 'The Impact of Gun Control (Bill C-51) on Homicide in Canada.' *Journal of Criminal Justice.* Vol. 29, pp. 287–94.

Leenaars, Antoon A., Ferenc Moksony, David Lester, and Susanne Wenckstern. 2003. 'The Impact of Gun Control (Bill C-51) on Suicide in Canada.' *Death Studies.* Vol. 27, pp. 103–24.

Lester, David. 2000. 'Gun Availability and the Use of Guns for Suicide and Homicide in Canada.' *Canadian Journal of Public Health.* Vol. 91, No. 3, May/June, pp. 186–87.

Lindgren, Göran. 2005. 'The Economic Costs of Civil War.' Paper prepared for the Ninth Annual International Conference on Economics and Security. Bristol, 23–25 June.

Londoño, Juan Luis and Rodrigo Guerrero. 1999. *Violencia en America Latina: epidemiologia y costos.* Documento de Trabajo R-375. Washington, DC: Inter-American Development Bank.

Lopez, Humberto. 2001. *The Cost of Armed Conflict in Central America.* Washington, DC: World Bank.

Ludwig, Jens and Philip Cook. 1999. *The Benefits of Reducing Gun Violence: Evidence from Contingent-valuation Data.* Working Paper Series No. 7166. Cambridge, MA: National Bureau of Economic Research. June.

Matzopoulos, Richard G., Megan Prinsloo, Alexander Butchart, Margie M. Peden, and Carl Lombard. Forthcoming. 'Estimating the South African Trauma Caseload.' *International Journal of Injury Control and Safety Promotion.*

Max, Wendy and Dorothy P. Rice. 1993. 'Shooting in the Dark: Estimating the Costs of Firearm Injuries.' *Health Affairs.* Vol. 12, pp. 171–85. <http://content.healthaffairs.org/cgi/reprint/12/4/171>

Miller, Ted. 1995. 'Costs Associated with Gunshot Wounds in Canada in 1991.' *Canadian Medical Association Journal.* Vol. 153, No. 9, pp. 1261–68.

——, Mark Cohen, and Brian Wiersema. 1996. *Victim Costs and Consequences: A New Look.* National Institute of Justice Report. Landover: National Institute of Justice.

—— and Mark A. Cohen. 1997. 'Costs of Gunshot and Cut/Stab Wounds in the United States, with Some Canadian Comparisons.' *Accident Analysis and Prevention.* Vol. 29, No. 3, pp. 329–41.

——, Deborah A. Fisher, and Mark A. Cohen. 2001. 'Costs of Juvenile Violence: Policy Implications.' *Pediatrics.* Vol. 107, No.1, pp. 1–7.

MSF (Médecins sans Frontières). 2004. *Accès aux soins de santé au Burundi: résultats de trois enquêtes épidémiologiques.* Bujumbura: MSF Belgique.

Netpress. 2005 'Le ministère ayant en charge la solidarité nationale fait libérer tous les convalescents indigents détenus dans les hôpitaux.' Bujumbura. 23 December. <http://www.tutsi.org/Act241205.htm>

Ordog, Gary, Jonathan Wasserberger, and Greg Ackroyd. 1995. 'Hospital Costs of Firearm Injuries.' *Journal of Trauma: Injury, Infection, and Critical Care*. Vol. 38, No. 2, pp. 291–98.

Paniagua, Ignacio, Emperatriz Crespin, Ademar Guardado, and Ana Mauricio. 2005. 'Wounds Caused by Firearms in El Salvador, 2003–2004: Epidemiological Issues.' *Medicine, Conflict and Survival*. Vol. 21, No. 3, pp. 191–98.

Pfizer Journal. 2001. *Responding to the Global Public Health Challenge of Violence*. Global Edition. Vol. 11, No. 1.

Phillips, Rozett. 1998. *The Economic Cost of Homicide to a South African City*. Unpublished MBA thesis. University of Cape Town, Graduate School of Business.

Rice, Dorothy P. 2000. 'Cost of Illness Studies: What Is Good about Them?' *Injury Prevention*. Vol. 6, pp. 177–79.

Richardson, John, Jr. and S. W. R. de A. Samarasinghe. 1991. 'Measuring the Economic Dimensions of Sri Lanka's Ethnic Conflict.' In S. W. R. de A. Samarasinghe and Reed Coughlan, eds. *Economic Dimensions of Ethnic Conflict*. London: St. Martin's Press.

Richmond, Therese, Rose Cheney, and C. William Schwab. 2005. 'The Global Burden of Non-conflict Related Firearm Deaths.' *Injury Prevention*. Vol. 11, pp. 348–52.

Schneider, Friedrich. 2002. 'Size and Measurement of the Informal Economy in 110 Countries around the World.' Paper presented at a Workshop of the Australian National Tax Centre. Australian National University, Canberra, 17 July.

Small Arms Survey. 2002. *Small Arms Survey 2002: Counting the Human Cost*. Oxford: Oxford University Press.

——. 2003. *Small Arms Survey 2003: Development Denied*. Oxford: Oxford University Press.

——. 2004. *Small Arms Survey 2004: Rights at Risk*. Oxford: Oxford University Press.

——. 2005. *Small Arms Survey 2005: Weapons at War*. Oxford: Oxford University Press.

Solagberu, Babatunde A. 2003. 'Epidemiology and Outcome of Gunshot Injuries in a Civilian Population in West Africa.' *European Journal of Trauma*. Vol. 2, pp. 92–96.

Spicer, Rebecca, Ted Miller, John Langley, and Shaun Stephenson. 2005. 'Comparison of Injury Case Fatality Rates in the United States and New Zealand.' *Injury Prevention*. Vol. 11, pp. 71–76.

Stewart, Frances, Cindy Huang, and Michael Wang. 2000. 'Internal Wars: An Empirical Overview of the Economic and Social Consequences.' In Frances Stewart and Valpy Fitzgerald, eds. *War and Underdevelopment, Vol. 1: The Economic and Social Consequences of Conflict*. Oxford: Oxford University Press, pp. 67–103.

Stewart, Frances and Frank Humphreys. 1997. 'Civil Conflict in Developing Countries over the Last Quarter of a Century: An Empirical Overview of Economic and Social Consequences.' *Oxford Development Studies*. Vol. 25, No. 1, pp. 11–41.

TRL (Transport Research Laboratory). 1995. *Costing Road Accidents in Developing Countries*. Overseas Road Note 10. Crowthorne: TRL.

UNDP (United Nations Development Programme). 2005. *Human Development Report 2005 Statistics*. <http://hdr.undp.org/statistics/>

Unsworth, Robert E. and James E. Neuman. 1993. 'Review of Existing Value of Morbidity Avoidance Estimates: Draft Valuation Document.' Memorandum to Jim DeMocker, Office of Policy Analysis and Review, Industrial Economics, Inc. 30 September.

Vassar, Mary and Kenneth Kizer. 1996. 'Hospitalizations for Firearm-related Injuries: A Population-based Study of 9562 Patients.' *Journal of the American Medical Association*. Vol. 275, No. 22, pp. 1734–39.

Walker, John. 1997. 'Estimates of the Costs of Crime in Australia 1996'. *Trends and Issues in Crime and Criminal Justice*. No. 72. Canberra: Australian Institute of Criminology. August.

Waters, Hugh, Adnan Ali Hyder, Yogesh Rajkotia, Suprotik Basu, and Alexander Butchart. 2005. 'The Costs of Interpersonal Violence: An International Review.' *Health Policy*. Vol. 73, pp. 303–15.

WHO (World Health Organization). 2002. *World Report on Violence and Health*. Geneva: WHO.

——. 2003. *International Statistical Classification of Diseases and Related Health Problems,* 10th revision. Geneva: WHO. <http://www3.who.int/icd/vol1htm2003/fr-icd.htm>

——. 2004a. *The Economic Dimensions of Interpersonal Violence*. Geneva: WHO.

——. 2004b. *World Report on Road Traffic Injury Prevention*. Geneva: WHO.

——. 2006. *Choosing Interventions that are Cost Effective (WHO-CHOICE): Country Specific Costs*. <http://www.who.int/choice/country/en/>

Wilkins, Kathryn. 2005. 'Deaths Involving Firearms.' *Health Reports*. Vol. 16, No. 4, June, pp. 37–41. Statistics Canada.

Zohoori, Namvar, Elizabeth Ward, Georgiana Gordon, Rainford Wilks, Deanna Ashley, and Terrence Forrester. 2002. 'Non-fatal Violence-related Injuries in Kingston, Jamaica: A Preventable Drain on Resources.' *Injury Control Safety Promotion*. Vol. 9, No. 4, December, pp. 255–62.

ACKNOWLEDGEMENTS

Principal author

Nicolas Florquin

Contributors

Katherine Aguirre, Gustavo Álvarez, Wendy Cukier, Alexis Huynh, Carolina Idarraga, Jeadran Malagón, Cristina Nascimento, Luciana Phebo, Jorge Restrepo, and Guilherme Werneck

A FARC member guards an area in La Macarena, southern Colombia, where rebels held soldiers captive. June 2001. © Scott Dalton/AP Photo

Colombia's Hydra

THE MANY FACES OF GUN VIOLENCE

9

INTRODUCTION[1]

Colombia has long been characterized as one of the most violent countries in the world. Violence arising from a protracted armed conflict and both organized and common crime has claimed the lives of almost half a million civilians and combatants since 1979—almost 17,600 per year—a human security crisis of extraordinary dimensions.[2] This chapter finds that while there is considerable heterogeneity in the nature of homicides over time and space in Colombia, there is a strong contributing factor: firearms. In fact, more than 80 per cent of all homicides in Colombia since the late 1970s have been perpetrated with guns. What is more, this percentage has steadily increased—from about 60 per cent in the 1980s to more than 85 per cent in 2002. By 2005 more than 15 per cent of all deaths by natural and external causes[3] were firearm-related.

This chapter offers the first comprehensive and evidence-based overview of the relationships between armed violence and firearms in the country. Drawing on a combination of data sources and extensive field-based research, the chapter presents the following findings:

- There have been nearly 39,000 violent deaths due to armed conflict since 1988. The yearly average is 2,221 violent deaths, many of them concentrated in rural areas.

- Colombia experienced a significant reduction in conflict-related civilian deaths in 2003 and 2004, followed by a pronounced reversal in the first half of 2005.

- There have been more than 475,000 firearm-related deaths as a result of crime and conflict violence since 1979, averaging 17,600 per year, with most deaths concentrated in urban centres.

- More than 80 per cent of all homicides are committed with firearms—with more than half of the variation in external death rates over time attributable to firearms.

- Most weapons in circulation are illegal and unregistered. The number of legally and illegally held weapons (excluding the state security forces) is estimated between 2.3 million and 3.9 million, an ownership rate of 5.05 to 8.42 per 100 inhabitants. Official statistics report only 1.53 legally held firearms per 100 inhabitants, a low rate in comparison with other Latin American countries.

- Illegal right-wing paramilitaries appear to have more modern and abundant weapons stocks than left-wing guerrillas. Paramilitaries are also party to a more lucrative and sustained source of funding.

- Men suffer more than 90 percent of all gun deaths. More than one-third of all firearm deaths are concentrated among men aged 20–29, with more than 340,000 potential years of life lost among men in 1985 alone.

- It appears that firearm control measures have yielded significant dividends in reducing violence in major cities such as Bogotá, Medellín, and Cali.

- Colombia's legal arms market is among the most transparent and tightly regulated in the world, despite uneven enforcement.
- The country exhibits a potentially unhealthy regulatory environment for firearms in which state-owned firms that produce and sell firearms also fall under the public entity that is responsible for arms control.

The causes and effects of Colombia's armed violence are complex;[4] this chapter presents their core features. The first section discusses the background and context within which conflict and criminal violence take place. Next, the chapter turns to the political economy of the legal arms industry, the dynamics of legal and illegal gun ownership, patterns of illegal production and trafficking in weapons, and the distribution and effects of conflict-related violence and criminal violence. It closes with a brief discussion of municipal arms control interventions and the nascent process of disarmament, demobilization, and reintegration (DDR) of the country's paramilitaries.

CONTEMPORARY AND HISTORICAL DIMENSIONS OF ARMED VIOLENCE

Colombia has been afflicted by a long-standing human security crisis, which includes kidnapping and forced displacement along with systemic violence.[5] In concentrating on lethal gun violence, this chapter presents new empirical insights generated by the Small Arms Survey. Lethal threats in Colombia are driven by a complex and interconnected array of armed groups and individuals. Since 1963, a major contributor to human insecurity remains the armed conflict, which has pitted the government against left-wing guerrilla groups, primarily the Revolutionary Armed Forces of Colombia (FARC) and the National Liberation Army (ELN), as well as the right-wing paramilitary groups such

Graph 9.1 **Homicides and firearm-related homicides, 1979–2002**

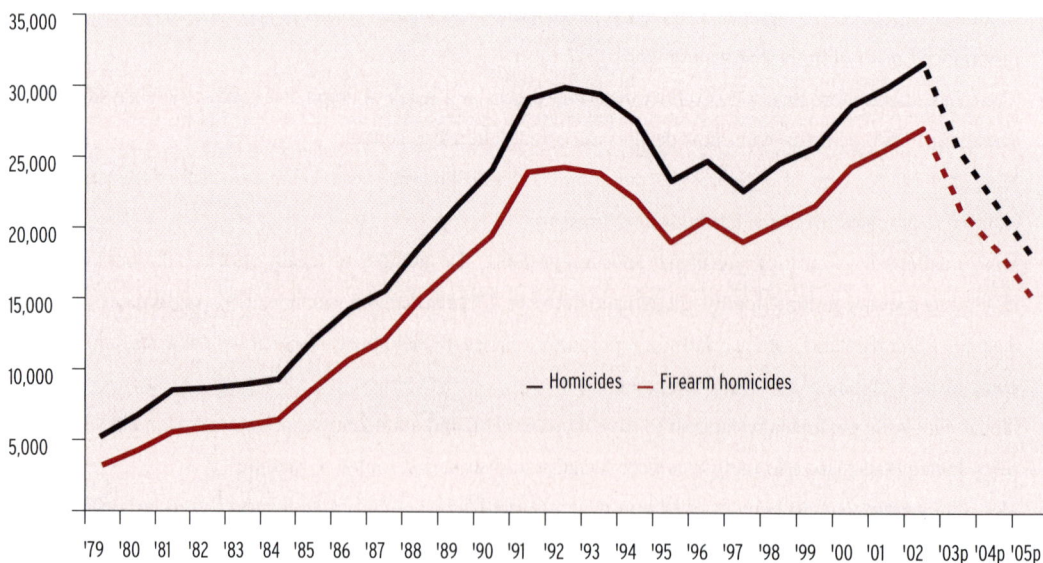

Note: 2003, 2004, and 2005 DANE figures drawn from National Police and INML.

Source: DANE; processed by CERAC

as the United Self-Defence Forces of Colombia (AUC). Another factor contributing to insecurity is deeply embedded organized and common criminal violence, much of it carried out by narco-traffickers, mafia gangs, and petty bandits.

While both conflict and criminal violence constitute very real threats to human security, international concern has focused disproportionately on the former. Colombia's armed conflict has been characterized as a 'low-intensity' contest for political power (Restrepo, Spagat, and Vargas, 2004, p. 398). Unlike other wars in Africa, South and Southeast Asia, or Europe, the conflict is not marked by evident regional, ethnic, or religious drivers. Nevertheless, the impact of Colombia's armed conflict on human welfare has been profound: more than 38,800 people have been killed directly in conflict since 1988. Recently, there have been dramatic oscillations in annual figures, with a substantial decrease in killings—particularly of civilians—from its peak in 2002. Nevertheless, the decrease was followed by a sharp rise in early 2005, mainly due to a spike in paramilitary violence. It should be noted that illegal paramilitaries register exceptionally high killed-to-injury ratios, indicating a high degree of intentionality in their attacks (Restrepo and Spagat, 2005a, p. 142).

While conflict-related violence has had a devastating impact on human security, criminal violence has exacted an even heavier toll. More than 400,000 civilians have been killed as a result of criminal violence since 1988, the vast majority of them shot dead with small arms and light weapons (see Graph 9.1). Indeed, longitudinal trends in homicide are virtually indistinguishable from firearm-related homicides.

> Criminal violence has exacted an even heavier toll on human security than conflict-related violence.

POLITICAL ECONOMY OF PRODUCTION, IMPORT, AND EXPORT

Before analysing the scale and distribution of armed violence arising from conflict and crime, it is important to review Colombia's arms production, as well as its exports and imports. Colombia is a significant producer of weapons and munitions. The country also legally imports arms of various calibres from at least 43 countries and is striving to generate greater revenues from weapons exports.[6] Moreover, Colombia has a massive illegal market for weapons, with a complex network of buyers and sellers—many of them driven by the armed conflict and narco-trafficking. It is useful to parse out the country's legal manufacturing capacities as well as the illegal dynamics of domestic arms acquisition in order to appreciate their relationship with human security.

As is the case in other countries, Colombia's domestic production of defence material and equipment has long been linked to intrinsic notions of national sovereignty, development, and security. Self-sufficiency in weapons and munitions production is frequently a stated goal of industrializing states, and Colombia is no exception. Indeed, the Colombian state has promoted the domestic production of firearms and explosives since the beginning of the twentieth century. The first production lines began operating in 1908, and domestic manufacturing grew steadily throughout the 1930s and 1940s, ensuring regular provision to the armed forces and police. The defence industry was itself consolidated into a single entity—INDUMIL (from the Spanish name *Industria Militar*) in 1954—which gained official monopolies on the production, import and export, and sales of firearms and explosives. Since its inception, INDUMIL has been a state-owned enterprise under the aegis of the Ministry of Defence (MOD), which acts as the regulator and keeper of the registry for all legally held firearms through the Office for Control and Trade of Arms and Explosives (OCCAE).[7]

Since the 1990s, INDUMIL has increased domestic production and diversified its product lines. This strategy is part of an import-substitution policy driven by three main factors. First, by Colombian standards, foreign weapons are expensive. Second, there has been a fear, particularly within domestic military circles, that the continued dependence on foreign weapons suppliers could expose Colombia to possible supply cut-offs and an unacceptable level of vulnerability.[8] Third, military planners have aspired with some success to capture some of the high profits generated by a domestic monopolization of military supply lines.

INDUMIL's primary function is to equip the armed forces and the National Police, though it also supplies legitimate domestic demand and is quietly nurturing an export capacity. INDUMIL has in fact achieved self-sufficiency in, among other items, the Galil rifle[9], revolvers, grenade launchers, and various types of ammunition.[10] Pistol production may also soon begin as part of a programme to upgrade the police force, which currently uses revolvers.[11] Secondary markets, including civilians and private security firms, are also being targeted for increased sales of revolvers.[12] As Colombia does not import significant quantities of revolvers, civilians, the armed forces, and the police are effectively a captive market for INDUMIL.

INDUMIL today is highly profitable. While imports have risen since 2001, INDUMIL has considerably expanded its production for export and domestic consumption (see Graph 9.2). To underpin this growth, the Colombian state has also initiated talks on the privatization of various aspects of production and the promotion of international partnerships. In 2003, for example, the defence minister indicated her interest in partially privatizing the company through

Graph 9.2 **Value of arms, munitions, parts, and accessories: imports, exports, and production, 1995–2004**

US DOLLARS (THOUSANDS. INDEX BASE 2000)
(IMPORTS AND ARMS PRODUCED)

US DOLLARS (THOUSANDS. INDEX BASE 2000)
(EXPORTS)

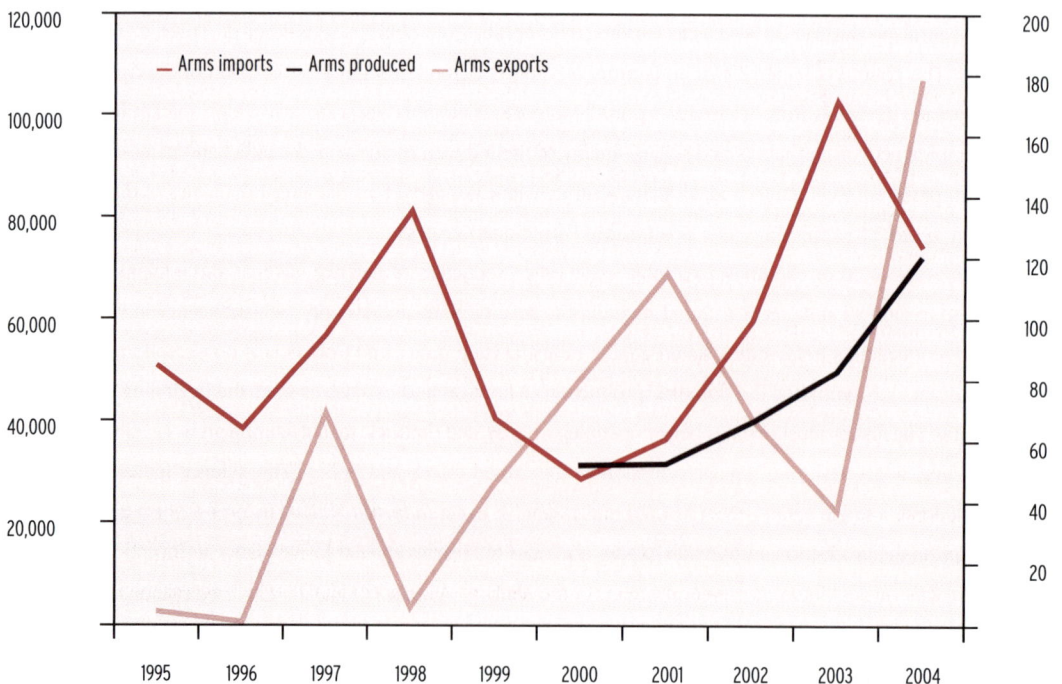

Source: National Tax Administration; INDUMIL;
deflator source: Bureau of Economic Analysis; processed by CERAC

a partnership with Spain, expecting to 'increase the production capacity of INDUMIL, especially in mortars and grenades'. Negotiations were reportedly suspended following a change of government in Spain in 2004; no official policy statement has been issued since.[13]

INDUMIL's monopoly is affirmed not just in the Firearm Control Law, but also in the Colombian Constitution itself (Article 223). INDUMIL is relatively transparent in comparison with other publicly owned defence companies. Unlike several of its Latin American counterparts, the firm is highly unlikely to have been selling or leaking arms to organized crime syndicates or to embargoed countries. Predictably, the strongly regulated legal gun market is also accompanied by an illegal parallel market that appears to deal principally in military-grade weapons for (organized) criminal use. This suggests that the regulation of the legal market is not so strict as to drive aspiring legitimate gun users into the parallel market.

The MOD oversees an assortment of public entities that participate in the supply, manufacture, and regulation of firearms. INDUMIL, the primary agency, is the sole outlet for legal firearms, ammunition, and explosives through its central office in Bogotá and some 30 retail outlets throughout the country—each of which is located within military garrisons. The MOD's OCCAE oversees all aspects of arms regulation and licensing to both individuals and corporate entities. The MOD thus oversees the production, import, and sales of firearms as well as the regulation of weapons sales. This potentially unhealthy regulatory environment is fraught with conflicts of interest. In particular, these current arrangements can create a situation whereby licensing requirements could be relaxed to enable certain weapons sales that should otherwise be blocked. While no systematic malfeasance has come to light, the separation of production and commercialization on the one hand and regulation on the other could avoid potential conflicts of interest and properly align incentives.

> Colombia has an ownership rate of 5.05 to 8.42 weapons per 100 inhabitants.

Patterns of ownership

In comparison with neighbouring countries, Colombia exhibits a low level of firearms ownership. The reason for this is twofold: the Colombian state enforces strong regulation on civilian arms possession and non-state conflict groups and organized crime exert tight control over the criminal market for guns. The Colombian constitution allows the licensing of a restricted type of firearm to civilians only if security needs are proven.[14] In that case the gun remains the legal property of the state. In areas where enforcement of gun regulation is weak, organized crime and conflict groups impose tight controls over firearm possession. Most weapons in circulation are nevertheless illegal and unregistered. The number of legally and illegally held weapons (excluding the state security forces) is estimated at 2.3 million to 3.9 million. With an estimated population of more than 46 million, the 2005 figure indicates an ownership rate of 5.05 to 8.42 per 100 inhabitants.

Legal possession

The OCCAE had issued a total of 706,210 firearm permits to civilians by mid-2005. This figure includes the 235,696 registered firearms issued in the 1994 firearm amnesty, when the new regulation entered into force.[15] The 2005 figure indicates a ratio of 1.53 legal arms per 100 civilians. But while the number of legally registered firearms is lower than those of its neighbours, it appears that Colombian civilians are arming themselves in greater numbers.

In Colombia, the combined firearms holdings of the state—the National Police, armed forces, and intelligence services—are roughly equivalent to those of civilians, a ratio that is high and unusual in the region. The Department of National Planning (DNP) currently registers 113,418 armed police officers in the country. A standard multiplier of

1.2 firearms per officer yields a range of 91,000 to 181,000 firearms in the hands of the National Police.[16] By way of comparison, the armed forces, according to the same source, presently include more than 249,190 personnel plus an estimated 60,700 reserve forces (IISS, 2005, pp. 329–30). The application of a conventional military small arms multiplier of 1.8 per recruit suggests that there are between 373,000 and 742,000 small arms and light weapons of various calibres in military armouries.[17] The Colombian armed forces are also shored up by more than 21,000 trained local battalions (*soldados de mi pueblo,* meaning 'soldiers from my village') with an estimated 1:1 ratio, thus raising total estimated state holdings to 486,000–944,000.

Unlike the National Police, the DAS (Administrative Department of Security)[18] does not publicly report its size, holdings, or budget, undermining any concerted effort to estimate its weapons stockpiles.[19] Nevertheless, the DAS recently announced an ambitious arms modernization plan, including the purchase of several thousand assault rifles, sub-machine guns (MP5), and 0.40 S&W pistols (*Revista Cambio,* 2005).

Illegal ownership[20]

Despite Colombia's comparatively strict regulatory regime, the country is home to many unregistered guns. Official military sources estimate that there are as many as 800,000 unregistered weapons in circulation. National police sources, on the other hand, contend that the number is at least three times greater—reaching 2.4 million. Neither of these figures can be easily verified.

The country's various non-state armed groups possess a wide range of weaponry. There are probably more than 12 different calibres and more than 42 brands distributed among the arsenals of non-state actors. But there are also conflicting estimates of the holdings of guerrilla and illegal paramilitary groups. The National Police estimates that

AUC fighters stand in formation at
La Dorada, Putumayo, February 2003.
© Jason P. Howe/WPN

the force strength of the FARC stands at around 12,500 (DNP, 2005).[21] Applying a standard multiplier of 1.6 military-style arms per combatant yields an upper threshold of 20,000 weapons of various types, including 9 mm pistols, Uzis, AKM-series assault rifles (since 1997), sniper rifles, various types of grenade launchers, and even man-portable air defence systems, or MANPADS. The ELN, for its part, is believed to consist of a much smaller force of around 3,600 active recruits (DNP, 2005). The application of an analogous multiplier of 1.6 arms per ELN combatant produces an estimate of about 5,900 weapons. The AUC, with more than 10,900 fighters (DNP, 2005), is believed to maintain a large stockpile despite the recent demobilization and disarmament process discussed in more detail below.[22] With at least 17,500 sophisticated and high-calibre firearms, including rifle and pistol silencers, they are among the most well armed of all non-state groups in the world today. Recent research and media reports confirm that the paramilitaries commonly use US-made R-15 assault rifles and M60 machine guns, as well as Israeli Galil rifles (AFP, 2004).

It is significant that the paramilitaries have acquired and deployed better weapons than the guerrillas. This distinction—one that is not yet properly appreciated in discussions of Colombia's armed conflict—implies that paramilitary groups are both wealthier and better connected internationally and domestically than guerrilla groups. In fact, while the paramilitaries have been actively procuring modern military technology on the international market, the guerrillas have settled for cheaper, and even home-made, weapons. The illegal drug business is to a large extent a mirror image of the gun trafficking business, with guns flowing into the country and drugs flowing out. The paramilitary gun premium, expensive to maintain, suggests that the paramilitaries participate more deeply in the lucrative narcotics trade than do the guerrillas. In particular, recent evidence from the demobilization process suggests especially strong paramilitary participation in the international transport and distribution phases of the drug business.[23] This inference reinforces a difficult-to-prove perception among some analysts that the paramilitaries are bigger narco-traffickers than the guerrillas.[24] It also demonstrates graphically and unequivocally the utility of examining violence through the lens of small arms.

> Paramilitaries have acquired and deployed better weapons than the guerrillas.

Command and control

An in-depth analysis of the command and control of non-state armed groups can yield insight into prospects for disarmament, demobilization, and reintegration in the event of a peace agreement or formal accord. Guerrilla deserters have revealed unusually strong levels of command and control over their weapons, in relation to both possession and use. The higher up a guerilla leader may be in a given organization, the more robust and sophisticated the weapons under his or her command. This applies to the FARC more than it does to the ELN, since the FARC closely resembles a classical army formation while the ELN consists principally of smaller cells.

Paramilitary groups, most of which conglomerated in 1997 under the umbrella organization AUC, do not have such a well-defined military structure. They are largely organized in regionally defined 'fronts' or 'blocks' that are administered by an assortment of paramilitary leaders, typically affiliated with the AUC. Moreover, the organization of individual blocks appears to vary over time as the paramilitaries exhibit high turnover rates among leaders and combatants alike. This is because their recruitment—unlike that of the guerrillas—is strongly tied to financial incentives and regular payments. Interviews with ex-paramilitary combatants reveal that their arsenals are well stocked with modern weapons and munitions.[25] Respondents gave little indication that paramilitary groups maintained the strict 'fire discipline' or hierarchical control over firepower that was reported among guerrilla groups. This relative carelessness with respect to firearms use is yet another indication that the paramilitaries have ample resources and international connections to maintain well-stocked arsenals.

ILLEGAL ARMS TRAFFICKING AND MANUFACTURING

The curbing of illegal production of and trafficking in small arms by guerrillas, paramilitaries, and narco-traffickers has been a top priority of successive Colombian administrations. In recent years, the US, Canadian, and Colombian governments have collaborated closely to improve their tracing of illegal arms flows, although this task remains immensely challenging.[26] In fact, arms smuggling into and out of Colombia has deep historical roots, particularly in the La Guajira peninsula bordering Venezuela and the Caribbean Sea (*El Pais,* 2004). Highly profitable exports of contraband goods, particularly cocaine, helped give rise to a roaring black market in small arms and light weapons (*O Globo,* 2005). Moreover, the country's extensive coastline and relatively porous frontiers with five countries, particularly the long tracts of isolated border areas, severely complicate the control of illegal weapons flows.

Although information on the scale and volume of Colombia's illegal arms trafficking is sparse, there are some clear patterns and trends.[27] For example, both the FARC and the ELN operate highly sophisticated procurement networks. Official arms seizure reports and media articles suggest that comparatively few weapons used by guerrilla groups originate directly from the stockpiles of the Colombian armed forces. It should be noted, however, that there are no publicly available statistics on state-owned weapons lost or stolen during combat. In fact, the army maintains that a mere ten per cent of guerrilla arms and ammunition retrieved by the armed forces were originally manufactured by INDUMIL (*El Tiempo,* 2005a). Though further investigation of this claim would be valuable, the guerrillas do appear to buy the overwhelming majority of their weapons from illegal dealers.

Guerrillas appear to buy the overwhelming majority of their weapons from illegal dealers.

The Colombian authorities have managed to intercept a number of weapons shipments, shedding some light on the foreign procurement practices of the guerrillas. Though information is scarce, it appears that weapons are purchased through a complex web of interactions to avoid directly implicating guerrilla leaders. To minimize the risk of capture or interdiction, consignments are regularly air-dropped or shipped to an agreed safe area, often within a conflict zone where the government exerts only limited control. From there, weapons are shifted to a 'zone of consolidation', and ultimately to the 'rear guard' of a guerrilla front. The Caribbean coast, particularly the Urabá Gulf corridor, is a primary entry point for FARC assault rifles and light machine guns, most of which originate in the Middle East and Eastern Europe and are transshipped via Central America.[28] Pistols and ammunition follow a variety routes, for example through the infamous triple border area of Paraguay, across the Brazilian border, and into Vaupés Department (*O Globo,* 2005). Armed factions fight regularly over various 'prized' trafficking routes, such as jungle areas along the Pacific coast that have traditionally served as major smuggling routes. Though vigorously denied by the Venezuelan authorities, Colombian military intelligence alleges that corrupt elements within the Venezuelan armed forces regularly provide FAL rifles and 7.62 ammunition to the FARC and ELN.[29] Similar types of arms are also believed to be transshipped from Peru (*El Peruano,* 2005), while press reports have revealed the smuggling of G3s, HK33s, and Galils across the border from Ecuador (*El Comercio,* 2003).

The primary paramilitary smuggling routes for AKM-series assault rifles and various types of machine guns include the Urabá Gulf and the Pacific port of Buenaventura—with most weapons allegedly originating in the United States and Central America. A key access and distribution point is the town of Apartadó in Antioquia, which itself lies relatively close to the Panamanian border in a fiercely contested area characterized by exceptionally inaccessible topography. Some paramilitary groups reportedly possess weapons that are identical to government stocks—particularly Galil rifles, light machine guns, and 5.56 ammunition. Compared to guerrilla groups, paramilitary groups seldom use explosive devices (such as mines) but have much better access to pistols, even if they are less frequently used.[30]

Colombia's illicit craft firearms industry operates on a fairly large scale, with two types of producers and suppliers. First, there are small-scale manufacturers of non-automatic firearms whose primary clients are petty criminals residing in urban centres. Second, since the mid-1990s, the FARC has ratcheted up production of sub-machine guns, mortars, mortar grenades, and hand grenades. Interviews with demobilized guerrillas and government-affiliated security officials have confirmed that the FARC domestically produces copies of the Ingram 9 mm sub-machine gun and a semi-auto-matic Beretta pistol. Other products include 60 mm and 120 mm mortars, hand grenades, and explosives. 'Cylinder' or 'canister' bombs are particularly insidious home-made weapons that are fabricated with gas containers into which the guerrillas often pack shrapnel and even rotten produce to infect their victims.[31] The production of these weapons requires a combination of common metalworking tools, vices and presses, and more expensive raw materials than do *changones,* or sawn-off shotguns.

Intelligence officers point out that blueprints and design plans for some of these more sophisticated items have often been illegally obtained from INDUMIL, retired civil servants, or engineers.[32] For example, mortars and mortar grenades appear to copy standard Colombian military mortars. Security personnel and analysts document the con-struction of other mortars and launch platforms, such as for gas canister bombs, which appear to have benefited from foreign expertise, including that of the IRA.[33] Finally, there is evidence of a sophisticated network of armouries, repair shops, and specialist armourers within each FARC unit, suggesting a concern with economizing, repairing, and maintaining their limited war material. The extent of autonomous production and maintenance activity points to some government success in choking off FARC financing and supply routes. It also underscores the depth of the problem; the FARC has developed many ingenious ways of acquiring or producing the weapons it requires. In con-trast, there is no evidence that paramilitary groups manufacture weapons to any significant degree. This also supports the assumption that they have sufficient resources and procurement capacity to satisfy their needs without resorting to improvisation.

Colombia has seen more than 38,800 conflict-related killings since 1988.

VIOLENCE LINKED TO ARMED CONFLICT

The spatial and temporal dynamics of Colombia's armed conflict are becoming better understood. Compiled data registers more than 38,800 conflict-related killings since 1988—with an average of 2,221 deaths per year.[34] Most of these killings appear to have been perpetrated in isolated rural areas (Restrepo and Spagat, 2005b, p. 15). The municipalities with population densities below 50 people per square kilometre suffer well over 65 per cent of all conflict killings despite accounting for less than 20 per cent of the total population. In fact, only ten per cent of conflict fatalities have been in municipalities with population densities exceeding 200 people per square kilometre where two-thirds of the Colombian population resides. But these general trends only reveal one dimension of the effects of conflict-related violence. This chapter now considers who is killed by whom, the categories of victims, the types of events in which killings occur, and the various categories of weapons used to carry out the killing. In so doing, the chapter renders a sharp distinction between 'clashes', in which two or more groups exchange fire, and 'attacks', defined as one-sided events with no effective resistance.[35]

The distinction between clashes and attacks is important to recognize: most victims in 'clashes' are combatants while the majority of those victimized during 'attacks' are civilians (see Graph 9.3). Importantly, the vast majority of conflict-related civilian killings are perpetrated by paramilitary groups in massacres characterized by high rates of

Graph 9.3 **Civilian and combatant casualties during clashes and attacks, 1 January 1988–30 June 2005**

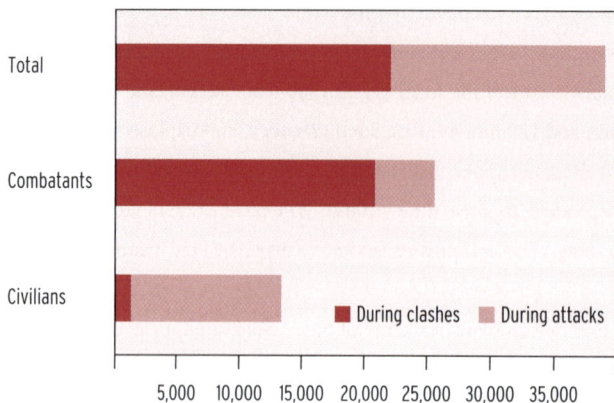

Total

Combatants

Civilians

■ During clashes ■ During attacks

5,000 10,000 15,000 20,000 25,000 30,000 35,000

Source: CERAC Colombian Conflict Database

intentional killing with few left injured.[36] The paramilitaries have long sought to kill civilians whom they suspect of supporting the guerrillas. Such killings surged in the late 1990s, fell significantly in 2003, but then jumped again in the first half of 2005 (see Graph 9.4).[37] The FARC and ELN guerrillas, in contrast, have pursued an alternative strategy of disrupting Colombian society, employing bombing campaigns, the storming and seizing of municipalities, and focused attacks directed against public and private infrastructure. In the process, the guerrillas have killed significantly fewer civilians than have the paramilitaries, though they have injured many more due largely to the indiscriminate nature of their bombings. Finally, government attacks are comparatively infrequent and account for a relatively modest proportion of overall civilian deaths, though they have nevertheless carried out a handful of aerial bombardments in which large numbers of civilians were killed.[38] This research reveals that paramilitaries are much more likely than the guerrillas or the government to fire a weapon in

Graph 9.4 **Civilian casualties during attacks by group, annualized quarterly, 1 January 1988–30 June 2005**

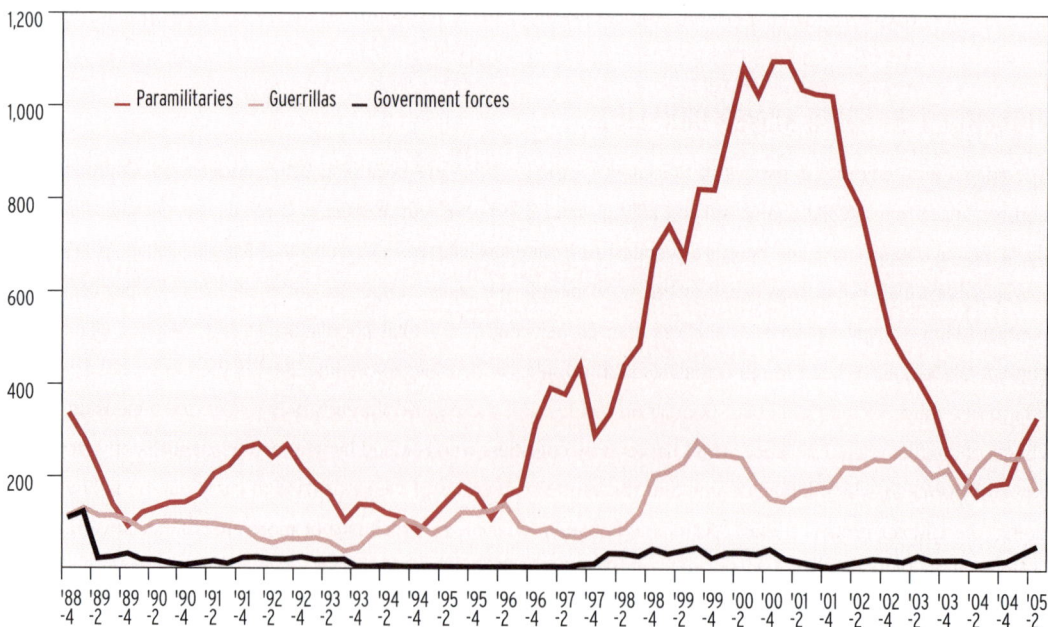

— Paramilitaries — Guerrillas — Government forces

1,200

1,000

800

600

400

200

'88 '89 '89 '90 '90 '91 '91 '92 '92 '93 '93 '94 '94 '95 '95 '96 '96 '97 '97 '98 '98 '99 '99 '00 '00 '01 '01 '02 '02 '03 '03 '04 '04 '05
-4 -2 -4 -2 -4 -2 -4 -2 -4 -2 -4 -2 -4 -2 -4 -2 -4 -2 -4 -2 -4 -2 -4 -2 -4 -2 -4 -2 -4 -2 -4 -2 -4 -2

Note: For each group (curve) in the figure, the number for every quarter is the sum of civilian killings for that group in the previous four quarters.

Source: CERAC Colombian Conflict Database

the course of their attacks. In fact, paramilitaries shoot in 70 per cent of their attacks compared to 11 and 10 per cent for the guerrillas and the government, respectively. This behaviour corresponds with the paramilitaries' apparent abundance of ammunition and weapons.[39]

Parsing out the role of specific weapon types in clashes and attacks by the state forces and non-state armed groups can generate insight into appropriate controls and restrictions. For 7,100 out of a total of 21,000 incidents, the type of weapon used was specifically documented. Of the remaining incidents, some 6,633 consisted of 'clashes' during which assault rifles were inevitably used. By adding these two categories, it is possible to discern weapons types in almost two-thirds of all reported conflict incidents since 1988.

All told, 14 specific categories of armament are deployed in conflict incidents. This varied set of categories ranges from home-made armoured vehicles, blunt objects, and sharp or bladed objects, to explosives and gas canisters. Firearms were the most widely documented instrument used in conflict. The most common types of weapons used include assault rifles and grenade launchers, followed by pistols, sub-machine guns, and mortars. Rifles appear to have been used in most of the high-impact incidents (see Table 9.1). Rifle incidents tend to produce more killings than injuries, whereas explosive incidents tend to generate the opposite trend.

Table 9.1 Relationship between weapon types and casualties in conflict incidents, 1 January 1988–30 June 2005

Casualty type	Casualty range	Firearms[1]	Assault rifle	Bladed weapons	Mines	Bombs and gas canisters	Other explosives[2]	Fire and bladed weapons	Combination[3]	Total
		Number of incidents								
Killed	0	537	1,225	3	255	133	2,506	2	98	4,759
	1–10	1,996	5,473	28	210	67	468	39	104	8,385
	11–25	88	123	0	1	11	21	8	14	266
	26–74	12	22	1	2	2	4	5	2	50
	75 or more	1	2	0	0	5	0	0	1	9
		Number of incidents								
Injured	0	1,990	5,269	29	130	128	2,336	48	102	10,032
	1–10	620	1,539	3	332	71	570	5	106	3,246
	11–25	17	31	0	5	14	62	1	10	140
	26–74	7	5	0	1	4	27	0	1	45
	75 or more	0	1	0	0	1	4	0	0	6

Notes:

1 Does not include assault rifles.

2 Includes explosives not enclosed in casing such as Anfo, C4, Dynamite, and Gelatinous dynamite among others but not bombs, mines, or gas canisters.

3 Two or more methods but not firearms or bladed weapons.

Source: CERAC Colombian Conflict Database

Graph 9.5 **Evolution of small arms use in armed conflict, 1 January 1988–30 June 2005**

RECORDED NUMBER OF INCIDENTS   RECORDED NUMBER OF INCIDENTS

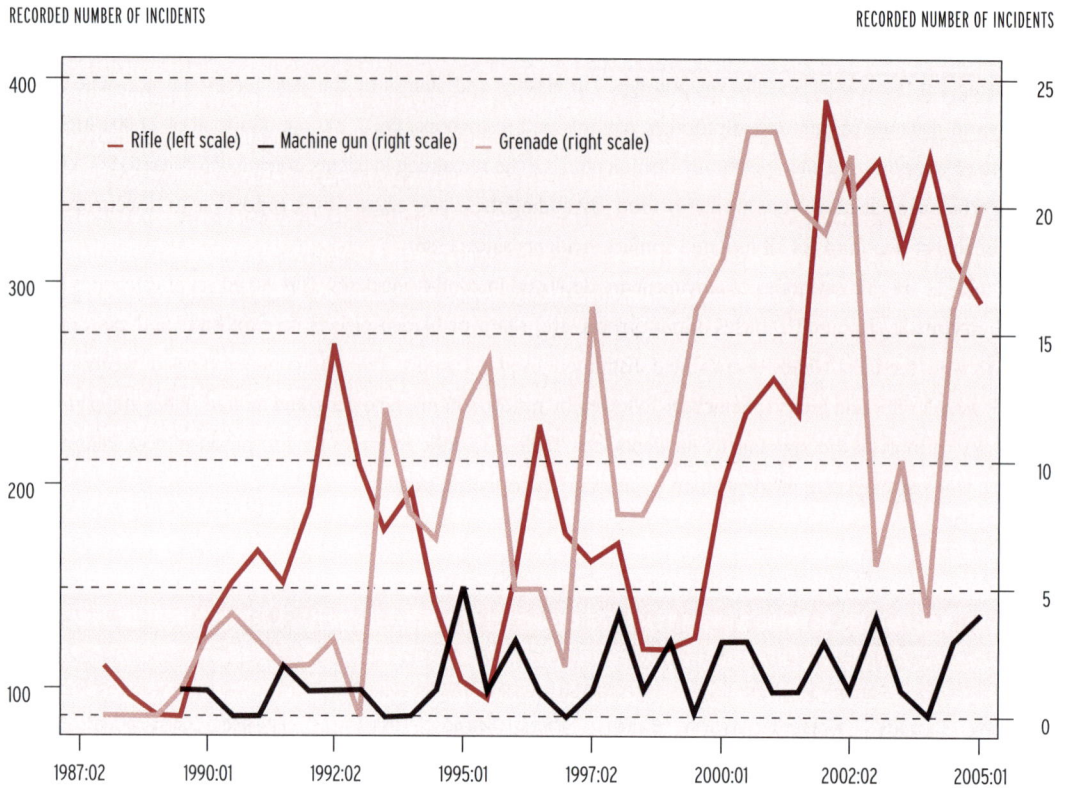

— Rifle (left scale)   — Machine gun (right scale)   — Grenade (right scale)

Source: CERAC Colombian Conflict Database

The specific types of weapons used in Colombia's armed conflict have evolved considerably over the past two decades. Graph 9.5 documents the number of times weapons of various types were reportedly used between 1988 and mid-2005. It shows a peak in grenade use during 1995, followed by a significant surge in rifle use between 1999 and 2001. This second peak coincides with the introduction by FARC of large numbers of Jordanian-sourced AKM assault rifles into the country (*El Tiempo*, 2004). This period also witnessed the consolidation of various paramilitary groups into the AUC, as described above. In addition, the graph reveals that there has been a considerable decline in rifle use since 2002: this steep reduction has coincided with a dramatic fall in conflict-related casualties and reported conflict incidents.[40]

The geographical distribution and types of firearms used also shed light on the dynamics of Colombia's armed conflict. Between 1988 and mid-2005, for example, light weapons were most commonly used in Antioquia, Cauca, and the regions of the Sierra Nevada, Catatumbo, and Magdalena Medio (see Map 9.1). Considerable levels of armed violence and population displacement accompanied gun and drug trafficking on the routes discussed above, particularly near the Venezuelan border and the ports of Buenaventura and Urabá. The geographical distribution of explosive use closely resembles that of firearms, though it exhibits a much higher concentration and frequency along the Venezuelan border where there are oil pipelines.[41]

Map 9.1 **Firearm use in conflict incidents by municipality, 1 January 1988–30 June 2005**

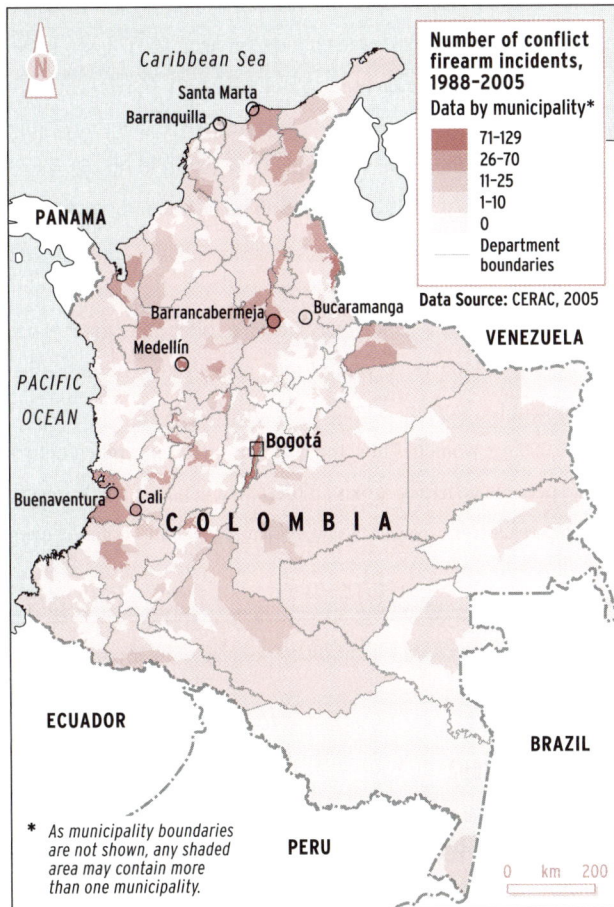

Number of conflict firearm incidents, 1988–2005
Data by municipality*

- 71–129
- 26–70
- 11–25
- 1–10
- 0
- Department boundaries

Data Source: CERAC, 2005

* As municipality boundaries are not shown, any shaded area may contain more than one municipality.

CRIMINAL VIOLENCE AND THE BURDEN OF FIREARMS

Although most international commentary on Colombia's human security crisis focuses on the armed conflict, organized and petty criminality is where guns exert their gravest toll. In contrast to conflict casualties, which are primarily rural (Restrepo and Spagat, 2005b, p. 15), criminal violence is a predominantly urban phenomenon. In fact, between 1979 and 2002, between 70 and 80 per cent of all firearm-related deaths occurred in urban areas.

There is considerable disagreement over the exact dimensions of Colombia's criminal violence. This debate is fuelled in part by separate data sets on criminal violence and firearm-related deaths, which are produced by three distinct authorities: the state statistical agency (DANE), the National Police Centre for Criminological Research (CIC), and the office of the medical examiner (INML). Despite protracted disagreements between agencies and with the human rights community, these three entities have reported remarkably similar trends since the mid-1990s so that the choice of data set is not as signifi-

cant as is often believed. The following section considers DANE data, which records some 475,000 people killed as a result of firearms use in homicides, suicides, accidents, and undetermined incidents between 1979 and 2005, some 11 per cent of all deaths over the period.[42]

While there is a lack of consensus over the magnitude of firearm-related deaths, there is unanimous agreement that the proportion of firearm-related deaths as compared to all naturally or externally occurring deaths is extremely high. In Colombia, firearm deaths are more than five times higher than in other comparator countries of Latin America, including Mexico, where firearm deaths constitute two per cent of all deaths.[43] There has also been considerable variation in this relationship over time, with firearm deaths rising from 3 per cent in 1979 to 15 per cent in 2002. In fact, the absolute number of firearm deaths increased an astonishing sevenfold during the same period—from 3,617 to 28,989—with the rate per 100,000 inhabitants increasing fourfold, from 16 to 66 (see Annexe 2). Firearm death rates peaked at 70 per 100,000 in 1991 at the height of the narco-trafficking era and declined dramatically over the last three years to 29 per 100,000 in 2005 (see Annexe 3).[44] Graph 9.1 highlights a parallel shift in homicides and firearm-related homicides; a similar relationship can be found in the total numbers of annual deaths in relation to firearm deaths, in terms of rates per 100,000. Thus, while there are some relatively constant causes of death (e.g. heart disease), variations in total numbers of annual deaths are essentially variations in gun deaths.

Firearms have thus played a central role in overall external causes of death. Of the more than 509,000 weapons-related deaths reported between 1979 and 2002, about 83 per cent were firearms-related (427,204), 16 per cent were attributable to bladed weapons, and less than one per cent were tied to explosives (see Annexe 4).[45] Perhaps more important, the percentage of all external deaths associated with firearms rose steeply until 1991 and has remained relatively constant since. Firearms thus currently account for more than 80 per cent of all homicides in Colombia, 36 per cent of all suicides, and more than 2 per cent of all accidents. In fact, firearm deaths account for fully 50 per cent of the variation over time in all external death rates.

It is also evident that criminal violence perpetrated with firearms is gender-specific. Men suffer more than 90 per cent of all gun deaths. In fact, firearms account for an astonishing 18 percent of total male deaths (of all causes, whether natural or external) compared to only 1.9 per cent for women. Annual male gun-death rates have ranged between 70 and 131 per 100,000 versus 5 to 10 per 100,000 for women. Thus, the absolute variation is much higher for men than for women, although they vary similarly in percentage terms. Annexe 5 presents a gender-specific review of the proportion of deaths from all causes, external causes, and firearms between 1980 and 2002 and demonstrates that women account for only seven per cent of firearm deaths.

Criminal violence perpetrated with firearms is also concentrated among specific demographic clusters of the population. It appears that men between 20 and 29 years of age account for almost 40 per cent of all firearm-related

Graph 9.6 **Demographic distribution of firearm death, 1979–2002**

PROPORTION OF TOTAL   ■ Firearms ■ Other external causes ■ Natural causes

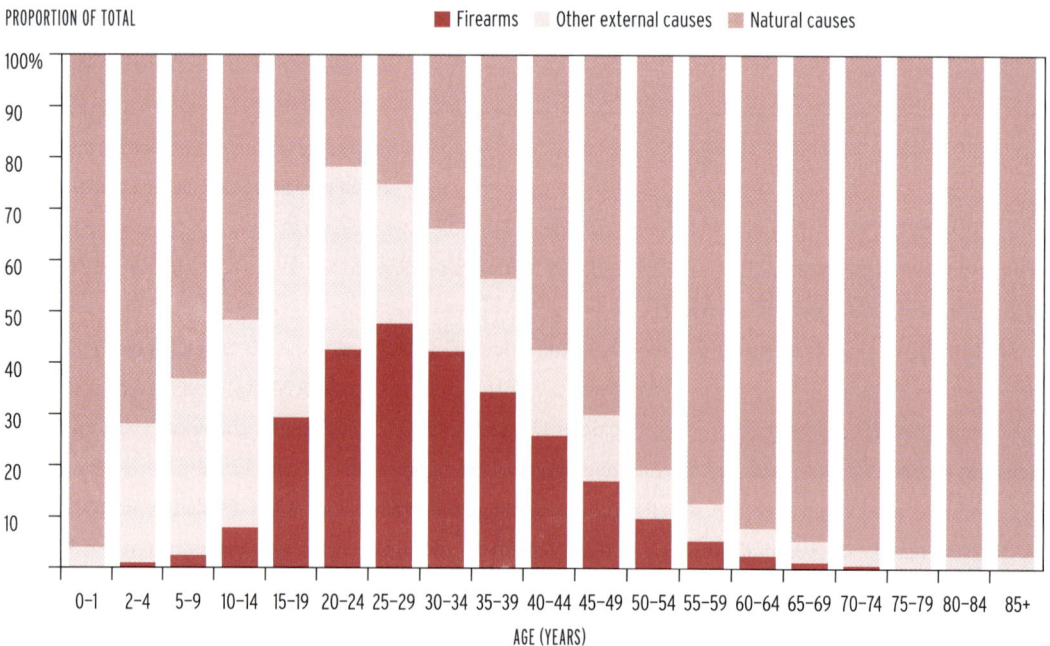

AGE (YEARS)

Source: DANE; processed by CERAC

Map 9.2 **Rates of firearm death by municipality, 1979–2002**

Map legend:

Firearm death rates per 100,000 inhabitants, 1979-2002
- 111-211
- 61-110
- 1-60
- 0
- No data
- Department boundaries

Data Source: DANE, 2005

As municipality boundaries are not shown, any shaded area may contain more than one municipality.

deaths. Less than four per cent of firearm deaths are of people younger than 14 or older than 65. Graph 9.6 provides an overview of the age distribution of firearm victims in relation to deaths from external and natural causes. It observes that firearm exposure rises sharply from around the age of 14. People between 16 and 39, particularly the 20 to 24 range, experience the highest proportions of deaths due to external causes in total deaths (77 per cent) and firearm deaths in total deaths (nearly 50 per cent). Further, the graph shows that before age 14 and after age 40, most deaths are due to natural causes, while the reverse is true within this age interval. Firearm deaths are the principal contributing factor in this dynamic.

Firearm violence is concentrated predominantly in densely populated areas. This stands in sharp contrast to conflict violence which is predominantly rural. The large cities of Medellín, Bogotá, Cali, and Barranquilla are particularly susceptible to firearm-related violence, accounting for almost one-third of all firearm deaths over the past three decades.[46] Much smaller cities such as Cúcuta, Bello, Itagüí, Pereira, and Manizales have also suffered unusually high firearm death rates.[47] Some conflict-affected areas such as north-western Antioquia (Urabá), Casanare, Arauca, Putumayo, and Meta also present very high rates of gun violence. Moreover, firearm death rates vary widely by municipality (see Map 9.2). For example, in the municipalities of Granada, San Luis, and Cocorná (east of Antioquia), and Vistahermosa (Meta) firearm death rates ranged between 515 and 640 per 100,000 inhabitants in 2002, compared to Majagual (Sucre), Uribía (La Guajira), and El Tambo (Nariño), with rates between 2.64 and 3.44 per 100,000 in the same year (see Annexe 6).[48]

The full impact of firearm violence on Colombia's society and economy is difficult to quantify precisely, but simple calculations hint at its enormous dimension. Young men between 15 and 35 are disproportionately affected by a wide margin, so the impact on both formal and informal productivity is extensive. This chapter finds that some 342,253 potential years of life were lost due to intentional firearm deaths in 1985 alone. Moreover, a male Colombian born in 1985 has a five per cent chance of dying from firearms over the course of his lifetime, as compared to a nine per cent chance of dying from any other external cause. Thus, under present conditions, male Colombians born in 2002 will lose some 40 months of life on average due to firearm violence.[49] The socio-economic consequences of this loss on the labour force are likely to be significant, coming just after these victims have finished their tax-supported educations. Moreover, it is plausible that some urban men have invested lightly in workforce preparation due to the fact that they are unlikely to have long lives during which to exploit such efforts. Clearly, reducing gun violence is an urgent public policy priority for Colombia and for the international community more generally.

MUNICIPAL ARMS REDUCTION AND DDR

There are arguably few more important policy interventions to reduce armed violence—both conflict and criminal-related—than arms control. But there is also a wide bandwidth of control policies, ranging from prohibitions on domestic production, imports, exports, and civilian controls to military and police enforcement measures, legislative reform, and changes in incarceration policy. Previous editions of the *Small Arms Survey* have reflected on the value of international, regional, and national regulatory frameworks for preventing armed violence. This chapter focuses instead on several examples of innovative municipal efforts to reduce gun violence at the local level, as well as aspects of the controversial DDR process under way since 2004, under which many of the paramilitary groups have demobilized.

Municipal arms reduction

Despite suffering one of the highest rates of armed violence in the world, Colombia has recently witnessed a dramatic decline in firearm-related deaths and victimization. This development warrants serious attention, although it is too early to identify a real trend. Analysts have tied declines in gun violence in Bogotá, Cali, and Medellín to identifiable municipal policy interventions that might be usefully applied elsewhere, inside and outside Colombia (Guerrero, 1999; Villaveces et al., 2000; Llorente et al., 2000).

Colombia has recently witnessed a dramatic decline in firearm-related deaths and victimization. In the mid-1990s, when national homicide rates were peaking, local authorities of large cities experimented with plans to reduce arms-related violence. Rodrigo Guerrero, then mayor of Cali, introduced temporary restrictions on the carrying of firearms, alcohol prohibitions, and other security guarantees such as road checks and increased police presence. The carrying restrictions were often introduced during festivals and on other occasions such as during weekends, prior to pay-day, or late at night (Guerrero, 1999). A similar tactic was employed by Jaime Castro, a former mayor of Bogotá (1992–94), when he invoked obscure legal mechanisms to suspend civilian carrying licenses in certain situations, particularly on election days.

But one of the most impressive examples of municipal arms control emerges from Bogotá. Antanas Mockus, concerned with spiraling levels of armed violence in the early 1990s, was elected mayor of Bogotá in 1995 with violence reduction as the central issue in his campaign. During his two terms (1995–96 and 2001–03) he responded to record-low levels of confidence in public security provision by introducing alternative approaches for citizens to protect themselves and their families. As part of a programme called 'security for everyone', and with the support of the Catholic Church and the police, he restricted carrying licenses during weekends and evenings throughout the city. Mockus introduced police roadblocks, both to reduce drunk driving and to undertake gun searches.

Opinion surveys have been conducted in order to investigate perceptions of safety and attitudes towards the preference to carry a firearm to feel secure. About 25 per cent of people surveyed by the local Secretary of Culture

Box 9.1 A snapshot of firearm homicides in Colombia, 2003–2005

There is a more up-to-date alternative to the DANE data applied elsewhere in this chapter. These include the national police crime database or SIEDCO (Sistema de Información Estadistico Delincuencial, Contravencional, y Operativo), which processes information from the 36 police departments throughout the country. Between 2003 and 2005, for example, SIEDCO recorded 61,299 homicides, of which 52,028 were reportedly caused by firearms. Table 9.2 details annual homicides, while also disaggregating these according to the type of weapon used in the incident. Firearms are far and away the leading cause of homicide, followed by bladed weapons.

SIEDCO also gathers information on the time and date of an incident. Graph 9.7 demonstrates that homicides peak at 8 p.m. with a follow-on peak at 10 a.m. Homicides also appear to peak on weekends, suggesting that there are potentially compelling benefits from restricting weekend carrying permits. SIEDCO also collected detailed data on almost 8,000 crimes and offences involving the use of firearms in the first trimester of 2005. Though an incomplete and imperfect sample, Table 9.3 provides a breakdown of these figures by type of weapon. While they reflect a modest sample, these statistics point to the dominant role played by revolvers, shotguns, and pistols. Police officials revealed that more than 95 per cent (7,594) of all the cases involved firearms without permits.

Table 9.2 Homicides by type of weapon in Colombia, 2003–05

Type	2003	2004	2005	2003–05
Firearm	20,058	17,208	14,510	51,776
Sharp object	2,257	2,368	2,486	7,111
Blunt object	309	310	287	906
Explosive material, smoke, fire and flames	347	197	317	861
Hanging, strangulation, and suffocation	32	67	107	206
Other	10	13	5	28
Poison	2	10	14	26
Total	**23,015**	**20,173**	**17,726**	**60,914**
Rate per 100,000 inhabitants				
Firearm homicides	44.99	37.97	31.54	38.34*
Total homicides	51.62	44.51	38.53	45.16*

Source: National Police–CIC (SIEDCO) and DANE; processed by CERAC

* Average rate 2003–05

Graph 9.7 A question of timing: homicides and firearm homicides, 2003–05

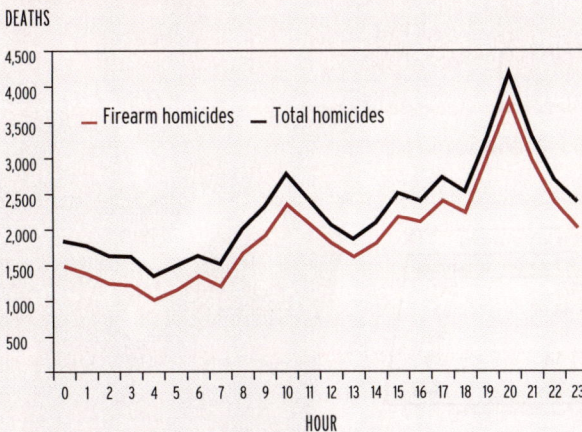

Source: National Police–CIC (SIEDCO); processed by CERAC

Table 9.3 Weapons used in crime and offences: January–April 2005

Category	Quantity
Revolver	4,317
Shotgun	2,243
Pistol	1,243
Assault rifle	78
Carbine	33
Sub-machine gun	19
Other firearm types	9
Mortar	4
Machine gun	2
Bazooka	2
Cohete	3
Grenade launcher	2
Total	**7,955**

Source: National Police–CIC (SIEDCO); processed by CERAC

in 2002 said that it was important to protect oneself with a firearm. In a subsequent survey, carried out in 2003, after the start of the implementation of Mockus's disarmament plan, the response fell to ten per cent. Finally, the Quality of Life Survey of 2003 revealed that disarmament campaigns were one of the measures that made people feel the safest: at least two-thirds of all respondents asserted that disarmament campaigns increased their perception of security.

Related research has explored the impact of active anti-gun policies and other security interventions—particularly those launched in the mid-1990s—on reducing firearm-related homicides in Bogotá. These studies have found a statistical dependency between specific arms control restrictions and reduced homicides on weekends. In other words, it appears that the temporary bans on carrying firearms, strongly enforced on weekends, yielded a positive effect (Aguirre et al., 2005, p. 26).

In the absence of strong national backing, the mayor of Bogotá drew on voluntary support and private funding to launch a buyback programme in 1995 and 1996. As there is no federal regulatory framework to support efforts by municipal institutions to collect firearms, the Church and its parishes were harnessed, with priests negotiating with gang leaders to turn in their weapons. The mayor's office simultaneously launched a series of measures that, while viewed by some as mere political stunts and political opportunism, may actually have contributed to altering the culture of violence in the city. These included an initiative of exchanging toy guns for poetry books, which is still actively endorsed by the police, and another programme of rendering spoons from guns. There have been no robust statistical studies of the impact of these distinct measures on reducing homicide, though few can dismiss the dramatic 26 per cent decline in Bogotá's homicide rate experienced during Mockus's tenure (1995 to 1996).[50]

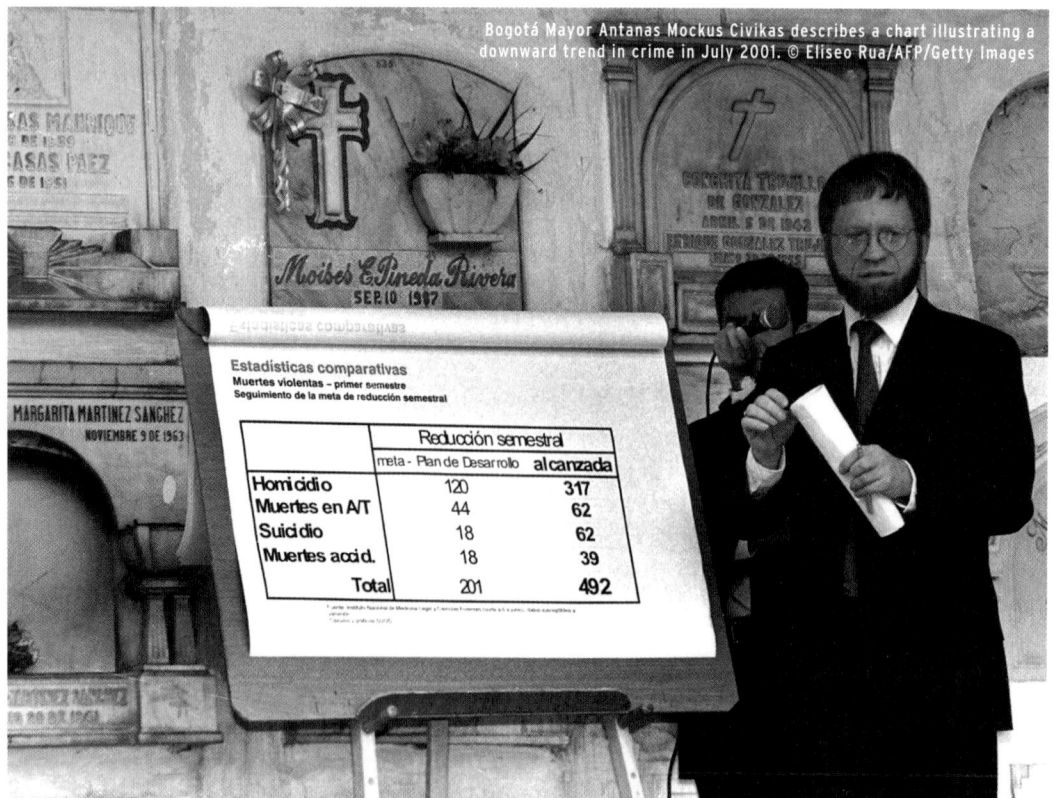

Bogotá Mayor Antanas Mockus Civikas describes a chart illustrating a downward trend in crime in July 2001. © Eliseo Rua/AFP/Getty Images

The physical collection of weapons represents only one feature of these various interventions. In fact, only a modest number of firearms were decommissioned in Bogotá during the 1990s. During the Mayor's programme, returned arms included 200 to 300 *changones* from 20 parishes, 2,300 firearms, and more than 800 grenades that were destroyed on Mother's Day towards the end of his first term. While fewer than 4,000 firearms have been destroyed in this programme during the past 10 years, the impact of these interventions cannot be measured solely by the number of weapons collected (*El Tiempo,* 2005d). The collection and destruction of firearms yielded important symbolic effects that were widely felt—'a flip side to Santander department exhibitionism and machismo', in the words of one local politician.

Disarmament, demobilization, and reintegration

A series of DDR processes have been applied to various rebel groups over the past few decades. In fact, successive Colombian administrations have eagerly pursued DDR since the 1950s, after the relatively positive DDR experience during and after the period known as *La Violencia.*[51] There have been nine separate 'collective' DDR programmes with distinct guerrilla groups since the 1970s. Since 1990, some 7,300 ex-guerrillas have been collectively disarmed, and 4,715 of them entered reintegration programmes.[52] Specific DDR frameworks have varied from negotiation to negotiation—and no formulaic or standardized approach has yet emerged.[53]

To be sure, DDR is an intrinsically complicated and ambitious exercise in any context (Small Arms Survey, 2005, ch. 10; Muggah, 2005). Colombia is a case in point: a vast array of operational agencies and departments are responsible for the execution of various components of DDR. For example, the Dirección General de Reinserción (Reinsertion Department) under the Ministry of the Interior is directly responsible for the DDR of individuals covered under various peace agreements.[54] The Instituto Colombiano de Bienestar Familiar (Colombian Institute of Family Welfare) ensures that the rights of youth and minors are respected during the DDR process. Meanwhile, the Ministry of Defence offers humanitarian assistance to demobilized people. The Fiscalía General de la Nación (National General Prosecutor's Office) is charged with defining the legal situation of adult ex-combatants. The costs of ensuring that these agencies are coordinated and effective are tremendous. All told, the government spent more than USD 94 million on DDR between 1998 and 2002.

By March 2006, more than 22,000 paramilitaries had turned in more than 17,000 weapons.

Despite the lack of evidence, it is widely believed that investments in DDR have reduced violence. It is true that properly demobilized groups are expected to stop killing people. However, DDR could also be blamed for the sharp rise in illegal paramilitary groups that began towards the end of the 1990s. In particular, the rise has been linked to outrage in some right-wing circles concerning the fact that left-wing guerrillas were demobilizing without penalties. The extent to which some of these formerly demobilized individuals resumed activities with armed groups, including the paramilitaries, has yet to be determined. While it is plausible that DDR has reduced violence over the years, there has not been a single study to carefully review the causal links between DDR and resumed livelihoods or reductions in armed violence or criminality.

The DDR process in Colombia is exceedingly controversial. In December 2002 most paramilitary groups initiated a ceasefire with the government, albeit one they have not always adhered to on the ground. This led to a suspension of hostilities with the government and paved the way for politically volatile negotiations in 2003. As early as January 2003, the first surrendered weapons were trickling in, many of them of high-quality but with serial numbers erased. By March 2006, more than 22,097 ex-combatants had turned in more than 17,600 weapons (Alto Comisionado para la Paz, 2005).[55] Despite involvement of the Organization of American States (OAS), the US and Canadian governments, the International Organization for Migration (IOM), and a host of others, the process has been subjected to withering criticism, especially from human rights organizations. Most critics argue that the process began before an adequate legal framework was put in place and also that the measures ultimately installed were too soft on the paramilitaries (HRW, 2005, p. 203).[56] Government officials have reacted to this criticism with frustration, pointing to the country's long history of impunity deals with left-wing groups, none of which provided any reparations for victims.

Though few of the DDR critics have addressed these apparent inconsistencies, there are nevertheless two compelling reasons for treating the paramilitaries differently from the guerrilla groups of the past. First, as this chapter has shown, the paramilitaries have been by far the most vicious faction in the Colombian armed conflict since the late 1990s. Second, in the present environment, weak punishment for paramilitaries can be challenged in the Colombian Constitutional Court, the Inter-American Court for Human Rights, and even the International Criminal Court.[57] Paramilitaries who have committed or ordered massacres will not voluntarily sign up for prison terms that are proportional to their crimes. In fact, the law may go too far in effectively capping sentences at 6.5 years. The paramilitaries are comparatively poor fighters whose negotiating position might soften up under military pressure (Restrepo and Spagat, 2005b, p. 67).

The government should respond more vigorously than it has to ceasefire violations committed by paramilitary groups currently sitting at the negotiating table. But it would be costly to pursue an exclusively military solution to paramilitarism and, therefore, some leniency is an inevitable component of any demobilization deal. The main problem is that the current Colombian administration has neither the time nor the resources to determine which particular crimes were committed by various paramilitaries, who the victims are, and what property has been stolen. Nevertheless, these issues can be addressed during the Justice and Peace law's implementation phase with sufficient resources, political will, and international engagement.

Bogotá, Medellín, and Cali have shown drastic reductions in homicide rates in recent years.

CONCLUSION

Colombia has been an extraordinarily violent place for a long time. This chapter has shown that firearms have played a central role in many of the country's challenges. Weapons of various calibres are the primary tool of those who murder, kidnap, and forcibly displace. Conflict violence is predominantly rural while criminal violence is mainly urban. Young males are the primary victims of gun violence. Left-wing guerrillas, right-wing illegal paramilitaries, drug dealers, and the government forces that oppose them are all well armed with relatively sophisticated weapons. Colombian civilians are also armed, though less so than their neighbours.

This chapter has found that easy availability of small arms and light weapons has been a major contributor to the onset, lethality, and scale of both criminal and conflict violence in Colombia. On average, more than half of the variation over time in external death rates is significantly explained by the variation in firearm death rates. Yet despite Colombia's severe problems there are real grounds for hope. In recent years there have been substantial reductions in homicides in several of the country's biggest cities. Some of the policy initiatives that have contributed to these improvements can be replicated elsewhere and extended, not just in Colombia.

The Colombian cities of Bogotá, Medellín, and Cali have shown drastic reductions in homicide rates in recent years, with gun control being central to their success. These experiences demonstrate that gun control policy can and had yielded a substantial impact even within an overall context of rampant violence. It should be noted, however,

National Police mascot Segurito (Safie) shows the press seized weapons in Medellín, December 2004.
© Gerardo Gomez/AFP/Getty Images

that the reductions in violence have been achieved principally in more densely populated areas, where the state has established a strong presence. In isolated areas, the state holds little sway and the conflict continues unbridled. Conflict violence, especially pertaining to civilians, had been greatly reduced in 2003 and 2004 but many of these gains were reversed in the first half of 2005. All in all, the country's major successes have been limited in scope.

Future progress depends largely on Colombia's paramilitaries, who have been the main perpetrators of conflict violence against civilians in recent years. As this chapter has shown, they are also more deeply involved in trading narcotics and weapons than the guerrillas, and are thus armed groups of major international significance. If they can be successfully demobilized and their criminal activity kept in check, the country could look forward to a future much less violent than its past. If, instead, the paramilitaries transition from a mixture of counterinsurgency and criminality into pure criminality, Colombian violence could reach new heights. Paramilitary DDR is, therefore, critical and should become a major focus of international attention. ◾

ANNEXE 1. TOP TEN COUNTRIES EXPORTING ARMS AND PARTS TO COLOMBIA, 1994–2005*

Country	Export value (USD)*
United States	195,887,011
South Africa	163,678,531
Israel	145,348,075
Brazil	50,364,170
Belgium	14,958,391
Italy	12,857,171
France	12,165,775
Czech Republic	11,159,777
United Kingdom	10,161,063
Spain	6,635,414
Other countries	51,677,648
Total top 10 countries	623,215,378
Total	**674,893,026**

* Cumulative value for January–October 2005

Source: National Tax Administration; processed by CERAC

ANNEXE 2. TOTAL, EXTERNAL, AND FIREARM DEATHS IN COLOMBIA, 1979–2005

Year	Total	External causes	Firearm	External causes/ total (%)	Firearm/total (%)	Firearm/external (%)
1979	110,400	15,680	3,617	14	3	23
1980	125,573	18,898	4,980	15	4	26
1981	139,505	22,084	6,552	16	5	30
1982	137,678	22,685	7,127	16	5	31
1983	140,292	23,767	7,343	17	5	31
1984	137,189	24,455	8,211	18	6	34
1985	153,947	29,218	11,505	19	7	39
1986	146,345	30,210	13,472	21	9	45
1987	151,957	32,179	14,780	21	10	46
1988	153,065	34,995	17,447	23	11	50
1989	154,694	36,228	18,947	23	12	52
1990	156,314	38,107	20,569	24	13	54
1991	162,063	43,066	24,941	27	15	58
1992	167,743	44,395	25,084	26	15	57

1993	168,647	44,621	24,572	26	15	55
1994	168,568	43,287	23,118	26	14	53
1995	169,896	41,532	21,313	24	13	51
1996	173,506	42,307	23,062	24	13	55
1997	170,753	41,087	22,222	24	13	54
1998	175,363	42,823	21,950	24	13	51
1999	183,553	43,959	23,320	24	13	53
2000	187,432	46,031	26,465	25	14	57
2001	191,513	47,175	27,618	25	14	59
2002	192,262	48,438	28,989	25	15	60
2003*	189,073		19,624	20	10	50
2004*	194,788		16,951	18	9	45
2005*			13,494			
Total 1979–2002	**3,818,258**	**857,227**	**427,204**	**22**	**11**	**50**
Change from 1979 to 2002	**74.2%**	**208.9%**	**701.5%**	**77.4**	**360.2**	**159.4**

Notes: * Projected

Source: DANE and National Police–CIC; processed by CERAC

ANNEXE 3. TOTAL, EXTERNAL, AND FIREARM DEATH RATES PER 100,000 INHABITANTS, 1979–2005

Year	Total	External causes	Firearms
1979	397	56	13
1980	441	66	18
1981	480	76	23
1982	463	76	24
1983	462	78	24
1984	442	79	26
1985	486	92	36
1986	453	94	42
1987	461	98	45
1988	455	104	52
1989	451	106	55
1990	447	109	59
1991	454	121	70

1992	461	122	69
1993	454	120	66
1994	445	114	61
1995	441	108	55
1996	442	108	59
1997	426	103	55
1998	430	105	54
1999	441	106	56
2000	443	109	63
2001	445	110	64
2002	439	111	66
2003*	424		44
2004*	430		37
2005*			29

Notes: * Projected

Source: DANE and National Police; processed by CERAC

ANNEXE 4. DISAGGREGATING WEAPON TYPES AND DEATHS IN COLOMBIA, 1979–2002

NUMBER OF DEATHS

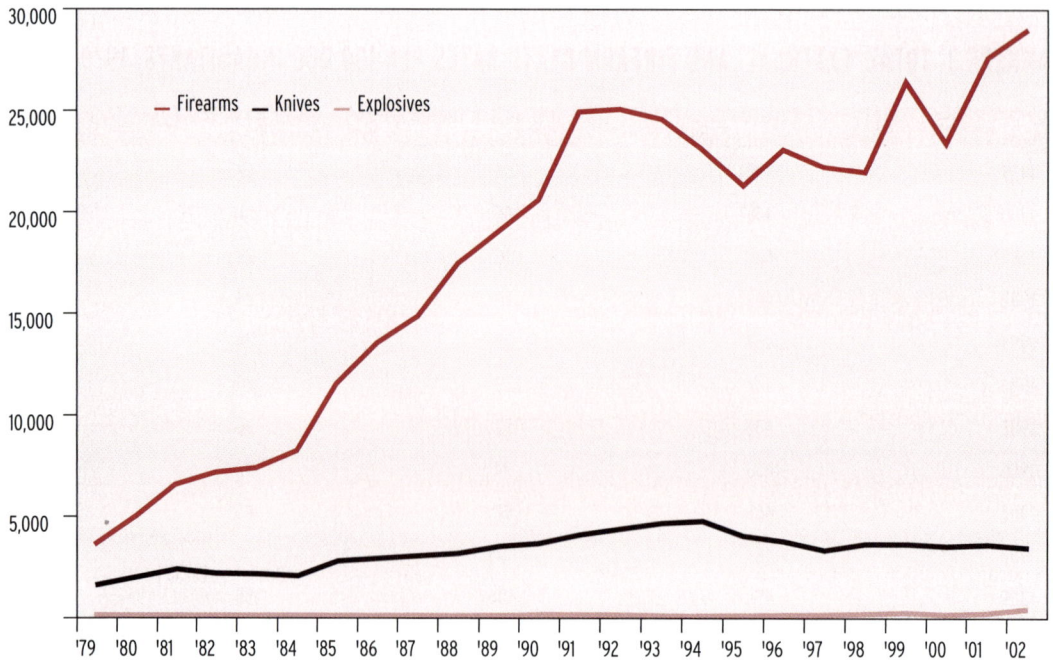

Source: DANE; processed by CERAC

ANNEXE 5. PROPORTION OF MALE AND FEMALE VICTIMS FOR ALL CAUSES, EXTERNAL CAUSES, AND FIREARMS, 1980–2002

Year	Total (%)		External causes (%)		Firearms (%)	
	Men	Women	Men	Women	Men	Women
1980	56	44	81	19	93	7
1981	56	44	83	17	93	7
1982	56	44	83	17	93	7
1983	56	44	83	17	94	6
1984	57	43	83	17	94	6
1985	57	43	84	16	94	6
1986	58	42	86	14	94	6
1987	59	41	86	14	94	6
1988	59	41	87	13	94	7
1989	60	40	87	13	93	7
1990	60	40	87	13	93	7
1991	61	39	88	12	93	7
1992	61	39	88	12	93	7
1993	61	39	87	13	93	7
1994	61	39	87	13	93	7
1995	60	40	86	14	93	7
1996	60	40	87	13	93	7
1997	60	40	87	13	93	7
1998	60	40	87	13	92	7
1999	60	40	86	14	93	7
2000	61	39	87	13	93	7
2001	60	40	87	13	92	7
2002	60	40	87	13	92	7
Total	**59**	**41**	**86**	**14**	**93**	**7**

Source: DANE; processed by CERAC

ANNEXE 6. FIREARM DEATH RATES BY MUNICIPALITY, 1985, 1993, 2002, AND AGGREGATE

1985				
Department	Municipality	Population	Firearm deaths	Rate per 100,000
Boyacá	La Victoria	3,017	15	497.18
Arauca	Cravo Norte	3,557	13	365.48
Antioquia	Salgar	22,652	73	322.27
Huila	Altamira	2,822	8	283.49

Guaviare	San José Del Guaviare	41,476	115	277.27
Meta	Puerto Lleras	9,576	25	261.07
Arauca	Puerto Rondón	1,985	5	251.89
Boyacá	Muzo	11,567	29	250.71
Arauca	Saravena	24,417	61	249.83
Risaralda	Balboa	8,953	22	245.73
Sucre	Corozal	47,329	1	2.11
Santander	Girón	53,547	1	1.87
Córdoba	Cereté	58,605	1	1.71
Tolima	Espinal	58,696	1	1.70
Atlántico	Soledad	170,854	2	1.17
	National totals	**31,658,715**	**11,505**	**36.34**
1993				
Meta	El Castillo	3,104	15	483.25
Antioquia	Apartadó	78,019	312	399.90
Casanare	Sácama	1,139	4	351.19
Santander	Sabana De Torres	20,000	68	340.00
Antioquia	Chigorodó	44,201	148	334.83
Boyacá	La Victoria	1,571	5	318.27
Cundinamarca	San Cayetano	5,464	17	311.13
Boyacá	Pauna	9,752	27	276.87
Antioquia	Medellín	1,834,881	5,000	272.50
Valle del Cauca	El Cairo	9,589	26	271.14
Córdoba	Lorica	120,961	2	1.65
Atlántico	Soledad	257,650	4	1.55
Atlántico	Sabanalarga	73,409	1	1.36
Bolívar	El Carmen De Bolívar	74,836	1	1.34
Atlántico	Malambo	75,807	1	1.32
	National totals	**37,127,295**	**24,607**	**66.28**
2002				
Antioquia	Granada	17,326	111	640.66
Meta	Vistahermosa	19,781	105	530.81
Antioquia	San Luis	16,445	87	529.04
Antioquia	Cocorná	21,552	111	515.03
Caquetá	El Paujil	16,833	83	493.08
Caquetá	Curíllo	14,700	70	476.19
Norte de Santander	Tibú	39,977	182	455.26
Meta	San Juan De Arama	10,426	47	450.80
Meta	Uribe	9,730	42	431.65

Caquetá	Solita	9,874	39	394.98
Nariño	Guaitarilla	28,184	1	3.55
Córdoba	Pueblo Nuevo	28,221	1	3.54
Nariño	El Tambo	29,044	1	3.44
La Guajira	Uribia	66,957	2	2.99
Sucre	Majagual	37,885	1	2.64
	National totals	**43,834,117**	**28,899**	**65.93**
Aggregate 1979–2002. Population of 2002				
Risaralda	Balboa	7,372	358	211.14
Antioquia	Valdivia	11,963	542	196.98
Antioquia	Salgar	18,110	815	195.66
Caldas	Viterbo	18,684	839	195.24
Boyacá	Muzo	16,445	727	192.21
Boyacá	La Victoria	1,311	57	189.04
Antioquia	Remedios	17,658	746	183.68
Antioquia	Apartadó	96,039	4,047	183.21
Antioquia	Olaya	2,686	110	178.06
Caquetá	Curillo	14,700	598	176.87
Bolívar	Hatillo De Loba	12,701	2	0.68
Nariño	Nariño	6,441	1	0.68
Chocó	Bajo Baudó	14,062	2	0.62
Chocó	Río Iro	7,184	1	0.61
Magdalena	Zapayán	8,944	1	0.49
	National totals	**43,834,117**	**427,204**	**42.37**

Source: DANE; processed by CERAC

LIST OF ABBREVIATIONS

AUC	Autodefensas Unidas de Colombia (United Self-Defence Forces of Colombia)	FARC	Fuerzas Armadas Revolucionarias de Colombia (Revolutionary Armed Forces of Colombia)
CERAC	Conflict Analysis Resource Center		
CIC	Centro de Investigaciones Criminológicas, Policía Nacional (National Police Centre for Criminological Research)	INDUMIL	Industria Militar (Military Industry)
		INML	Instituto Nacional de Medicina Legal y Ciencias Forenses (National Institute of the Medical Examiner and Forensic Sciences)
DANE	Departamento Administrativo Nacional de Estadística (National Administrative Department of Statistics)		
		MOD	Ministry of Defence
DAS	Departamento Administrativo de Seguridad (Administrative Department of Security)	OAS	Organization of American States
		OCCAE	Oficina de Control y Comercio de Armas

DDR	disarmament, demobilization, and reintegration		y Explosivos (Office for Control and Trade of Arms and Explosives)
DNP	Departamento Nacional de Planeación (Department of National Planning)	SIEDCO	Sistema de Información Estadistico Delincuencial, Contravencional, y
ELN	Ejército de Liberación Nacional (National Liberation Army)		Operativo (Statistical Information System of Delinquency, Offences, and Operations)

ENDNOTES

1   This chapter was produced by the Conflict Analysis Resource Center (CERAC) and the Small Arms Survey. References to CERAC's work relate to its conflict database on Colombia or to its research on violence and conflict in Colombia, which was undertaken with the support of the Small Arms Survey. Unless otherwise noted, information reflects details revealed during structured author interviews administered in October and November 2005 in Bogotá, with strong cooperation provided by the Vice Ministry of Defence. The information gathered during these interviews was triangulated with debriefings by intelligence officers of the armed forces.

2   Roughly 75 per cent of these deaths are criminal homicides. An examination of the complex interplay between criminality and armed conflict lies beyond the scope of this study.

3   External causes of morbidity and mortality include accidents, intentional self-harm, assault, events of undetermined intent, legal intervention and wartime operations, and complications during medical and surgical care.

4   A comprehensive assessment of armed violence in Colombia by CERAC and the Small Arms Survey is forthcoming and will be available in English and Spanish.

5   This chapter uses the concept of human security as a human welfare lens through which to analyse violence.

6   For a list of Colombia's main arms suppliers and the corresponding quantities, see Annexe 1.

7   This particular arrangement is rare in Latin America, where most arms industries have been entirely or partially privatized. Full or partial privatization is not necessarily an improvement in terms of the institutional arrangement for arms control.

8   On human rights grounds, Colombia has long been subject to a de facto arms embargo from many producers of high-quality arms. For example, the European Union does not sell arms to the Colombian forces and the United States, one of their main suppliers, has conditioned certain arms sales on human rights performance. As a result, Colombia is dependent on South Africa, Israel, Singapore, and a few European countries for arms supplies. Colombia is now actively seeking new suppliers, including China (*El Tiempo*, 2005b).

9   The armed forces adopted the Israeli Galil in 1992, and INDUMIL began production in 1994. INDUMIL estimates that 12,000–40,000 are produced annually. Previous assault rifles included the Heckler & Koch G3, imported from Germany.

10   In the late 1990s INDUMIL began a major shift towards self-sufficiency. By 2002 this goal was fulfilled in the production of Galil rifles, for which it currently produces some 30 million rounds of 5.56 military-grade ammunition each year. INDUMIL is self-sufficient in the production of three types of revolvers (i.e. 38L, 32L, and 38S), various kinds of ammunition (i.e. 38L, 9 mm, 32L, 7.62, and shotgun ammunition), mortars, and hand grenades. INDUMIL is also striving to become a leading producer of high-grade explosives and explosive services for the mining and oil sectors by the end of 2006. It has significant exports to Central America.

11   Such an upgrade would probably necessitate decommissioning existing weapons to prevent their leakage into the open market, although neither the national police nor INDUMIL have yet addressed this issue officially. Increased Colombian production of high-grade pistols would bring Colombia into competition with Venezuela, which recently announced its own domestic pistol production (*El Tiempo*, 2005c).

12   Primary products include the Llama INDUMIL Martial .32 long and .38 special, as well as the Llama INDUMIL Scorpio and Cassidy. INDUMIL is believed to have produced between 4,000 and 12,000 revolvers per year since 1999.

13   See, for example, *El Espectador* (2003).

14   The Colombian Constitution (Article 223) states: 'Nobody may own or carry weapons without a permit issued by the competent authorities.' The specific rules about firearms appear in legislative decree 2535 of 17 December 1993.

15   Of the new permits issued since 1994, 80.91 per cent went to Colombian citizens, 18.96 per cent to foreign residents, and 0.13 per cent to private security companies. Permits are divided into 'holding' and 'carrying' varieties. Colombian civilians appear more inclined to acquire carrying licenses for personal protection, while foreigners and companies overwhelmingly acquire holding licenses.

16   This range results from multiplying the number of estimated police officers by the standard multiplier and then establishing a margin of error of 33 per cent. It should also be noted that in the case of Colombia, these police-issued firearms are subject to rigorous command and control

procedures. Police-issued weapons are carried only during the officers' shifts and handed in during off-duty periods. Nevertheless, police are entitled to up to two personal firearms under the law and many avail themselves of this right. Based on author interviews with police and army officers, November 2005.

17   As with the police stockpile estimate above, this range is determined by applying a standard multiplier and ascribing a margin of error of 33 per cent.

18   DAS is a security institution with judicial police, intelligence, and immigration functions that is roughly equivalent to a combination of the FBI and the CIA in the United States.

19   There are indications that DAS uses much more sophisticated weapons than the police: in 2005, the United States provided USD 4 million for DAS to purchase 1,500 M16s and 3,000 pistols for a new special task force. See, for example, *Revista Semana* (2005).

20   All the figures of personnel strength of the illegal non-state armed groups were confirmed during interviews with intelligence officers of the Colombian army. Author interviews with police and army officers, November 2005.

21   Senior demobilized FARC commanders interviewed for this study maintain that the FARC has some 42 active 'fronts' and several mobile columns and units.

22   This is the latest official figure. The current figure may have changed as paramilitary DDR is an ongoing process.

23   Several high-ranking paramilitary commanders are currently being tried for their alleged involvement in narco-trafficking in Colombia and the United States. See Fundación Ideas para la Paz (2005a) for a description of the prosecution of these leaders. The leader of the paramilitaries, Diego Fernando Murillo, is requested by the District Court of New York on drug trafficking charges and his extradition was petitioned by the US government according to the Nota Verbal No. 1733, US, 26-07-04.

24   See, for example, Pizarro (2004, ch. IV), who deals with the impact of narcotics traffic on the conflict groups. Carlos Castaño, the former leader of the largest paramilitary group, the AUC, claimed in an interview that his organization received up to 70 per cent of its funding from narcotic-related activities (AP, 2000).

25   Author interviews, October and November 2005.

26   See, for example, http://www.usdoj.gov/criminal/icitap/TextColombia.html.

27   Moreover, the director of the forensic laboratories of the judicial police (Dirección Central de Policía Judicial, or DIJIN) claims that some 70 per cent of all firearm-related offences are committed with illegal arms.

28   For a documented case of smuggling, see the report of the Organization of American States (OAS) on the cache of assault rifles for the para-militaries that was intercepted in Panama through Nicaragua (OAS, 2003). See also AP (2006).

29   Intelligence and police sources, as well as several press reports, claim that FAL rifles are often seized during counter-guerrilla operations. See also Schroeder (2004) and García-Peña (1999, p. 3).

30   The CERAC database reveals that guerrillas use explosive devices 12 times more often than the paramilitaries. Furthermore, author interviews with demobilized guerrillas and paramilitaries show that the guerrillas use pistols more frequently than the paramilitaries, issuing them to rank-and-file members as well as their leaders.

31   In one incident, in Bojayá-Chocó in May 2002, the FARC used such cylinders to bomb a church in which people had taken shelter from guerrilla–paramilitary clashes, killing 119 and wounding 90.

32   Author interviews with DAS detectives, October 2005.

33   Author interview with security personnel and analysts, November 2005. There has been much speculation on an IRA–FARC link. Colombian army officers claim that there are strong similarities between IRA and FARC tactics and techniques in the use of explosives. The *Sunday Times* (2005) provides an account of this subject.

34   For initial data, see Restrepo et al. (2004, p. 407).

35   Attacks cover a wide range of events such as massacres, bombings, mine explosions, economic sabotage (e.g. attacks on oil pipelines or electricity grids), incursions, and aerial bombardments.

36   This paragraph draws on Restrepo and Spagat (2005a, p. 135).

37   In this latest outbreak, the paramilitaries generally killed people one or two at a time rather than in larger numbers, as they had before.

38   The government tends to clash with guerrillas rather than engage in attacks such as those staged by the guerrillas and paramilitaries. Relatively few civilians are harmed during these clashes.

39   The low percentages for the guerrillas and the government are largely explained by the types of attacks. Shots are not usually fired during an aerial bombardment or during an attack on economic infrastructure, for example.

40   Grenade use follows a similar pattern until 2004.

41   The conflict in Arauca tends to be waged in small towns with heavy explosive use by guerrillas.

42   Between 1979 and 2002, there were 857,227 deaths from 'external causes', i.e. not from natural diseases. More than 430,000 people died from external causes without a firearm being involved.

43   The rest of this section draws only on DANE data, which runs through 2002.

44   In Colombia the firearm homicide rate reached a peak of 66 per 100,000 in 2002 (DANE). See, for example, Godnick et al. (2002).

45   For more than 95 per cent of all reported firearm deaths between 1979 and 2002 (406,855), no specific type of weapon was identified. In approximately 2 per cent of these cases (8,762), the data specifies 'handguns', while in 3 per cent of them (12,533) it specifies 'long guns'.

46   Medellín experienced 12.6 per cent of all firearm deaths, Bogotá 12 per cent, Cali 6.5 per cent, and Barranquilla 2.19 per cent.

47   The projected populations of these cities for 2005 are: Medellín (2,093,624), Bogotá (7,185,889), Cali (2,423,381), Barranquilla (1,386,895), Cúcuta (742,689), Bello (400,291), Pereira (521,684), Itagüí (288,207), and Manizales (382,193).

48   The areas east of Antioquia and south of the Meta department are major clusters of conflict violence.

49   In contrast, female Colombians born in 2002 will lose only four months on average.

50   See, for example, Aguirre et al. (2005).

51   *La Violencia* refers to a period marked by vicious fighting between liberal and conservative party supporters from 1948 to 1952, although lower-level violence continued for some years afterwards. A large disarmament and demobilization programme of the liberal guerrillas and conservative militias coincided with the start of the agreement that put an end to the infighting known as Frente Nacional (1958); the programme was also carried out during the previous military government (1953–57).

52   Guerrilla groups and their numbers are as follows: M-19 (900), PRT (200), EPL (2,000), Movimiento Armado Quintín Lame (157), Comandos Ernesto Rojas (25), CRS (433), Milicias de Medellín (650), Frente Francisco Garnica (150), COAR (200). See, for example, Guáqueta (2005).

53   See Fundación Ideas para la Paz (2005b) for a detailed discussion of this evolution.

54   For more information, see Decree 2546 of 1999.

55   Of the total number returned, there were 13,333 long-range weapons, 2,460 handguns, 1,161 machine guns and mortars, and 8,550 grenades. At least 2,000,000 rounds of ammunition were also surrendered.

56   In its *World Report 2005,* Human Rights Watch asserts that '[a] significant obstacle to a full and effective paramilitary demobilization is the lack of a legal framework to govern the demobilization process and the benefits to be provided to those who demobilize. A draft bill initially proposed by the administration of President Alvaro Uribe in 2003 would have allowed cooperative paramilitary leaders responsible for atrocities to go virtually unpunished. After an international and domestic outcry, the proposed law was modified. However, a new version of the bill circulated in April 2004 still contains serious flaws—a failure to provide for thorough investigations of paramilitary crimes and illegal assets, and a loophole allowing those convicted of atrocities to entirely avoid incarceration—that make the effective demobilization and dismantling of paramilitary structures unlikely' (HRW, 2005, p. 203).

57   The legal framework provided by the Justice and Peace law limits the jail sentence to a maximum of eight years. Sentences can be served in penal rural colonies.

BIBLIOGRAPHY

AFP (Agence France-Presse) 2004. 'El bloque paramilitar más poderoso de Colombia entrega las armas el viernes.' 9 December.

Aguirre, Katherine, et al. 2005. *Assessing the Effect of Policy Interventions on Small Arms Demand in Bogotá, Colombia.* Unpublished background paper. Bogotá: CERAC.

Alto Comisionado para la Paz (High Commissioner for Peace). 2005. Demobilization page.
    <http://www.altocomisionadoparalapaz.gov.co/desmovilizaciones/2004/index_resumen.htm>

AP (Associated Press). 2000. 'Colombian Paramilitary Chief Shows Face, Presents New Image.' 2 March.

——. 2006. 'El Salvador: investigarán tráfico de armas a Colombia.' 20 January.

DNP (Departamento Nacional de Planeación). 2005. *Visión Colombia II Centenario: 2019. Propuesta para la discusión.* Bogotá: Editorial Planeta Colombia S.A.

El Comercio. 2003. 'Otro cargamento de balas fue detectado en un barco.' 9 February.

El Espectador. 2003. 'La oferta española.' 13 April.

El Pais. 2004. 'M-19 cambió drogas por armas.' 6 October. <http://elpais-cali.terra.com.co/paisonline/notas/Octubre062004/A206N2.html>

El Peruano. 2005. 'Montesinos al banquillo por violación a DD HH.' 20 July.

El Tiempo. 2004. 'Juicio a Vladimiro Montesinos revela operativo con el que se armó la guerrilla de las Farc en 1998.' 24 January.

——. 2005a. 'El 10 por ciento de las municiones de las FARC y de los "paras" tienen el sello de Indumil.' 12 June.

——. 2005b. 'Gobierno de Colombia tiene la mira en China para la compra de armas nuevas.' 8 August.

——. 2005c. 'Armas, los primeros productos de la "revolución" de Hugo Chávez en Venezuela.' 7 November.

——. 2005d. 'El sábado hadrá jornado de desarme en Bogotá.' 22 November.

Fundación Ideas para la Paz. 2005a. 'Instancias del Gobierno Nacional encargadas o relacionadas con los procesos de desmovilización y reinserción de los grupos ilegales armados: 1982–2004.' *Boletín Informativo No. 6.* <http://www.ideaspaz.org/proyecto03/boletines/boletin06.htm>

——. 2005b. 'Juego de espejos.' *Siguiendo el conflicto: hechos y análisis de la semana*. 14 October.
   <http://www.ideaspaz.org/new_site/secciones/publicaciones/download_boletines/boletin_conflicto34.pdf>

García-Peña, Daniel. 1999. *War, Peace and Light Weapons in Colombia : A Case Study*. Geneva: Geneva Forum.
   <http://www.geneva-forum.org/Reports/salw_vol1/19991129.pdf>

Godnick, William, Robert Muggah, and Camilla Waszink. 2002. *Stray Bullets: The Impact of Small Arms Misuse in Central America*. Occasional Paper 5.
   Geneva: Small Arms Survey.

Guáqueta, Alexandra. 2005. 'Desmovilización y reinserción en El Salvador: Lecciones para Colombia', *Informes FIP 1*. Bogotá: Fundación Ideas para
   la Paz. <http://www.ideaspaz.org/new_site/secciones/publicaciones/download_documentos/desmovilizacion.pdf>

Guerrero, Rodrigo. 1999. *Programa Desarrollo, Seguridad y Paz, DESEPAZ de la Ciudad de Cali*. Rio de Janeiro: Inter-American Development Bank.
   <http://www.iadb.org/IDBDocs.cfm?docnum=362232>

HRW (Human Rights Watch). 2005. *World Report 2005*. 'Human Rights Overview: Colombia.' New York: HRW.
   <http://hrw.org/english/docs/2005/01/13/colomb9847.htm>

IISS (International Institute for Strategic Studies). 2005. *Military Balance 2005–2006*. Oxford: Routledge.

Krause, Keith, et al. 2005. 'Colombia's Chimaera: Reflections on Human Security and Armed Violence.' *Human Security Bulletin*. Vancouver: Canadian
   Consortium on Human Security (CCHS). <http://www.humansecuritybulletin.info/page230.htm>

Llorente, María, Jairo Núñez, and Mauricio Rubio. 2000. *Efectos de los controles al consumo de alcohol y al porte de armas de fuego en la violencia
   homicida*. Bogotá: Mayor's Office. <http://www.suivd.gov.co/investigaciones/Alcohol%20&%20Armas.htm>

Muggah, Robert. 2005. 'No Magic Bullet: A Critical Perspective on Disarmament, Demobilization and Reintegration (DDR) and Weapons Reduction in
   Post-conflict Contexts.' *The Round Table*. Vol. 94, No. 379, pp. 239–52. April.

O Globo. 2005. 'Traficantes brasileiros se aliam às FARC no Paraguai.' 10 April.

OAS (Organization of American States). 2003. *Informe de la Secretaría General de la Organización de los Estados Americanos sobre el desvío de armas
   nicaragüenses a las Autodefensas Unidas de Colombia*. January. <http://www.oas.org/OASpage/NI-COarmas/NI-COEsp3687.htm>

Pizarro Leongómez, Eduardo. 2004. *Una democracia asediada: balance y perspectivas del conflicto armado en Colombia*. Bogotá: Norma.

Restrepo, Jorge, et al. 2005. 'Colombia's Chimaera: Reflections on Human Security and Armed Violence.'
   <http://www.humansecuritybulletin.info/page230.htm>

Restrepo, Jorge and Michael Spagat. 2005a. 'Colombia's Tipping Point?' *Survival*. Issue 2, Summer 2005, pp. 131–52.

——. 2005b. *The Colombian Conflict: Where Is It Heading?* Bogotá: CERAC. <http://www.cerac.org.co/pdf/CSISPresentationwithtext-V10_Low.pdf>

Restrepo, Jorge, Michael Spagat, and Juan Vargas. 2004. 'The Dynamics of the Colombian Civil Conflict: A New Dataset.' *Homo Oeconomicus*. Vol. 21 (2),
   pp. 396–429.

Revista Cambio. 2005. 'Secretos.' 7 February.

Schroeder, Matt. 2004. *Small Arms, Terrorism and the OAS Firearms Convention*. FAS Occasional Paper No. 1.
   <http://www.fas.org/asmp/campaigns/smallarms/OAS_Firearms_Convention.html#report>

Small Arms Survey. 2005. *Small Arms Survey: Weapons at War*. Oxford: Oxford University Press.

Sunday Times. 2005. 'FARC rebel "admits IRA trained him."' 15 May.

Villaveces, Andrés, et al. 2000. 'Effect of a Ban on Carrying Firearms on Homicide rates in two Colombian Cities.' *Journal of the American Medical
   Association*. Vol. 283, No. 9. 1 March.

ACKNOWLEDGEMENTS

Principal authors

Katherine Aguirre, Robert Muggah, Jorge A. Restrepo, and Michael Spagat

Contributors

Andrés Ballesteros, Óscar Becerra, Brodie Ferguson, Simón Mesa, Nicolás Suárez, and CERAC

Nigerian rebels of the Niger Delta People's
Volunteer Force patrol the creeks of the Niger
Delta near Port Harcourt, September 2004.
© George Esiri/Reuters

A Constant Threat
ARMED GROUPS IN WEST AFRICA

10

INTRODUCTION

With these militias, the government unleashed a monster. It now has to be fed.
—United Nations official in Darfur, Sudan[1]

By referring to militias as 'monsters' the United Nations official in Darfur highlighted a central challenge facing the world today: armed groups, typically unaccountable and often unpredictable, will fight to sustain themselves, even when that entails changes in the group's nature or objectives. The number of conflicts involving one or more armed groups has eclipsed those involving only states. In 2004, all 19 of the world's 'major armed conflicts' (those causing more than 1,000 battle-related deaths in a year) could be characterized as 'intra-state', meaning they involved one or more armed groups (SIPRI, 2005, pp. 121–33).

State security forces have heavy weapons, armoured vehicles, and aircraft at their disposal. Armed groups, on the other hand, are generally equipped mainly with small arms and light weapons.

This chapter provides an overview of the dynamics of small arms acquisition, management, and control by armed groups in the Economic Community of West African States (ECOWAS)[2] region, where the proliferation of armed groups is a phenomenon of particular concern.[3] It then looks into control and reduction measures that can help to minimize the deleterious effects of conflict and prevent future fighting. What measures can be taken to address the problem of armed groups in West Africa? Normative moral persuasion,[4] attempting to influence armed groups' actions through an appeal to norms of behaviour, is an important but insufficient means of engagement. Experience from West Africa suggests that providing incentives (job creation, security sector reform, and so on), affecting, in particular, the demand for weapons, is more effective.

Examples from South-East Asia, Northern Ireland, and Colombia point to the truly global nature of the challenges posed by armed groups and complicate the picture of motives and methods among these diverse actors (LORD'S RESISTANCE ARMY). It is hoped the study's findings will be useful to actors in West Africa and beyond.

The main findings of the chapter include the following:

- Many West African armed groups' motivations, allegiances, and scope of activities shift over time.
- There is a robust market for small arms and light weapons in West Africa to which armed groups have easy access.
- Recirculation of small arms stockpiles among conflict areas (including leakage from government stocks, corruption, and craft production) remains substantial.
- Reducing supplies of ammunition is an underutilized means of limiting the deleterious effects of small arms conflict.

- Weapons collection programmes in West Africa have had some success, but the quality of weapons collected is questionable.

- A lack of alternative employment opportunities may drive demobilized ex-fighters to return to fighting in an armed group.

- Arms reduction efforts linked to development incentives show promise.

- Supply-side interventions alone will not curb groups' access to small arms, and efforts to reform the security sector are needed to address groups' willingness to arm.

ARMED GROUPS: AN OVERVIEW OF THE ISSUES

'Armed group': a definition

There is no single, universally accepted definition of 'armed group'. Labels with political connotations (such as 'freedom fighters', 'terrorists', and 'revolutionaries') are often used to channel debate. The ideological charge of such labels risks simplifying a group's nature and concealing changes in its goals or tactics. The advantage of a neutral definition lies in its ability to incorporate and describe those changes.

Several widely used definitions focus only on groups outside of state control. The International Council on Human Rights Policy (ICHRP), for example, defines 'armed groups' as entities that 'are *armed and use force to achieve their objectives and are not under state control*' (ICHRP, 2000, p. 5, emphasis in original). Claude Bruderlein of the Harvard School of Public Health provides a similar definition. He stipulates that armed groups have 'a basic command structure'; use 'violence to achieve political ends'; and are independent 'from state control' (Bruderlein, 2000, p. 8). Some definitions, such as the one employed by Geneva Call, an NGO that campaigns for armed groups to sign pledges to adhere to international laws, prefer the term 'non-state actors' to 'armed groups'. By adding the stipulation that such actors operate 'outside of state control', however, this explanation falls into the same pattern of exclusion as the ICHRP's and Bruderlein's (GC, 2002).

The Small Arms Survey, however, favours a broader definition that includes groups linked to the state. This chapter focuses more on the effect these groups have on human security than on political considerations. Accordingly, even ostensibly 'pro-state' armed groups are of interest to our study. The chapter therefore employs a definition proposed by Pablo Policzer (2005): 'Non-state armed groups are challengers to the state's monopoly of legitimate coercive force.' As Policzer explains, 'this definition suggests focusing on the relationships between groups, states, and populations that determine the degrees of coercive monopolization. Such relationships are fluid, but the definition need not be' (pp. 8, 9).

A typology of armed groups

A typology assists us in identifying trends and changes, both at the level of individual groups and at the level of national or regional relations. Three main parameters characterize armed groups: their relationship to the state; the scope of their operations; and their motivations for taking up arms. Interpreting a group's motivations is an empirically fraught endeavour because it would require a comparison of a group's stated aims with its actions, which may differ from its aims or change as incentives arise. Instead, our typology, based on Policzer's definition, will use only the first two parameters: (*a*) the relationship of armed groups to the state, and (*b*) the geographic scope of the threat to

Table 10.1   A typology of armed groups in the ECOWAS region, active at some point during the period 2000–05

No.	Country	Armed group	Year formed	Type when formed	Subsequent type(s)
1	Côte d'Ivoire	**New Forces**	2003	REB-N	REB-N
2	Côte d'Ivoire	**Group of Patriots for Peace (GPP/CPP/FLN)**	2002	PGM-N	PGM-N
3	Côte d'Ivoire	**Front for the Security of the Centre-West (FSCO)**	2002	PGM-SN	PGM-SN
4	Côte d'Ivoire	**Front for the Liberation of the Great West (FLGO)**	2002	PGM-SN	PGM-SN
5	Côte d'Ivoire	**Young Patriots**	2000s	PGM-N	PGM-N
6	The Gambia	**Green Boys**	1999	PGM-SN	PGM-SN
7	Guinea	Young Volunteers	2000	PGM-N	VG-N*
8	Guinea	Movement of the Democratic Forces of Guinea (RFDG)	1996	REB-N	REB-N
9	Liberia	Liberians United for Reconciliation and Democracy (LURD)	2000	REB-N	VGN-SN*
10	Liberia	Movement for Democracy in Liberia (MODEL)	2003	REB-N	VGN-SN*
11	Liberia	Government of Liberia militias	1997	PGM-N	PGM-N*
12	Nigeria	**Bakassi Boys**	1999	VG-SN	PGM-SN/VG-SN*
13	Nigeria	**O'odua People's Congress (OPC)**	1994	VG-N	VG-N
14	Nigeria	**Arewa People's Congress (APC)**	1999	VG-N	VG-N
15	Nigeria	**Egbesu Boys of Africa (EBA)**	1998	VG-SN	VG-SN
16	Nigeria	**Niger Delta Volunteer Force (NDVF)**	1998	VG-SN	VG-SN
17	Nigeria	**Niger Delta Vigilante (NDV)**	2000	VG-SN	PGM-SN*
18	Nigeria	**Niger Delta People's Volunteer Force (NDPVF)**	2003	VG-SN	REB-SN VG-SN*
19	Nigeria	**Hisbah Groups**	1990s	VG-SN	VG-SN PGM-SN*
20	Nigeria	**Federated Niger Delta Ijaw Communities (FNDIC)**	1997	VG-SN	VG-SN
21	Nigeria	**Al-Sunna Wal Jamma**	2002	REB-N VG-N	REB-N VG-N
22	Nigeria	**Zamfara State Vigilante Service (ZSVS)**	1999	PGM-SN VG-SN	PGM-SN VG-SN
23	Senegal	Movement of the Democratic Forces of Casamance (MFDC)	1982	REB-SN	REB-SN
24	Sierra Leone	Revolutionary United Front (RUF)	1991	REB-N	REB-N
25	Sierra Leone	Armed Forces Revolutionary Council (AFRC)	1997	REB-N	PGM-N*
26	Sierra Leone	Civil Defence Force (CDF)	1990s	PGM-N	VG-SN*
27	Sierra Leone	West Side Boys (WSB)	2000	VG-SN	VG-SN

Key: **PGM-N**: pro-government militia with national ambitions; **PGM-SN**: pro-government militia with sub-national ambitions; **REB-N**: rebel group with national ambitions; **REB-SN**: rebel group with sub-national ambitions; **VG-N**: vigilante group with national ambitions; **VG-SN**: vigilante group with sub-national ambitions; **BOLD**: active in 2005

* Change in type

Source: Florquin and Berman (2005)

One-third of West African armed groups active at some point in 2000–05 changed type following their formation.

government authority (i.e. whether the group seeks to challenge the state's monopoly of force on a national or sub-national scale). Since groups can evolve along both these parameters, even groups that are not created in opposition to the state (such as paramilitaries or local defence forces) exercise coercive powers over a part of national territory and thus challenge the state's monopoly over the provision of public order.

Characterizing their relationship to the state yields three categories of groups. Groups that support the government or regime in power will be termed 'pro-government militias'. Those that oppose the government or regime in power will be classified as 'rebels'. Groups that operate without an explicitly pro- or anti-state agenda will be termed 'vigilante groups'. We can further refine the typology by dividing these groups into those that seek to exercise force nationally—throughout the entire country—and those whose ambitions are sub-national—that is, regional or local.

In the ECOWAS region, applying this typology enables an analysis of developments among particular groups as well as within the broader region. One-third of West African armed groups active at some point in 2000–05 changed type following their formation, with the shift most often concerning their relationship to the state.

Motivations for joining armed groups

In order to understand the dynamics of weapons acquisition and use in armed groups, it helps to probe the reasons why such groups form and take up arms. Some groups arise for political reasons, to either overthrow or protect a government. Other groups are primarily driven by economic motives. For example, arms may provide the necessary clout to control extraction of a natural resource or impose taxes on a local population. Some groups are motivated—or claim to be motivated—by ethnic or religious grievances. Commanders who draw on these 'selective incentives'

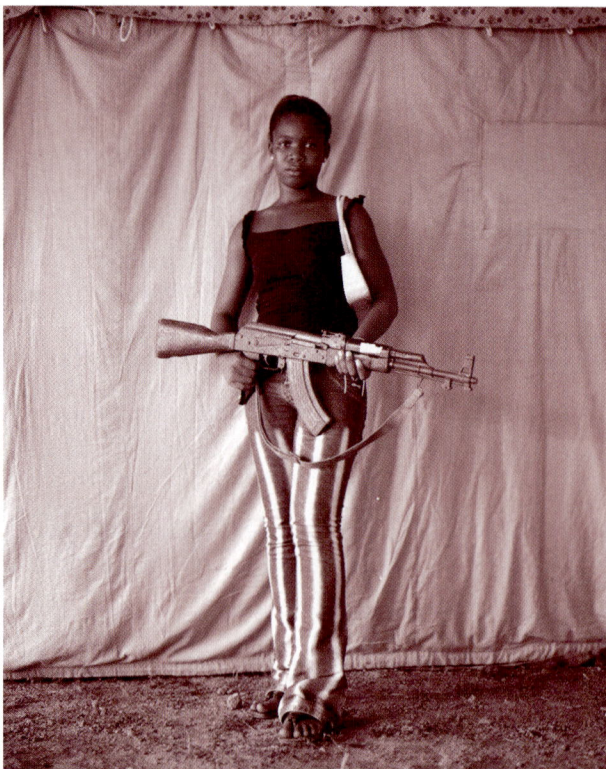

(Weinstein, 2004, pp. 9–10) to enlist recruits may thereby enable greater cohesion among fighters. Additional motivations, however, such as the material spoils that accompany force, or the political power that comes with inclusion in an eventual transitional government, render even relatively unified groups unpredictable.

Some armed groups may, at least initially, be sources of security rather than insecurity for their communities. William Reno (2003) has shown how, in some cases, armed groups can provide protection from a predatory state. Vigilante groups may come into existence because members see a market for their enforcement services in an area where state security organs are absent or ineffective. For

A LURD fighter poses with her weapon at a UN disarmament point in May 2004. Originally from Sierra Leone, she had been caught up in the war in Liberia.
© Tim A. Hetherington/Panos Pictures

instance, the Bakassi Boys began as a vigilante group in Abia state, hired by local traders to provide security in the markets. They subsequently spread to provide similar services in Anambra and Imo states as well, and gained the endorsement of the three states' governors for their success in staving off robberies. In Nigeria, the police are a federal service, which state governors cannot control and often feel ill-served by.[5] However, group control and leadership among the Bakassi Boys have waned, and their stated purpose of provision of security has often become a screen to justify extreme measures, such as extra-judicial killings and arbitrary arrest. On 4 August 2005 in Aba, Abia state, they arrested 37 people at the Orie Olabiam electronics market and locked them in a ten-feet-by-ten-feet cell; the following morning, 27 of the captives were found dead (Orji, 2005, p. 4). Attitudes towards them have shifted from respect to fear and disdain, and Abia is the only state in which the Boys have managed to maintain their role as purveyors of 'jungle justice' (immediate and summary trials and executions).[6]

Box 10.1 Children associated with fighting forces in the Mano River Union

Estimates of the total number of children associated with fighting forces (CAFF) serving worldwide range up to 300,000.[7] The Mano River Union (MRU)–Guinea, Liberia, and Sierra Leone–has been particularly plagued. Examining how these children were recruited, whether forced or voluntary, illuminates some of the motivations behind joining, as do the roles small arms played in their recruitment and their terms of service.[8]

Recruitment methods varied widely within the region.[9] Though most of the children conscripted in Sierra Leone were forced into service at the point of a gun, this was not the case in Liberia or Guinea. However, in all three countries combined, 40 per cent of the children appeared fatalistic, and many went along and joined simply because 'that was the only thing to do' (Wille, 2005, p. 191).

Figure 10.1 Self-reported motivations for joining armed groups

NUMBER OF RESPONDENTS

Legend: Sierra Leone, Liberia, Guinea

Access to firearms and the power associated with being a member of an armed group seemed important to some of the children, as reflected in sentiments such as the following: 'When you are in possession of arms you can loot and get anything you want. You can say or do anything to anybody without fear.'[10] The majority (91 per cent) of CAFF interviewed reported that they had access to firearms, particularly Kalashnikovs, which are lighter than many other assault rifles.

The widespread availability of weapons and ammunition in the MRU helps explain why armed groups there could afford to enlist large numbers of children (who may waste ammunition or be poor shots) without jeopardizing the groups' effectiveness.

Studying the recruitment of child soldiers in the MRU and the tasks they carry out shows that motivations vary, both in terms of trends among armed groups and among individual children. From the perspectives of the leadership of the armed groups, recruiting children enabled the adults to build on their own power bases as well as divest themselves of tasks they did not want to carry out.

Table 10.2 CAFF access to firearms

	CAFF with access to firearms	Percentage of total CAFF interviewed
Guinea	82	90
Liberia	94	94
Sierra Leone	72	91
Total	**248**	**92**

Source: Wille (2005, pp. 180–221)

At the level of individual fighters, multiple motivations probably enter into their decision to enlist in an armed group. In some cases, conscription may be forced, itself carried out at the point of a gun. In West Africa, the dearth of employment opportunities drives some to join an armed group. Exacerbating the dire poverty of the region, the population distribution leans heavily towards the young: in 2000, 45 per cent of West Africa's population was below the age of 15 (UN, 2004). Limited opportunities for schooling or other vocational training mean that many people have few avenues towards sustenance, let alone enrichment; even those who do obtain an education often lack means of supporting themselves. University students, who received arms and a payment of 5,000 CFA francs (approximately USD 9) from the government for their 'service', perpetrated extensive violence and vandalism in Abidjan, Côte d'Ivoire, in November 2004 (ICG, 2005a, pp. 5, 10). Material incentives, whether provided upfront or promised, thus become powerful recruitment tools.

Fluidity in armed groups' command structures

The command structures of armed groups in West Africa vary greatly. For instance, Yan Daba and Almajiri, two Nigerian vigilante groups, have no centralized structure but rather exist as a loose network of cells, each with its own leader (Orji, 2005). Groups that have an 'activist' stance are more likely to be well-disciplined than groups that are more 'predatory'. However, the lack of accountability on the part of armed groups means that the extent of leaders' control over subordinates' actions may shift alongside changes of objectives, sources of support, or access to weapons and ammunition.

In northern Mali, Arab and Tuareg rebels fought a rebellion for greater autonomy from the central government. The conflict lasted from 1990 to 1996, and initially rebels actively sought public support for their cause. To do so, they located their bases outside of civilian areas so the Malian army would not harass the population, and punished fighters who mistreated civilians (Lecocq, 2002, p. 235). However, as the years of conflict wore on, the rebels began to splinter and with that development came a subsequent breakdown in discipline: by 1994, reports stated that various rebel groups had engaged in killings of civilians across tribal lines (2002, pp. 275–76, citing various press and Amnesty International reports[11]).

Militias created to support standing governments provide some compelling examples of the unpredictability of armed groups. For instance, President Lansana Conté of Guinea established the Young Volunteers to counter attacks by Liberian rebels, the Revolutionary United Front (RUF), and the Movement of the Democratic Forces of Guinea (Rassemblement des forces démocratiques de Guinée, RFDG) in 2000–01.[12] The Guinean military was in charge of recruitment (with assistance provided by local leaders such as the mayor of the town of N'Zérékoré) and provided the new fighters with weapons (ICG, 2003b, p. 18). The Guinean military and government also served as the leaders of the Young Volunteers. When the conflict ebbed and the need for the militia became less pressing, some were integrated into the army or into marching bands, while others returned to civilian life; however, as of 2003, more than a third (3,879 of 9,000) had yet to be demobilized (UNOCHA, 2003, p. 56). Some of these young fighters have organized themselves into gangs to intimidate and rob the population (UNOCHA, 2003, p. 56). Though the government adopted tough anti-crime measures to curb such banditry,[13] militias continue to menace the population, particularly in the Forest Region bordering Liberia and Côte d'Ivoire; few of them have relinquished their guns (ICG, 2005a, p. 22).

Armed groups in West Africa are unpredictable, and even those that initially build public security can come to undermine it. 'Original "ideology" tends to evaporate rapidly in the face of temptations to engage in financial or political aggrandizement—usually at the point of a gun' (Florquin and Berman, 2005, p. 386). Indeed, some commanders

Armed groups in West Africa are unpredictable.

> ### Box 10.2   Armed groups in South-East Asia: primed and purposeful
>
> In contrast to West Africa, where unemployed youth can quickly become 'guns for hire', in South-East Asia many people who join armed groups (with the exception of opportunistic criminal gangs) are committed to a particular political end.
>
> In Indonesia, three armed groups have engaged in struggles for independence or greater autonomy. The Armed Forces for the National Liberation of East Timor spent 24 years in jungle warfare principally armed with decades-old US- and Portuguese-made firearms, craft-produced small arms, and what it could capture from the Indonesian army and its local militias. International intervention and a referendum in 1999 resulted in Indonesia granting independence to the half-island state of East Timor. The West Papua Independence Movement (Organisasi Papua Merdeka, OPM) has been fighting for the independence of West Papua since the mid-1960s. It is poorly armed and has been harshly repressed by the Indonesian military. Currently, OPM hopes for a negotiated settlement. The Free Aceh Movement (Gerakan Aceh Merdeka, GAM) and its armed wing, the Tentara Nasional Aceh (TNA), were founded in 1976. After nearly 30 years of conflict, the GAM and the Indonesian government signed a peace accord in August 2005, which provided for limited autonomy for Aceh.[14]
>
> In the Philippines, 'Moros' (an appropriation of the Portuguese slang for 'Muslim') have been fighting for the creation of an Islamic state in the southern region of Mindanao since the 1960s, citing centuries of grievance and the perceived illegitimacy of the island's conquest by the Spanish in the 16th century and the United States in the 19th century. The Moro Islamic Liberation Front (MILF) has considerable local support and controls territory in Mindanao. The group has an extensive network of camps and training facilities throughout the islands.
>
> Thailand and Myanmar also host armed groups that have opposed their governments for nearly half a century. These conflicts have now involved several generations of fighters struggling for the same political goals as those who preceded them.
>
> **Source:** Koorey (2005)

see personal enrichment as a legitimate aspect of war. Not allowing subordinates to loot would sap the leaders' legitimacy in the fighters' eyes, a state of affairs that reveals the tenuous command structures in many armed groups. As one Sierra Leonean commander, a veteran of three wars and five armed groups, told Human Rights Watch,

> Anywhere you have rebel war you're entitled to get money . . . . My boys were looting a lot at the port as well. A commander can't know all their secrets. After all, they're the ones who made me a commander. You have to let them do it or they could blow you off. (HRW, 2005a, p. 21)

Since many fighters cycle through the various conflicts in a region, any individual armed group's efforts to instil discipline can be undermined by bad habits already ingrained. Veterans who had served in multiple armed groups singled out Liberians United for Reconciliation and Democracy (LURD) for its leadership's attempts to limit abuses by their ranks, but fighters' adherence to the rules was mixed (HRW, 2005a, pp. 35–36). In addition, in some instances the employment of foreign fighters in an internal conflict has led to an increase in attacks on civilians, reflecting the callous tactics of mercenaries, who have little at stake, within either the armed group or the communities that serve as battlegrounds (HRW, 2005a, p. 31).

SOURCES OF FIREARMS FOR ARMED GROUPS

In West Africa, armed groups' weapon supplies are most often a result of recirculation of stocks already within the region (Florquin and Berman, 2005). This section will first examine sources of weapons from within the country ('domestic') and then turn to external sources.

Domestic

Theft or seizure of official government-owned weapons is a primary source of armament for armed groups, whether via a strategic raid, the spoils of battle, or looting accompanying the chaos of war. Because stockpile management is rarely transparent, it becomes difficult to ascertain exactly what has been taken and to trace its whereabouts.

Guinea-Bissau's recent history provides a case in point. In 1998 Ansumane Mané, the Chief of Staff of the Armed Forces, formed a junta to overthrow the president. The ensuing conflict involved troops from both of the neighbouring countries, Senegal and Guinea (IISS, 1998). Allegations of Mané's support for a Senegalese armed group, the Movement for the Democratic Forces of Casamance (Mouvement des forces démocratiques de la Casamance, MFDC), was a main reason for his dispute with the central government (Evans, 2004, p.5; MALAO, 2003, p. 42). Weapons used by armed groups during the fighting came from three main sources: those looted from government stockpiles during the 1974 war for independence from Portugal, which fighters kept as symbols of their triumph; those distributed to civilians by the government in 1998–99 (mostly Kalashnikov assault rifles of Ukrainian and Bulgarian origin, and handguns) in an effort to create informal pro-government forces in communities; and those seized by the junta from government stockpiles in 1998–99.[15] Thus, all three of these means of weapons acquisition originate in government stocks.

Similarly, in Côte d'Ivoire many rebel group weapons came from captured government armouries. The Patriotic Movement of Côte d'Ivoire (Mouvement patriotique du Côte d'Ivoire, MPCI), for one, claimed to have taken government stocks from Bouaké and Korhogo, as well as to have unearthed weapons hidden by deserting government soldiers in 2000 (ICG, 2003a, p. 11). A few hundred kilometres away in Sierra Leone, armed groups filled their weapons

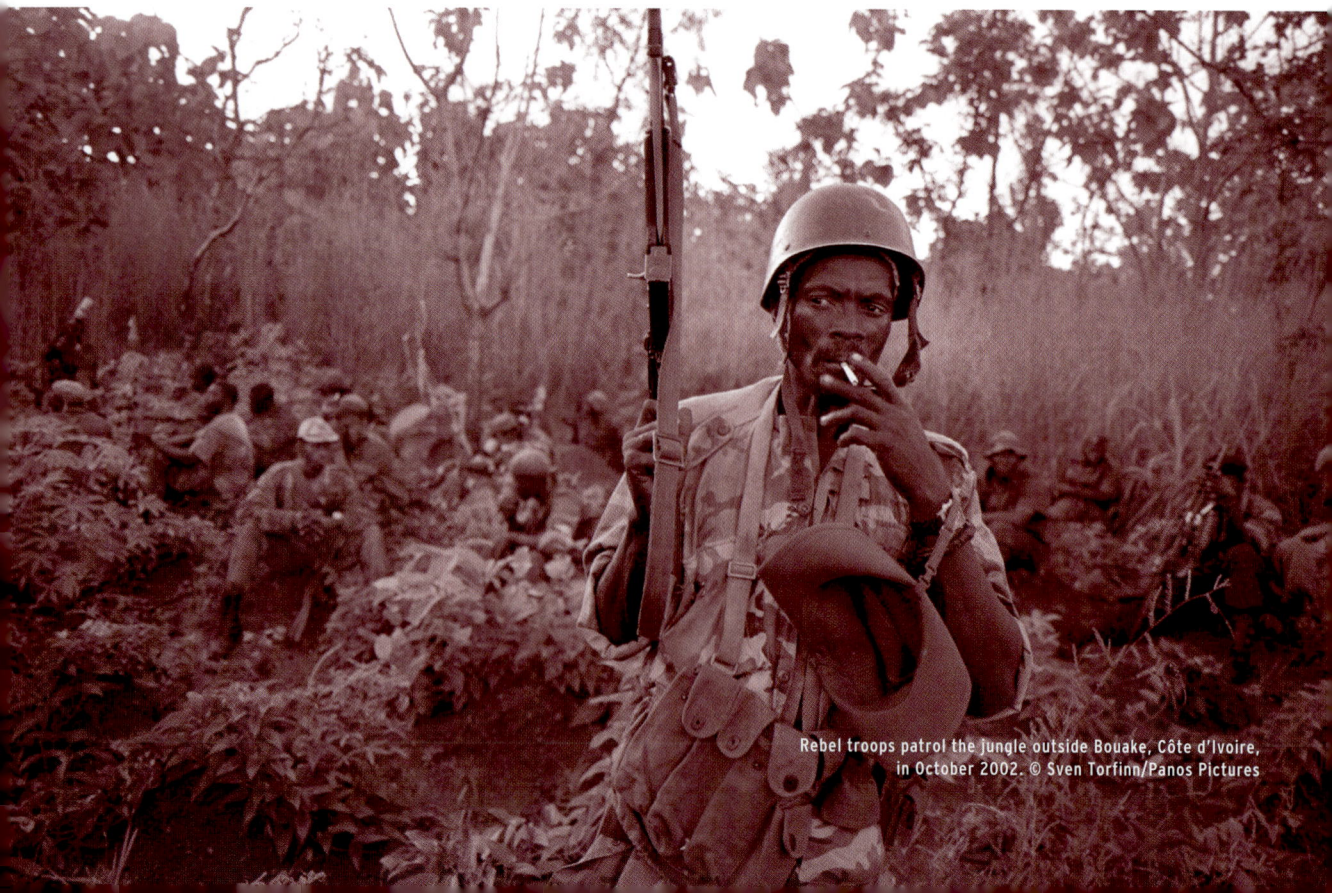

Rebel troops patrol the jungle outside Bouaké, Côte d'Ivoire, in October 2002. © Sven Torfinn/Panos Pictures

stocks from the holdings of the Sierra Leonean armed forces. In addition, the RUF captured hundreds of weapons from the Guinean, Kenyan, and Zambian troops who were taking part in the United Nations Mission in Sierra Leone. Also in Sierra Leone, the Armed Forces Revolutionary Council targeted the regional peacekeeping force, the ECOWAS Monitoring Group (ECOMOG), seizing weapons from its Malian troops (Berman, 2000, pp. 6–7). These few examples are illustrative of the much wider problem of seizure and theft of government stockpiles.[16]

In West Africa, the practice of governments creating and arming militias to shore up their hold on power or to counter rebel groups operating within their borders is widespread. Because these transfers of weapons from government stocks are of questionable legality, detailed records might not be kept, making it difficult to ascertain the strength and firepower of these pro-government militias.

The governments of Côte d'Ivoire, the Gambia, Guinea, and Liberia have all created armed groups they expected would defend their hold on power. Since 2000, the government of Côte d'Ivoire has increasingly used militia groups to bolster its standing, suppress political activism, and fight the rebellion that arose in 2002. These pro-government militias engage in extortion, crime, and harassment of civilians, but enjoy impunity due to their relationship with the government. The government's ability to control even the groups' membership appears limited, however, as some militia members are common criminals. One UN official told a Human Rights Watch researcher, 'Half of the militias could be political bully boys and half freelancers out to make money. It is hard to distinguish between them' (HRW, 2005b, p. 20).

Government creation of and support for armed groups should not be undertaken lightly, as these groups are unpredictable and can quickly become uncontrollable, as the above example from Guinea illustrates. They have a tendency to outlive the period during which they were deemed politically useful and will reinvent themselves as the incentive arises. In the end, it is the local communities that suffer from these 'armed and aimless' young men's robberies and vigilantism (Florquin and Berman, 2005, p. 386).

Corruption is another important source of weapons for armed groups. Soldiers and officers' illegal sales of weaponry were a 'main source of arms supply' during the Malian rebellion (Florquin and Pézard, 2005, p. 53). Returning peacekeepers may also be prone to corruption. In Nigeria, for instance, many small arms in circulation were introduced by soldiers who had served in peacekeeping missions abroad; police officers and government soldiers have also been reported to sell their weapons on the black market (Obasi, 2002, pp. 76–77), or rent their firearms for short periods of time (Best and Von Kemedi, 2005, p. 36).

Another important domestic source of small arms throughout the West African region is craft production, a practice that dates back to the introduction of iron working to the region several hundred years ago. Craftsmen in Ghana are particularly known for their skill. The government ban on the practice has not stopped the craft but rather resulted in sophisticated and secretive networks of gunsmiths throughout the country. The weapons produced are those for which ammunition (largely imported from abroad) is available on the open market. Currently, the most popular and prevalent types of ammunition include 12-bore shotgun shells and .410 cartridges. Reportedly, gunsmiths possess the skill to copy imported Kalashnikov assault rifles. Though gunsmiths in Ghana produce other implements, such as agricultural tools, guns are by far their most lucrative product (Aning, 2005, pp. 80–81). Individual sellers deal guns made in Ghana throughout the region, and reports indicate that Ghanaian gunsmiths have been invited to neighbouring Côte d'Ivoire to demonstrate their skills (Aning, 2005, p. 93). Craft-produced weapons have constituted a part of armed groups' arsenals in at least the following West African countries: Guinea, Niger, Nigeria, Senegal, and Sierra Leone (Florquin and Berman, 2005, pp. 281, 321, 342, 362, 373).

Corruption is an important source of weapons for armed groups.

Regional and global

Shipments from suppliers from outside the region continue to take place despite the 1998 ECOWAS Moratorium (ECOWAS, 1998) and various UN embargoes on specific countries and regions. Illicit arms transfers from outside West Africa generally involve at least one of two strategies to avoid detection: false end-user certificates and use of a go-between country as a transhipment point.

Though targeting suppliers in other countries is important, looking at the regional dimensions of arms acquisition is arguably more relevant to West Africa. In Liberia, the two main rebel groups, LURD and the Movement for Democracy in Liberia, received most of their weapons from neighbouring states. The governments of Guinea[17] and Côte d'Ivoire, respectively, imputed strategic value in arming these challengers to the government of President Charles Taylor. Some of the 81 mm mortar rounds LURD used in its June–July 2003 attacks on Monrovia were reportedly shipped from Iran to Guinea and then smuggled to LURD (HRW, 2003, p.15). In addition, LURD received arms and ammunition (such as United Arab Emirates-made mortar ammunition) directly from Guinean stockpiles (UNSC, 2002, para. 94; HRW, 2003, pp. 18–25).

The West African region is replete with instances of countries facilitating the armament of their neighbours' opponents. The government of Guinea-Bissau, for instance, reportedly provided weapons to the MFDC in Senegal. Côte d'Ivoire became a particularly tangled strategic web, with the Ivorian Popular Movement for the Great West (Mouvement populaire ivoirien du Grand Ouest) and the Movement for Justice and Peace (Mouvement pour la justice et la paix) armed and supported by the Charles Taylor government in Liberia, and the MPCI backed militarily by Burkina-Faso. Most of these continually recirculating weapons were originally manufactured in far-off factories and shipped as legal state-to-state transfers.

Complementing regional leaders' provision of small arms to various armed groups is the steady trickle of weapons known as the 'ant trade', when arms bought, often legally, in one country are smuggled in small quantities into another country. Widespread insecurity fuels demand for these weapons, delivered via West Africa's highly porous, largely unpatrolled borders. This 'ant trade' is hard to document, as shipments are rarely intercepted. Available evidence suggests, however, that it is a significant source, sometimes linked to larger-scale trafficking operations. For instance, Warri, the capital of Delta state in Nigeria, is known as a centre of such trade, with smugglers from Guinea-Bissau, Gabon, and Cameroon reportedly purchasing arms from speedboats off the coast. Many of these guns then circulate throughout the region (Best and Von Kemedi, 2005, p. 25). In many instances, illegally bunkered crude oil (crude oil that has been siphoned illicitly) is bartered for the weapons (Orji, 2005, p. 8).

Box 10.3 Armed groups and MANPADS

Armed groups, whether pro-government militias, rebels, or vigilantes, are largely thought to be entirely reliant on small arms such as Kalashnikovs and assault rifles. Yet several armed groups in West Africa (MPCI, RFDG, LURD, RUF) reportedly possessed man-portable air defence systems (MANPADS), and only a few have been recovered (Florquin and Berman, 2005, p. 387). The Algerian Salafist Group for Preaching and Combat (Groupe Salafiste pour la prédication et le combat, GSPC) allegedly acquired surface-to-air missiles during a stay in Mali (Smith, 2004). The arsenals of seven West African countries (Burkina Faso, Cape Verde, Côte d'Ivoire, Ghana, Guinea, Guinea-Bissau, and Nigeria) contain these weapons (IISS, 2004), rendering effective stockpile management crucial. These weapons are highly lethal, capable of attacks on both military and civil aviation.

WEAPONS MANAGEMENT AND CONTROL WITHIN ARMED GROUPS

Depending on the group's territorial objectives, relationship to the government, and motivations, weapons control strategies vary widely. However, certain trends in weapon management are apparent when one takes the traits of armed groups into account.

During the 1990–96 rebellion in northern Mali, both weapons and ammunition were scarce. Some researchers contend that the insurgency was launched with a single AK-47 (Humphreys and ag Mohamed, 2003, p. 3). On the rebels' side, the early months of the rebellion were taken up with obtaining weapons, ammunition, and supplies—largely from government stockpiles. Even after they had stocked their arsenal with pistols, assault rifles, light and heavy machine guns, RPGs, and mortars,[18] leaders' control over the weapons' use was strict. Except during states of alert, when each man was issued a weapon for defence, the weapons were kept under tight supervision. All weapons were the property of the movement as a whole, and one man in each base was given responsibility for accounting for their whereabouts.

Perhaps even more important than weapons control was control of ammunition. If ammunition was scarce, only those fighters with the best aim were given ammunition and were dispatched on a specific raid, such as to a police station, to obtain more. Moreover, when fighters could open fire was regulated: shooting in the air, for instance, was wasteful, an infraction punishable by temporary isolation (and being prohibited from going on mission), head shaving (considered a sign of shame), or even the infliction of severe pain (Florquin and Pézard, 2005, pp. 54–56).

Weapon and ammunition scarcity was not the only reason the Malian insurgents enforced strict control over their use. The Malian rebels had the focused political aim of greater autonomy, and weapon control was one facet of a broader strategy of employing strict rules of engagement and soliciting the civilian support they saw as critical to their mission.

Nevertheless, the specific policies and procedures the movement adopted early on to maximize the effectiveness of its (relatively) limited supplies clearly reflect the constraints imposed by its equipment. In such cases, one can point to a correlation between the availability of weapons or ammunition and the degree of control exerted over them (LORD'S RESISTANCE ARMY).

The case of Liberia provides further evidence of such a relationship. The Liberian war was largely seen as an economic war of plunder rather than a conflict engendered by political ideals. Members of armed groups in Liberia engaged in looting and widespread terrorization of the population, as documented extensively by a number of human rights organizations and other observers.[19] As described in the above section on sources of small arms, Liberian armed groups had access to deep reserves of both weapons and ammunition.

During their shelling of Monrovia in July 2003, LURD fighters were 'often untrained in mortar use and were seen to fire without making any effort to distinguish targets', with civilians bearing the brunt of this indiscriminate violence (HRW, 2003, p. 10).

One indicator of weapons availability within an armed group is whether arms are given to enlisted CAFF, who are often believed to be poor shots. In one study of former Liberian child soldiers, 94 per cent reported that they had access to a firearm, the majority of which were AK-47 assault rifles (Wille, 2005, pp. 198, 200). Despite the leaders' apparent laissez-faire attitude towards their subordinates' actions, control over weapons management could be exerted when scarcity rendered it necessary.[20]

Weapons availability and the degree of control exerted over them are sometimes correlated.

INFLUENCING WEAPONS SUPPLY AND DEMAND

The previous sections document and flesh out the various strategies and characteristics of armed groups in relation to the small arms and light weapons they employ. This section investigates the strategies that are being used to control and reduce the stocks of weapons and ammunition supplies obtainable by armed groups. These strategies may target the supply of weapons and the demand for them. The two categories are not unrelated. Because many people in West Africa conclude they have good reason (such as a breakdown in public security, or unemployment) to acquire or retain weapons, supply-side efforts must be complemented by initiatives that address the motives underpinning weapons demand. In both cases, strategies involve creating incentives, both positive and negative, to influence actions.

Pure 'supply-side' measures

The first lesson that emerges from an analysis grounded in West Africa is that supply-side measures for weapons control are insufficient given that the region is already awash in guns (Florquin and Berman, 2005). The arms embargo imposed on both sides of the Ivorian conflict in November 2004 (and strengthened in February 2005) had important political value but, even if customs officials and UN inspectors had the capacity to enforce it rigorously (which they currently lack), UN officials recognize that the fighters already possess sufficient armament for the conflict to continue long into the future (HRW, 2005b, pp. 32–33). The laws of supply and demand ensure that these weapons will be recirculated from conflict region to conflict region as demand arises. Most of the armed groups currently active in West Africa were formed after 1998, and importation of weaponry to the area has continued in the years since.[21]

In terms of supply-side measures, a more promising tactic would be to target the provision of ammunition. Most countries in West Africa do not produce their own ammunition and are thus dependent upon imports. If ammunition is not stored properly, it quickly deteriorates (Small Arms Survey, 2005, p. 17). Without ammunition, weapons quickly lose their value.

LURD, preparing its 2003 assault on Monrovia, had an ample supply of British-made mortars from 1973 (HRW, 2003, p. 18), but had to postpone the attack while waiting for a new supply of mortar rounds. The craft weapons producers in Ghana do not bother to produce ammunition as it can easily be purchased new on the market. Instead, they produce weapons of the same calibre as the already available ammunition.

Box 10.4 Ammunition production in Tanzania

On 23 June 2005, the Belgian government decided–after some wavering–not to license a proposed Belgian-financed ammunition factory in Mwanza, Tanzania, on the shores of Lake Victoria (Small Arms Survey, 2005, p. 14). Tanzania claimed to want only to upgrade an existing facility; the Belgian government decided, based on equipment that had been imported, that the renovations in fact constituted an entirely new factory. Further, the Belgian government determined that Tanzania did not have an effective strategy in place for assuring that the ammunition did not end up in the hands of illegal users. International outcry over the planned development of the ammunition factory–including the threat of the stoppage of aid from the European Union to Tanzania–probably also contributed to the decision to scuttle the project (Mpinganjira, 2005). In addition, research by the Belgium-based Groupe de recherche et d'Information sur la paix et la sécurité (GRIP) has repeatedly shown that Mwanza is a transhipment point for tonnes of weaponry, arriving on flights from Belgium, Latvia, the Czech Republic, Qatar, Slovakia, Israel, and Ukraine[22] (Bilali, 2005).

Security sector reform

Reform of the formal security sector has implications for both the supply of small arms to non-state armed groups and these groups' demand for such weapons. As described above, weak oversight and control of official weapon stocks provides a steady source of weapons for many armed groups. At the same time, such groups often take up arms in response to the state's failure to provide public security. This section will briefly present examples of security sector reform (SSR) and show how they involve issues of both supply and demand (CAMBODIA).

At present, many militaries in West Africa lack the capacity to protect their stores of arms when they come under attack by armed groups. Moreover, government accounting for the weapons in their possession is currently weak in many countries. This gives rise to a number of challenges. Monitoring and control of groups armed by governments becomes difficult and, once these groups have outlived their intended purpose, it is hard to bring them under control or to ensure that all the weapons that have been handed out are returned.

Stockpile management is also undermined by corruption in many West African state security forces (Florquin and Berman, 2005, pp. 387–88). With oversight of stockpiles minimal, it becomes easier for armed groups to purchase weapons from corrupt officials. In addition, the failure of many armed forces to provide a basic standard of living for conscripts increases the temptation to engage in corrupt acts, and promotes the kind of disgruntlement that can lead to mutinies, coups, or related unrest. For example, the poorly remunerated military in Guinea used roadblocks to generate personal income. Though President Conté banned the roadblocks in 2003, by 2005 they had reappeared, showing how state security forces have the potential to operate as armed groups if economic incentives are sufficiently strong (ICG, 2005b, p. 16).

Targeted skills training for the security sector can also be helpful in areas plagued by conflict. Refugee-populated areas in Guinea, home to some 450,000 people, became militarized beginning in 2000. Many of the refugees were combatants from regional conflicts, and some used this foreign soil, and its effectively unpoliced refugee camps, to regroup (Milner, 2005). A training programme launched by Canada in 2003 involving the deployment of several Canadian mounted police to teach Guinean guards policing strategies and ethics, though far from solving all of the camps' problems, nevertheless resulted in greatly improved security for their inhabitants.[23]

Targeted skills training for the security sector can improve human security.

Removing the tools and incentives for war

Given that significant numbers of small arms will remain in circulation for the foreseeable future, it becomes critical to consider why people take up arms and to develop incentives that create peaceful alternatives (Regehr, 2003, p. 4). Disarmament, demobilization, and reintegration (DDR) programmes have been developed as one mechanism for transitioning former fighters back into productive and peaceful civilian lives. In West Africa, a lack of employment opportunities is one reason why people join armed groups, and so DDR policy-makers hope that, by trading in the fighting lifestyle for cash or training assistance, their demand for arms will decline.

In the ECOWAS region, DDR initiatives have been carried out in Liberia, Mali, Niger, and Sierra Leone, and programmes are planned for Côte d'Ivoire and Senegal. 'Weapons for Development' projects, voluntary gun collection projects, and other small-arms control initiatives have reached Ghana, Mali, Nigeria, and Sierra Leone. The results of these programmes, however, have been mixed at best. Between 1998 and 2004 more than 200,000 small arms were collected in the West African region, with at least 70,000 of these subsequently destroyed. However, the quality of the weapons destroyed appears to have been quite poor, which suggests that better-quality models are still circulating throughout the region (Florquin and Berman, 2005, pp. 388–98). With the large number of programmes already

Box 10.5 Northern Ireland

In West Africa, economic incentives, such as the reintegration assistance that is part of DDR programmes, have been used to persuade people to lay down their arms. In Northern Ireland the incentives were political in nature.

On 28 July 2005 the Irish Republican Army (IRA) announced its intention to 'verifiably put its arms beyond use'. This promised an end to what had been one of the main stumbling blocks on the road towards peace in Northern Ireland: 'decommissioning' of paramilitary arms (disarmament). Two months later, on 26 September, the Independent International Commission on Decommissioning reported that it had observed and verified a 'full range of ammunition, rifles, machine guns, mortars, missiles, handguns, explosives, explosive substances and other arms' representing 'the totality of the IRA's arsenal', which were then destroyed by being covered with concrete (IICD, 2005, pp. 1–2). As of November 2005, the Loyalist paramilitary groups had yet to decommission, though some appeared to be taking steps towards doing so (BBC, 2005).

The debate on decommissioning as a prerequisite for political negotiations dates back to 31 August 1994, when the IRA announced a ceasefire. Weapons repeatedly stymied progress towards peace as they took on a symbolic and political value that far exceeded their physical worth, both in society and within the political process. The Westminster elections in June 2001 proved to be a turning point. Sinn Féin, the political party associated with the IRA, won a majority of the Nationalist votes and thus saw that it could rely on political processes to achieve its goals (Hauswedell and Brown, 2002, pp. 11, 12).

A lack of success at the polls alienated the Loyalists from the peace process. Two leading paramilitary groups, the Ulster Volunteer Force and the Ulster Defence Association, jockey for supremacy and cannot commit to disarmament without appearing to have surrendered. Following verification of IRA decommissioning by international monitors in September 2005, the Reverend Ian Paisley, leader of the Democratic Unionist Party, remained sceptical of IRA intentions.

Sinn Féin's president, Gerry Adams, explained the move toward decommissioning: 'I think that violence is a response to particular political conditions. When there was no alternative to armed struggle, I defended the use of armed struggle' (Solomon, 2005, p. 13). In the Northern Ireland case, mediators put political pressure on the IRA to lay down their weapons while also including them, and taking their concerns seriously, in the peace process.

A man and child stand on a railing in front of a large Loyalist mural in north Belfast, Northern Ireland, October 2001.
© Adam Butler/AP Photo

completed, and several envisaged for coming years, it is important to look at the factors underpinning the success or failure of these DDR efforts.

Mali underwent a disarmament programme from 1995 to late 1997. In all, some 12,000 ex-armed group members benefited from the effort, which relied on the voluntary return of arms.[24] At the Timbuktu Flame of Peace ceremony in 1996, 3,000 weapons were set ablaze in front of 10,000 spectators, an event both symbolic and practical that has been repeated elsewhere in the years since. However, former combatants from all sides of the conflict argue that only a minority of the weapons destroyed were actually used in the rebellion; many people turned in old, barely usable rifles, and others who had not fought turned in some armament just so they could qualify for the reintegration assistance. In the end, some of the weapons used in the Malian conflict made their way to other conflict zones.[25] The borders with neighbouring Mauritania and Côte d'Ivoire are particularly permeable for the purposes of arms trafficking. Within Mali itself, the peace process did not initially bring an end to insecurity (armed banditry was on the rise), and many ex-combatants and local leaders opted to retain their weapons as a 'wait-and-see' tactic. On a positive note, Mali's DDR programme was apparently successful in a way few such endeavours can claim to be: the reintegration component, which consisted of three payments of 100,000 CFA francs (approximately USD 200) to start small businesses, which were in some cases supplemented with micro-credit loans, worked. According to UNESCO, by 2000, three years after the end of the programme, 90 per cent of reintegrated former combatants continued to earn a living from the employment they entered as a result of the reintegration initiative (Boukhari, 2000).

The Liberian DDR experience is particularly instructive. In December 2003, less than five months after a peace agreement ended Liberia's civil war that had lasted for more than a decade, the United Nations Mission in Liberia (UNMIL) began a DDR programme in the country. The programme was rushed through without proper planning, and the first few weeks of its operation were marked by violence as thousands more ex-combatants presented themselves than organizers had anticipated.

After this inauspicious beginning, the disarmament and demobilization phase of the programme officially ended on 31 October 2004. All told, 102,193 people went through the screening process, more than two and a half times the expected number (38,000). Partly, this disparity can be explained by the fact that armed factions failed to provide UNMIL with any lists or tallies of the number of fighters to expect, and the DDR team thus had to make do with best estimates.[26] In addition, the screening process was insufficient: it was carried out by local NGO staff, who lacked the training and authority to identify and disqualify from the programme those who, rather than being ex-combatants, simply showed up with an old firearm or some ammunition (only 150 rounds was required for entry) so as to qualify for the promised reintegration benefits (Nichols, 2005). Programme participants turned in 27,804 weapons, many of which were of poor quality, and more than ten per cent of which were old shotguns. Faction leaders' promises to turn in larger weapons, such as those used during the August 2003 siege of Monrovia, went unfulfilled (NCDDRR, 2004). Demobilization consisted of five days spent at a cantonment site, a duration too short to engender any substantive behavioural or attitudinal change. Commanders were in effect in control of the process. For instance, when rioting broke out among disgruntled ex-combatants in Tubmanburg, UNMIL officials had to call in a LURD commander to calm the men down.[27] In both Sierra Leone and Liberia, commanders restricted access to the DDR programme to those fighters (and civilians) who would return a portion of the benefits to their superiors (HRW, 2005a, p. 50).

What lessons can be drawn from the Liberian experience of disarmament and demobilization? First, it was an enormous and complicated undertaking, and so the fact that the process occurred without any major setbacks aside from the December 2003 troubles is fortunate. Planning was insufficient, and policies were unclear during the duration of the programme. In addition, there was a lack of regional coordination. Part-way through the process, Côte d'Ivoire announced its forthcoming DDR programme, which would offer more lucrative reintegration assistance than the Liberian one, raising speculation that Liberian ex-combatants would wait and hand in their weapons across the border. Because programme officials in Liberia gave insufficient credence to the volatile regional dynamics, the moment they started offering cash for weapons they created a market in guns and ammunition. Because the screening process was flawed and several times the expected number of people participated in the disarmament–demobilization phase, funds for the reintegration portion were inadequate, leaving many frustrated ex-combatants without expected assistance (Nichols, 2005).

A former LURD combatant surrenders weapons at a disarmament point set up by the Pakistani contingent of the United Nations Mission in Liberia in April 2004.
© Tim A. Hetherington/Panos Pictures

Box 10.6 An investigation of amnesties for armed groups: Colombia

In recognition of the fact that, without some concessions, armed conflict has the potential to drag on interminably, recent years have seen a number of amnesty laws designed to entice combatants to lay down their guns. For instance, both Uganda and Algeria have passed laws providing clemency for ex-armed group members. One of the longest-running conflicts in the Western hemisphere has been in Colombia (COLOMBIA) and, with the aim of bringing hostilities to an end, the government has negotiated an amnesty for the approximately 20,000 paramilitaries who operate throughout the country in the loose confederation known as the United Self-Defence Forces of Colombia (Autodefensas Unidas de Colombia, AUC). Tied to disarmament and demobilization, the Colombian Justice and Peace Act offers concessions that may in the end undermine long-term peace. In particular, the Colombian law does not address the issue of dismantling the paramilitary networks, which, like armed-group structures in West Africa, have a tendency to transform themselves (in this case into criminal narcotic syndicates) in such a way that their continued survival is assured.

The Colombian government helped form the paramilitaries more than 20 years ago to combat the leftist guerrillas that have been fighting the government since the 1960s. Though the paramilitaries were officially deemed illegal in 1989, they have continued to benefit from government military logistical support (Isacson, 2005, p. 1). The negotiation of an amnesty, which critics have labelled a 'negotiation among friends', has taken several years, largely because of disagreements over the so-called 'framework law', which provides an outline of the demobilization and reintegration process, a planning step the international donors who will fund the estimated USD 250 million project insisted upon. The 'framework law' eventually agreed upon by the two sides has drawn criticism from many observers, who argue that its narrow timelines for any prosecutions for the estimated thousands of civilian murders perpetrated by the paramilitaries in effect means none will be brought to justice.

Given the weak 'framework law' it is likely the paramilitary groups will remain intact, though they will officially go through a demobilization process. Since the turn of the present century many have shifted from resembling typical paramilitaries, with uniforms and heavy weapons, to Italian-style mafias, that is, regional crime syndicates powered by drug money and corrupt politics (Isacson, 2005, pp. 4, 7).

Neither the Justice and Peace Act nor the associated 'framework law' addresses this change in paramilitary organization. Both measures operate at the level of the individual—providing for some reintegration assistance—while failing to address the issues of the group's structure and how to eliminate them from society. As one 'demobilized' paramilitary told Human Rights Watch researchers, 'The demobilization . . . is a farce. It's a way of quieting down the system and returning again, starting over from another side' (HRW, 2005d, p. 1). This failure could be ominous given that the paramilitary groups, whatever their current incarnation, have vast resources at their disposal in the form of land they have commandeered and the profits from narcotics trafficking. As of mid-August 2005, 8,804 paramilitaries had been disarmed, with 5,843 weapons recovered (MAPP-OEA, 2005, p. 6).

Many of Colombia's citizens have accepted the amnesty in the hope that it will bring some peace and end the persecution they have faced from these armed groups (Forero, 2005). Whether it will bring an end to the armed groups, however, remains to be seen. Though amnesty laws achieve some good by providing an avenue for combatants to lay down their arms, it is important that they nevertheless address all the dimensions of combatants' potentialities for violence.

DDR initiatives are an important part of a transition from conflict to peace, as they represent best efforts to deal with the large number of armed ex-fighters who find themselves unemployed and back among their communities. However, several paradoxes immediately emerge as such programmes are being planned. First, with DDR, the international community is essentially rewarding those who took up arms with cash, training, or other assistance, while the civilian population—which often suffered at the hands of these same armed groups—is left with nothing.[28] Commanders have been known to recruit fighters using the prospect of eligibility for an eventual DDR programme as a carrot (HRW, 2005a, pp. 24, 49). Second, voluntary DDR programmes inevitably interfere with whatever small arms and ammunition market already existed, often artificially raising prices or creating new markets. Third, DDR programmes often neglect the close links many armed groups have with local communities. The bond between commander and fighter may prove stronger than that of a traditional army (HRW, 2005a), raising the spectre of future conflict if conditions change.

Box 10.7 An investigation of disarmament in Rivers State

Violence has plagued Rivers State, Nigeria, since the 2003 elections, when militias and cult groups arose to influence the outcome. The dialogue that began in September 2004 between the Nigerian government and leaders of the two largest militia groups, the Niger Delta People's Volunteer Force (NDPVF) and the Niger Delta Vigilante (NDV), began amidst an atmosphere of hope: rather than the violence that usually characterized both sides' tactics, discussion flowed into solid commitments. The leaders of the NDPVF and the NDV agreed to disband their militias and to disarm totally while maintaining a strict ceasefire. Under the terms of the disarmament agreement, militia members would receive cash for turning in weapons voluntarily; however, if they refused, they would face legal sanctions. The state government, led by Dr Peter Odili, offered USD 1,800 for an assault rifle; it is unknown how this figure was reached, but it appears inflated, and it quickly boosted the small arms market. In practice there was some variation in the amounts disbursed. The total number of weapons collected between October and December 2004, when the programme was in operation, is not known. The government figures and the totals claimed by the militia leaders do not match; the lack of transparency as guns were collected—with only a portion destroyed—means that the real numbers will probably never be known. The effects, however, are clear.

For one thing, the large amounts of cash being offered meant that an arms trade quickly sprang up: militia leaders and cultists sourced arms from elsewhere and sold them back to the government, pocketing a sizeable profit. Alhaji Asari Dokubo, the leader of the NDPVF, allegedly collected and sold two rocket launchers in this manner. Some government members of the disarmament programme, too, profited from their involvement, using it as a way to strengthen their own armed groups.

In addition to augmenting the local arms trade, the cash incentives created rifts within the armed groups, as members accused their leaders of keeping all the money for themselves rather than sharing it among the rank and file. Some of these disgruntled fighters took their grievances to the government, demanding their own payments under threat of increased violence, while others, notably Boma George of the NDPVF, formed breakaway groups and confronted their former leaders.

By removing arms from communities, the disarmament programme was supposed to bring peace. However, the fragile ceasefire lasted only one month, and violence returned to Rivers State in early November 2004. In the period since, gang-related violence has been continual. Murder, rape, robbery, and extortion are common occurrences. Government arrests of some leaders resulted in further splintering of the armed groups, with periodic battles for supremacy among them. Fighters remain well-armed, with Kalashnikovs, grenades, and other armaments. In the end, then, the disarmament programme offered few positive results; instead, the various armed groups were strengthened, as were some government officials and their militia, by the surging arms trade, and violence continues to engulf the region.

Source: NDPEHRD (2005)

DDR programmes aim to transition combatants back to civilian life, taking away their guns and replacing them with more productive tools. Such programmes can be an important way to calm the immediate post-conflict situation, but they are only the first step in assuring long-term peace: civilian demand for weapons, too, must be addressed. Arms for development (AfD) programmes aspire to do just that. Some community-based voluntary disarmament programmes have had success, notably those that take account of the needs and inputs of particular communities.

Sierra Leone's AfD programme is often cited as a model. In addition to initiatives aimed at improving border security and countering illegal trafficking, each chiefdom established a Project Management Committee to oversee the collection of arms and development projects. These local leaders, chosen through input from all community members, created 'dropping centres' where residents could drop off their weapons, which were then registered and taken to a secure stockpile (the residents are able to apply for the return of the weapon and a proper license at a later date). Once all the weapons in the community have been turned in, the police (monitored by UNDP) conduct a verification search of a random sample of 30 per cent of the village's households. If no weapons are found, the community is issued a 'gun-free' certificate and can then determine the development projects it would like to implement, with the support of UNDP, such as a health centre, school, or market. By mid-2005, 18 chiefdoms had been declared 'weapons free', and the programme was on track to reach 25 by the end of the year (UNDP, 2005). Though it is too early to tell whether there has been a decrease in small arms violence nationwide, there have been no gun incidents in the chiefdoms involved with the programme. Also, 'communities in arms-free chiefdoms are more inclined to report sight of small arms to the police as such behaviour is no longer regarded as acceptable by the communities' (Werthein and Widmer, 2005, p. 3).

The success of such programmes points to the need for dialogue in communities about steps that can be taken to stop the spread of guns and violence. In West Africa, where many ex-combatants report that they were driven to join an armed group due to the impossibility of finding alternative employment, it seems that a broadening of economic opportunity would lead to a reduction in violence. The large numbers of young people, especially men, with some kind of connection to an armed group, whether dormant or active, and the continuing youth bulge mean that creating incentives linked to lowering weapons supply and demand will be of utmost importance in the coming years (ANGRY YOUNG MEN).

INFLUENCING BEHAVIOUR

This section looks at some of the strategies employed by humanitarian organizations and others on the ground to curb weapon misuse by armed groups and to bring them into compliance with international humanitarian law (IHL) and other international norms.

In their 2004 study *Roots of Behaviour in War,* Munõz-Rojas and Frésard of the International Committee of the Red Cross (ICRC) found that combatants generally have an understanding of IHL. Although they may or may not refer to IHL as such, it is widely believed that civilians should be spared the effects of violent conflict. However, when it comes to action, combatants are motivated by peer pressure and adherence to authority rather than principles. Normative moral persuasion on its own, the authors found, has little effect on conduct, because combatants generally act in accordance with what they are told and what those around them are doing. Their conduct is not based on particular principles, even if they acknowledge and believe such principles to be important.

Combatants are motivated more by peer pressure than principles.

Human Rights Watch conducted interviews with ex-combatants in Côte d'Ivoire, Guinea, Liberia, and Sierra Leone that provide further evidence of the importance of commanders' orders:

> [T]he degree of effective command and control, and discipline maintained in the different armed groups played a key role in the kind and frequency of violations observed and perpetrated by the interviewees while fighting outside their own countries. (HRW, 2005a, p.31)

In their interviews, fighters singled out LURD for its attempts to instil some respect for civilians, but the broader culture of impunity meant that these efforts were 'inconsistent and often unsuccessful' (2005a, p.31). Indeed, the leadership of all of the armed groups represented within the HRW report tacitly approved of or explicitly ordered such war crimes as attacks on civilians, looting, and pillaging. Ending the culture of impunity that currently benefits the leadership of many armed groups could convince them of the importance of reining in their subordinates and curbing attacks on the civilian population.

Another strategy for engaging with armed groups involves working with the policy-makers and diplomats who, though far from any particular conflict, can exert an influence on its course by 'naming and shaming'. That is, by drawing international attention to an armed group's abuses and raising the spectre of sanctions or other punishment, a change in that armed group's methods of warfare can be engendered. Naming and shaming has been seen to be quite effective in some instances, but it has limitations. In particular, in order to induce a change in behaviour it must be tied to rewards and punishments. Also, 'only some groups care about their international reputations and/or international legitimacy, while others often lack the organizational capacity to even know they are being shamed' (Armed Groups Project, 2003, p. 3).

An extensive, institutionalized measure to implement a naming and shaming policy for those groups that employ child soldiers was adopted by the UN Security Council on 26 July 2005. Security Council Resolution 1612 will establish a 'monitoring and reporting system' that will target those parties, both armed groups and governments, who exploit children in conflict. Those offenders who fail to develop and carry out 'action plans' to address their abuses of children will be subject to 'targeted and graduated' UN sanctions, such as 'a ban on the export and supply of small arms and light weapons and of other military equipment and on military assistance' (UNSC, 2005, paras. 7, 9).

While it remains to be seen how its mechanisms for punishment will be deployed in practice, on paper the UN resolution constitutes a significant advance in international efforts to curb the use and abuse of children in war. By punishing the groups that are 'named and shamed', the resolution aims at deterring the offending behaviour. Still, punishments such as arms embargoes or account freezing must be fully monitored or else risk irrelevance to the leaders using child soldiers; the violation of past arms embargoes in West Africa suggests some commanders in the region see more benefit in continuing conflict than adhering to UN resolutions.

By working to reduce violence through targeting behaviour during conflict, 'naming and shaming' is an important tool for influencing armed groups, particularly when combined with punitive measures designed to coerce compliance, as in the resolution on CAFF. However, the weakness of normative moral persuasion in West Africa lies in the difficulty of enforcing consequences in settings characterized by widespread impunity. Though some groups, such as the Malian rebels, employ their own rules of engagement, 'spoilers' can mask criminality with the chaos of conflict in a region where state security forces lack the capacity to provide public order, undermining any single group's norms of behaviour. In this context, incentives influencing weapon demand outside of conflict, such as development assistance to 'gun-free' communities, appear a more promising means of arresting the cycle of violence that has beset the West African region with the proliferation of armed groups.

CONCLUSION

Armed groups—'challengers to the state's monopoly of legitimate coercive force'—constitute a persistent threat worldwide. In West Africa, the challenges arising from the diversity of such groups, from pro-government militias to rebel groups and vigilantes, is matched only by their unpredictability, with allegiances, geographic ambitions, and motives likely to change over time. In such an impoverished region, even groups that are not self-serving (motivated by real ideologies and grievances) can quickly become 'predatory', feeding on the communities they once called home. The great variety of armed groups makes blanket approaches to their control problematic.

Controlling the supply of the weapons armed groups rely upon—namely, small arms and light weapons—is one important means of reining them in. As government-owned stocks are a key source of weapons for many armed groups, more rigorous stockpile management is one obvious priority. Moreover, in a region already awash with small arms, controlling the (largely external) supply of ammunition is at least as important.

Using normative moral persuasion, the international community often seeks to influence the behaviour of armed groups during conflict for the purpose of protecting civilians. The weak command and control structures of many West African armed groups, however, tend to undermine the persuasive power of norms. Incentives, targeting fighters as individuals as well as their leaders, that address the underlying motivations for armed violence seem more likely to engender changes in fighters' actions. In the West African context, given armed groups' nature and motivations, those incentives will more often be economic in nature than political.

As yet the promise of incentive-based strategies remains unfulfilled. In order to reduce demand for weapons, future disarmament efforts need to develop meaningful ways of generating post-conflict employment, both for ex-combatants and for the members of the communities they live in. Successfully tackling the problem of armed groups requires in-depth analysis of specific local and regional dynamics, an arduous process for which funding and time are often limited. Nevertheless, this challenge will define human security in West Africa in the coming years. ◢

LIST OF ABBREVIATIONS

AfD	arms for development	**LURD**	Liberians United for Reconciliation and Democracy
AUC	Autodefensas Unidas de Colombia		
CAFF	children associated with fighting forces	**MANPADS**	man-portable air defence system
CFA	Communauté financière d'Afrique	**MFDC**	Mouvement des forces démocratiques
DDR	disarmament, demobilization, and reintegration	**MILF**	Moro Islamic Liberation Front de la Casamance
ECOMOG	ECOWAS Monitoring Group	**MPCI**	Mouvement patriotique du Côte d'Ivoire
ECOWAS	Economic Community of West African States	**MRU**	Manu River Union
		NDPVF	Niger Delta People's Vigilante Force
GAM	Gerakan Aceh Merdeka	**NDV**	Niger Delta Vigilante
ICHRP	International Council on Human Rights Policy	**OPM**	Organisasi Papua Merdeka
		RPG	rocket-propelled grenade
ICRC	International Committee of the Red Cross	**RUF**	Revolutionary United Forces
		SSR	security sector reform
IHL	international humanitarian law	**UNMIL**	United Nations Mission in Liberia
IRA	Irish Republican Army	**USD**	United States dollars

ENDNOTES

1   Quoted in Moorehead (2005, p. 57).

2   ECOWAS comprises 15 member countries: Benin, Burkina Faso, Cape Verde, Côte d'Ivoire, The Gambia, Ghana, Guinea, Guinea-Bissau, Liberia, Mali, Niger, Nigeria, Senegal, Sierra Leone, and Togo.

3   See also Nicolas Florquin and Eric G. Berman, eds., 2005, *Armed and Aimless: Armed Groups, Guns, and Human Security in the ECOWAS Region*. Geneva: Small Arms Survey, upon which this chapter is based.

4   The author thanks Pablo Policzer for contributing the term 'normative moral persuasion' in his review of this chapter.

5   For instance, the federal government withdrew the police from Anambra state in 2004, apparently due to displeasure with the state's governor, Chris Ngige (Onyeozili, 2005, pp. 48–49). Human Rights Watch has documented the pervasiveness of torture and lack of respect for the law on the part of the Nigerian police (HRW, 2005c; 2006, pp. 146–47). Chevigny (1995) shows that in the Americas a culture of police violence and impunity has given rise to violent vigilantism, raising important questions for West Africa.

6   Written correspondence with Dickson Orji, Country Director, Nigeria Action Network on Small Arms, 9 November 2005.

7   The Coalition to Stop the Use of Child Soldiers uses the figure 300,000 (http://www.child-soldiers.org/childsoldiers/some-facts), while the former UN Special Representative for Children and Armed Conflict's count comes to 250,000 (Otunnu, 2005). Some question the empirical basis of such statistics and believe the actual number of CAFF to be much lower (see Human Security Centre, 2005, ch. 3).

8   Normative moral persuasion might be more effective at stopping children from joining armed groups than it is for adults. One study highlights the role parents play in keeping their children from joining armed groups, including the importance of 'changing attitudes to stop children wanting to join the fighters' (SC, 2005, p. 11).

9   This data is drawn from interviews with 270 CAFF: 100 in Liberia, 91 in Guinea, and 79 in Sierra Leone (Wille, 2005).

10   Interview 118, cited in Wille (2005, p. 191).

11   See also Florquin and Pézard (2005, p. 57); Humphreys and ag Mohamed (2003, p. 27).

12   Confidential interviews with Guinean military officials, Ministry of Social Affairs representatives, and UNICEF officials, Conakry, March 2004, cited in Milner (2005, p. 280).

13   Confidential interview with UN officials, Conakry, March 2004, cited in Milner (2005, p. 283).

14   By 19 December 2005, GAM had handed over all 840 weapons it agreed to surrender under the terms of the peace agreement. On 27 December, it officially demobilized the TNA. The Indonesian government fulfilled all its obligations under the peace accord by 5 January 2006, marking the end of armed conflict in Aceh (UNORC, 2006, p. 1).

15   Written correspondence with Robin Edward Poulton, International Consultant and Senior Research Fellow at the United Nations Institute for Disarmament Research, 5 May 2004, cited in Florquin and Berman (2005, p. 289).

16   For more information, see Florquin and Berman (2005, p. 387).

17   In Guinea's case, the transfers occurred through a single 'mid-level functionary in the agriculture ministry, who worked in conjunction with European arms dealers, [and] was responsible for issuing 80 per cent of the documentation for illicit arms fuelling West Africa's regional war in the 1990s' (ICG, 2005b, p. 16).

18   Focus group discussions with Malian ex-combatants, Bamako, 2–3 September 2004, cited in Florquin and Pézard (2005, p. 52).

19   Government of Liberia soldiers were equally, or more, culpable of atrocities against civilians (HRW, 2005a).

20   For instance, LURD postponed its final attack on Monrovia in July 2003 due to insufficient stores of mortar rounds, ordering fighters to retreat while accumulating additional armament (HRW, 2003, p. 6).

21   See, for instance, Global Witness (2004b).

22   The arms trade on Lake Victoria was the subject of an award-winning 2004 documentary film, *Darwin's Nightmare*.

23   Meetings with refugee communities in Lainé and Kouankan camps, 2 October and 4 October 2004, cited in Milner (2005, p. 159).

24   Phone interview with Colonel Sirakoro Sangaré, President of the Malian National Commission on Small Arms, 22 March 2005, cited in Florquin and Pézard (2005, p. 59).

25   Focus group discussion with Malian ex-combatants, Bamako, 2–3 September 2004, cited in Florquin and Pézard (2005, p. 59).

26   The disarmament committee in neighbouring Côte d'Ivoire does not seem to have learned from the mistake of underestimating ex-combatants. The National Commission for Disarmament, Demobilization, and Reintegration has estimated militia membership at 10,000, prompting the UN to caution that the figure was 'very likely to be underestimated' (HRW, 2005b, pp. 14–15).

27   Interview with UNDP DDRR official in Monrovia, 21 September 2004, cited in Nichols (2005, p. 127).

28   'In the east [of Liberia] near the border with Côte d'Ivoire, the residents of Zwedru see former fighters zipping around town on shiny new scooters and setting up small stalls with their disarmament money, while they struggle to feed their own families' (IRIN, 2005).

BIBLIOGRAPHY

Aning, Emmanuel Kwesi. 2005. 'The Anatomy of Ghana's Secret Arms Industry.' In Nicolas Florquin and Eric G. Berman, eds. *Armed and Aimless: Armed Groups, Guns, and Human Security in the ECOWAS Region*. Geneva: Small Arms Survey, pp. 78–107.

Armed Groups Project. 2003. *Curbing Human Rights Violations by Non-State Armed Groups*. Conference Summary and Report. Vancouver: University of British Columbia. 13–15 November.

Berman, Eric G. 2000. *Re-armament in Sierra Leone: One Year After the Lomé Peace Agreement.* Occasional Paper No. 1. Geneva: Small Arms Survey.

Best, Shadrack Gaya and Dimieari Von Kemedi. 2005. 'Armed Groups and Conflict in Rivers and Plateau States, Nigeria.' In Nicolas Florquin and Eric
G. Berman, eds. *Armed and Aimless: Armed Groups, Guns, and Human Security in the ECOWAS Region.* Geneva: Small Arms Survey, pp. 12–45.

Bilali, Charles Nasibu. 2005. 'La Persistance du traffic d'armes vers la RDC et le Burundi.' Groupe de Recherche et d'Information sur la Paix et la
Sécurité (GRIP). 29 April. <http://www.grip.org/bdg/g4571.html>

Boukhari, Sophie. 2000. 'Mali: A Flickering Flame.' *UNESCO Courier.* January. <http://www.unesco.org/cpp/uk/projects/zapata.htm>

BBC (British Broadcasting Corporation). 2005. 'UDA Meets Decommissioning Body.' 27 October.
<http://news.bbc.co.uk/1/hi/northern_ireland/4383760.stm>

Bruderlein, Claude. 2000. 'The Role of Non-State Actors in Building Human Security.' Paper prepared for the Centre for Humanitarian Dialogue.
Geneva, May.

Chevigny, Paul. 1995. *Edge of the Knife: Police Violence in the Americas.* New York: New Press.

ECOWAS (Economic Community of West African States). 1998. *Declaration of a Moratorium on Importation, Exportation and Manufacture of Light
Weapons in West Africa.* Abuja, 31 October. Reproduced in UN document A/53/763 – S/1998/1194 (Annexe) of 18 December.

Evans, Martin. 2004. *Sénégal: Mouvement des forces démocratiques de la Casamance (MFDC).* Africa Programme Armed Non-State Actors Project
Briefing Paper no. 2. London: Royal Institute of International Affairs.

Florquin, Nicolas and Eric G. Berman. 2005. 'Part II: Armed Groups and Small Arms in ECOWAS Member States, 1998–2004.' In Nicolas Florquin and
Eric G. Berman, eds. *Armed and Aimless: Armed Groups, Guns, and Human Security in the ECOWAS Region.* Geneva: Small Arms Survey, pp.
223–383.

Florquin, Nicolas and Stéphanie Pézard. 2005. 'Insurgency, Disarmament, and Insecurity in Northern Mali, 1990–2004.' In Nicolas Florquin and Eric
G. Berman, eds. *Armed and Aimless: Armed Groups, Guns, and Human Security in the ECOWAS Region.* Geneva: Small Arms Survey, pp. 46–77.

Forero, Juan. 2005. 'Colombia Unearthing Plight of its Disappeared.' *New York Times.* 10 August.

Frésard, Jean-Jacques. 2004. *The Roots of Behaviour in War: A Survey of the Literature.* Geneva: International Committee of the Red Cross. October.

GC (Geneva Call). 2002. 'About GC.' <http://www.genevacall.org/about/about.htm>

Global Witness. 2003. *The Usual Suspects: Liberia's Weapons and Mercenaries in Côte d'Ivoire and Sierra Leone.* London: Global Witness.

——. 2004a. *Liberia: Back to the Future.* Washington, DC: Global Witness.

——. 2004b. 'Dangerous Liaisons: The Continued Relationship Between Liberia's Natural Recourse Industries, Arms Trafficking and Regional
Insecurity.' Briefing document. December. London: Global Witness.

Hauswedell, Corinna and Kris Brown. 2002. *Burying the Hatchet: The Decommissioning of Paramilitary Arms in Northern Ireland.* Brief 22. Bonn:
Bonn International Center for Conversion.

HRW (Human Rights Watch). 2003. *Weapons Sanctions, Military Supplies, and Human Suffering: Illegal Arms Flows to Liberia and the June–July 2003
Shelling of Monrovia.* Briefing Paper. New York: HRW. 3 November.

——. 2005a. *Youth, Poverty and Blood: The Lethal Legacy of West Africa's Regional Warriors.* Vol. 17, No. 5(A). March. New York: HRW.
<http://hrw.org/reports/2005/westafrica0405/>

——. 2005b. *Country on a Precipice: The Precarious State of Human Rights and Civilian Protection in Côte d'Ivoire.* Vol. 17, No. 6(A). May. New York:
HRW. <http://hrw.org/reports/2005/cdi0505/>

——. 2005c. *'Rest in Pieces': Police Torture and Deaths in Custody in Nigeria.* Vol. 17, No. 11(A). July. New York: HRW.
<http://hrw.org/reports/2005/nigeria0705/>

——. 2005d. *Smoke and Mirrors: Colombia's Demobilization of Paramilitary Groups.* Vol. 17, No. 3. August. New York: HRW.
<http://hrw.org/reports/2005/colombia0805/>

——. 2006. *Human Rights Watch World Report 2006.* January. New York: HRW.
<http://www.hrw.org/wr2k6/index.htm>

Human Security Centre. 2005. *Human Security Report 2005: War and Peace in the 21st Century.* Vancouver: Oxford University Press.

Humphreys, Macartan and Habaye ag Mohamed. 2003. 'Senegal and Mali.' Paper presented at the World Bank/PRIO Civil Conflict Workshop, Oslo,
January. <http://www.columbia.edu/~mh2245/papers1/sen_mali.pdf>

ICHRP (International Council on Human Rights Policy). 2000. *Ends and Means: Human Rights Approaches to Armed Groups.* Versoix, Switzerland:
ICHRP. Accessed 15 September 2005. <http://www.reliefweb.int/library/documents/2001/EndsandMeans.pdf>

ICRC (International Committee of the Red Cross). 2004. *Improving Compliance with International Humanitarian Law.* Background paper prepared for
the Informal High-Level Expert Meeting on Current Challenges to International Humanitarian Law. Cambridge, 25–27 June.

ICG (International Crisis Group). 2003a. *Côte d'Ivoire: The War is Not Yet Over.* Africa Report No. 72. Freetown/Brussels: ICG. 28 November.

——. 2003b. *Guinée: Incertitudes autour d'une fin de règne.* Africa Report No. 74. Freetown/Brussels: ICG. 19 December.

——. 2005a. *Côte d'Ivoire: Le pire est peut-être à venir.* Africa Report No. 90. Dakar/Brussels: ICG. 24 March.

——. 2005b. *Stopping Guinea's Slide*. Africa Report No. 94. Dakar/Brussels: ICG. 12 June.

IICD (Independent International Commission on Decommissioning). 2005. *Report of the Independent International Commission on Decommissioning*. Dublin/Belfast: IICD. 26 September. <http://www.nio.gov.uk/iicd_report_26_sept_2005.pdf>

IISS (International Institute for Strategic Studies). 1998. *The Military Balance 1998–1999*. London: Oxford University Press.

——. 2004. *The Military Balance 2004–2005*. Oxford: Oxford University Press.

IRIN (United Nations Integrated Regional Information Networks). 2005. 'Liberia: "What About Us?" Ask Those Who Didn't Fight or Flee.' 27 June. Accessed 13 September 2005. <http://www.irinnews.org/report.asp?ReportID=47845&SelectRegion=West_Africa&SelectCountry=LIBERIA>

Isacson, Adam. 2005. *Peace—or 'Paramilitarization?'* International Policy Report. Washington, DC: Center for International Policy. July. Accessed August 2005. <http://ciponline.org/colombia/0507ipr.htm>

Koorey, Stephanie. 2005. 'Primed and Purposeful: Armed Groups in South-East Asia.' Unpublished paper prepared for Small Arms Survey. September.

Lecocq, Baz. 2002. *'That Desert is Our Country.' Tuareg Rebellions and Competing Nationalisms in Contemporary Mali (1946–1996)*. Doctoral thesis. Amsterdam: University of Amsterdam. November.

MALAO (Mouvement contre les Armes Légères en Afrique Occidentale). 2003. *Les armes légères au Sénégal*. Dakar: MALAO and OXFAM. September.

MAPP–OEA (Mission to Support the Peace Process in Colombia). 2005. *Fifth Quarterly Report*. 5 October. Brasilia: Organization of American States.

Milner, James. 2005. 'The Militarization and Demilitarization of Refugee Camps in Guinea.' In Nicolas Florquin and Eric G. Berman, eds. *Armed and Aimless: Armed Groups, Guns, and Human Security in the ECOWAS Region*. Geneva: Small Arms Survey, pp. 144–79.

Moorehead, Caroline. 2005. 'Letter From Darfur.' *New York Review of Books*, Vol. 52, No. 13. 11 August, pp. 55–57.

Mpinganjira, Ernest. 2005. 'Illegal Arms Trade.' *East African Standard* (Nairobi). 3 July. <http://www.eastandard.net/archives/cl/hm_news/news.php?articleid=24248&date=3/07/2005>

Muñoz-Rojas, Daniel and Jean-Jacques Frésard. 2004. *The Roots of Behaviour in War: Understanding and Preventing IHL Violations*. Geneva: ICRC. October.

NCDDRR (National Commission for Disarmament, Demobilisation, Rehabilitation and Reintegration). 2004. 'Warring Groups Pledge to Submit Heavy Weapons'. 15 July. Press release. Monrovia: NCDDRR.

NDPEHRD (Niger Delta Project for Environment, Human Rights and Development). 2004. *Rivers State (Niger Delta, Nigeria): A Harvest of Guns*. Small Arms Project Brief Report. Ogale-Nchia: NDPEHRD.

——. 2005. *Rivers State (Niger Delta, Nigeria): The Big Disarmament Gamble; The Comeback of Small Arms and Light Weapons*. Small Arms Project Report No. 3. Ogale-Nchia: NDPEHRD.

Nichols, Ryan. 2005. 'Disarming Liberia: Progress and Pitfalls.' In Nicolas Florquin and Eric G. Berman, eds. *Armed and Aimless: Armed Groups, Guns, and Human Security in the ECOWAS Region*. Geneva: Small Arms Survey, pp. 108–43.

Obasi, Nnamdi K. 2002. *Small Arms Proliferation & Disarmament in West Africa: Progress and Prospects of the ECOWAS Moratorium*. Abuja: Apophyl Publications.

Onyeozili, Emmanuel C. 2005. 'Obstacles to Effective Policing in Nigeria.' *African Journal of Criminology and Justice Studies*, Vol. 1, No.1. April, pp. 32–54.

Orji, Dickson. 2005. 'Nigeria Report.' Unpublished paper commissioned by Small Arms Survey.

Otunnu, Olara. 2005. 'Ending Wars Against Children.' *International Herald Tribune* (Paris). 6 August. <http://www.iht.com/articles/2005/08/05/news/edotunnu.php>

Policzer, Pablo. 2002. 'Human Rights and Armed Groups: Toward a New Policy Architecture.' <http://www.genevacall.org/resources/testi-reference-materials/testi-other-documents/policzer-jul02.pdf>

——. 2005. 'Neither Terrorists Nor Freedom Fighters.' Paper presented at the International Studies Association Conference. Honolulu, 3–5 March.

Regehr, Ernie. 2003. *Reducing the Demand for Small Arms and Light Weapons: Priorities for the International Community*. Working Paper 04-2. Waterloo, Ontario: Project Ploughshares.

Reno, William. 2003. 'Sovereign Predators and Armed Non-State Protectors?' Paper presented at Curbing Human Rights Violations by Non-State Armed Groups, The Armed Groups Project annual conference. Vancouver, 13–15 November.

Republic of Guinea. 2004. *Rapport national sur l'application du Programme d'action des Nations unies en vue de prévenir, combattre et éliminer le commerce illicite des armes légères sous tous ses aspects*. Conakry: Ministère à la Présidence Chargé de la Défense Nationale. 26 April. Courtesy of Cissé Mahmoud, Sécretaire Permanent de la Commission nationale de lutte contre le commerce illicite des armes légères.

SC (Save the Children). 2005. *Fighting Back: Child and Community-Led Strategies to Avoid Children's Recruitment into Armed Forces and Groups in West Africa*. London: Save the Children.

Small Arms Survey. 2005. 'Rounding Out the Gun: Ammunition.' In *Small Arms Survey 2005: Weapons at War.* Geneva: Oxford University Press, pp. 9–37.

Smith, Craig S. 2004. 'U.S. Training African Forces to Uproot Terrorists.' *New York Times.* 11 May.

SIPRI (Stockholm International Peace Research Institute). 2005. *SIPRI Yearbook 2005.* Stockholm: Oxford University Press.

Solomon, Deborah. 2005. 'Questions for Gerry Adams.' *New York Times Magazine.* 14 August, p. 13.

UN. 2004. *World Population Prospects: The 2004 Revision.* <http://esa.un.org/unpp/index.asp?panel=3>

UNDP (United Nations Development Programme). 2005. *Arms for Development Programme—Draft Progress Report: Second Quarter.* Freetown, Sierra Leone: UNDP.

UNOCHA (United Nations Office for the Coordination of Humanitarian Affairs). 2003. *Consolidated Appeals Process (CAP): Guinea 2004.* New York and Geneva: UNOCHA.

UNORC (Office of the UN Recovery Coordinator for Aceh and Nias). 2006. *Humanitarian and Recovery Update—Aceh and Nias.* 16 December 2005–15 January 2006. Banda Aceh, Indonesia: UNORC.

UNSC (United Nations Security Council). 2000. *Report of the Panel of Experts appointed pursuant to Security Council resolution 1306* (2000), paragraph 19, in relation to Sierra Leone. S/2000/1195. December.

——. 2002. *Report of the Panel of Experts concerning Liberia.* S/2003/498 of 24 October.

——. 2003. *Report of the Panel of Experts concerning Liberia.* S/2003/498 of 24 April.

——. 2005. Resolution 1612, adopted on 26 July. S/RES/1612 (2005).

Weinstein, Jeremy. 2004. 'Resources and the Information Problem in Rebel Recruitment.' Stanford University. October. <https://upload.mcgill.ca/rgchr/jcr-weinstein.pdf>

Werthein, Lucas and Mireille Widmer. 2005. *Sierra Leone: From DDR to Comprehensive Disarmament.* Background Paper No. 2. Geneva: Centre for Humanitarian Dialogue. <www.hdcentre.org/datastore/Small%20arms/ Rio/Sierra%20Leone%20Paper%20FINAL.pdf>

Wille, Christina. 2005. 'Children Associated With Fighting Forces (CAFF) and Small Arms in the Mano River Union (MRU).' In Nicolas Florquin and Eric G. Berman, eds. *Armed and Aimless: Armed Groups, Guns, and Human Security in the ECOWAS Region.* Geneva: Small Arms Survey, pp. 180–221.

ACKNOWLEDGEMENTS

Principal author

Louisa N. Lombard

Contributors

Eric G. Berman, Nicolas Florquin, and Glenn McDonald

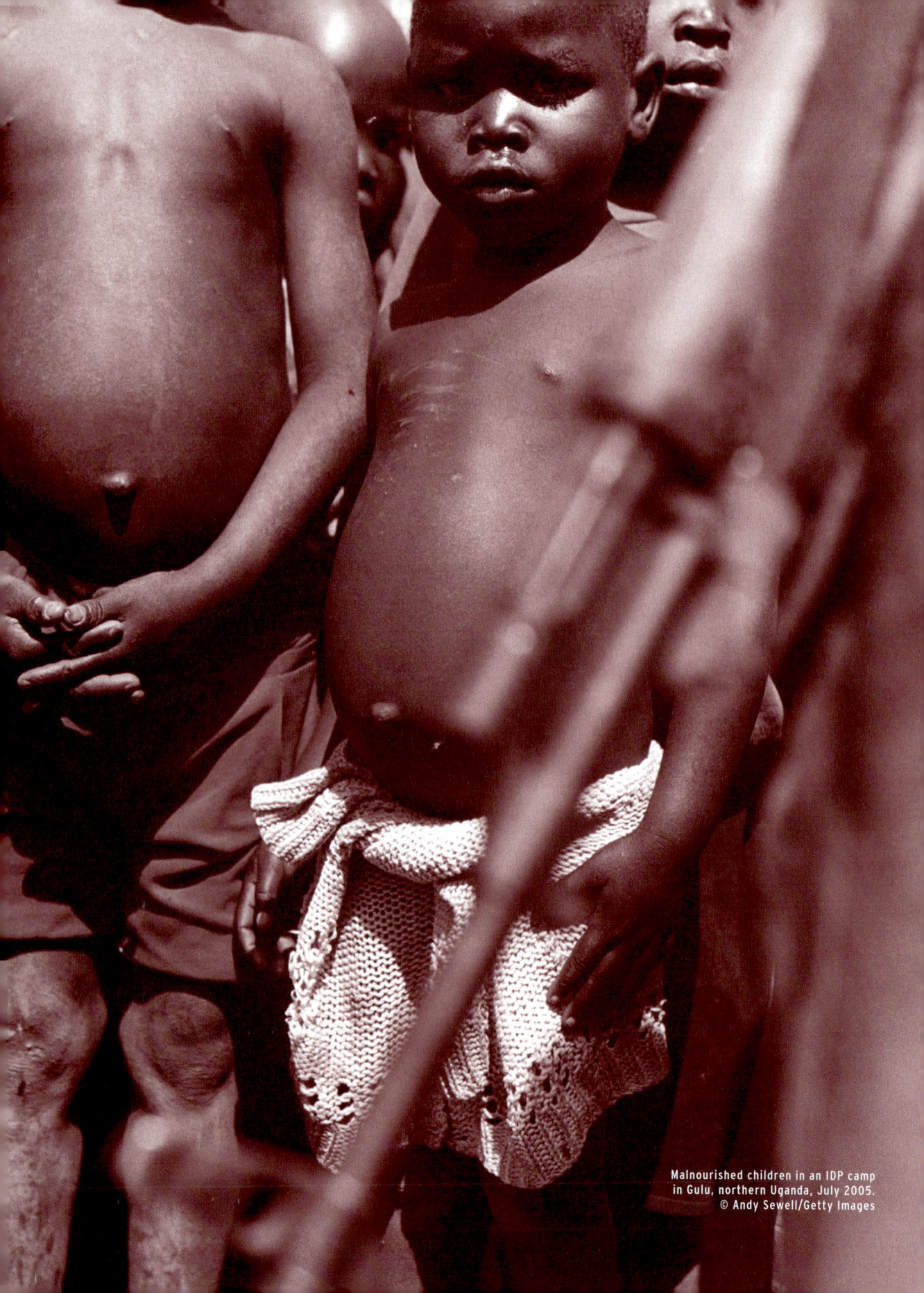

Malnourished children in an IDP camp
in Gulu, northern Uganda, July 2005.
© Andy Sewell/Getty Images

Fuelling Fear
THE LORD'S RESISTANCE ARMY AND SMALL ARMS

11

INTRODUCTION[1]

The Lord's Resistance Army (LRA) is a non-state armed group that forces children to fight a war using small arms. The war is directed, for the most part, against the civilian population of northern Uganda. The LRA commits massacres and atrocities, and abducts children, forcing them with extreme violence to become soldiers. An estimated 25,000 to 30,000 children have been abducted since 1987. Some have escaped; others have died through violence, disease, hunger, and exhaustion. Children now constitute between 80 and 90 per cent of the estimated 500 to 1,000 remaining LRA fighters.

In the past 19 years, the fighting has killed thousands of people and displaced close to 1.3 million within northern Uganda. Many more are maimed and tortured in displays of the LRA's strength. Although people are often attacked with knives and agricultural implements, small arms remain the fundamental facilitators of violence. They are used to corral people, preventing them from running away.

The Ugandan army has not been able to defeat the LRA militarily. It fights the LRA with armoured personnel carriers, aircraft, and around 40,000 troops, thereby curtailing some of the LRA's activities and disrupting its supply lines. But, although the LRA has declined in numbers, has few resources, and has difficulty moving equipment, it is able to continue fighting, killing, and abducting. Because it is well equipped with small arms, it is able to attack the local population and the Ugandan army in both Uganda and Sudan.

Small arms are the most suitable weapons for the LRA's operations, and the group consequently uses few larger weapons. The supply and maintenance of small arms is therefore a crucial gauge of the LRA's capacity to continue fighting. The following are among the main findings of this chapter:

- The LRA depends on small arms to conduct its operations.
- The LRA requires few resources other than small arms and people to use them.
- Children are easily captured, indoctrinated, and trained to use small arms.
- Small arms facilitate a deliberate policy of terrorizing the civilian population.
- The LRA acquires small arms constantly and keeps them in good repair.
- Weapons have been cached throughout northern Uganda and southern Sudan.
- Plentiful arms stocks mean the LRA is far from finished as a fighting force.

The chapter concludes that the current military solution to the conflict adds to the small arms problem in northern Uganda. The conflict has led to high levels of armament among the civilian population in the region, levels exacerbated by government and military policies of arming sections of society against the LRA and other armed groups. This contributes to a cycle of small arms, insecurity, and further armament, of which the conflict with the LRA is but a part.

A TROUBLED HISTORY

Ugandan politics is characterized by a deep north–south divide that is the legacy of colonial rule. The crisis with the LRA is rooted in this divide and also deepens it.

Northerners, and particularly the Acholi, constituted the majority of the army under British rule, while southern peoples were favoured in the administration of government. This order prevailed following independence in 1962. Northerners and southerners came to view each other as favoured in one way or another. Successive oppressive regimes in the 1970s and early 1980s, including those of Idi Amin (1971–79) and Milton Obote (1962–71; 1980–85), deepened this divide. The north, and particularly the largely northern Uganda National Liberation Army (UNLA), was increasingly viewed as repressive.

The final years of struggle against the regime of Milton Obote and his short-lived successors witnessed extreme persecution of the civilian population in the south. The UNLA is estimated to have killed some 300,000 people—a fact that remains fresh in the memories of many southerners (ICG, 2004, p. 2). When the National Resistance Army (NRA), led by Yoweri Museveni, overran Kampala in January 1986 to set up the current political order,[2] many UNLA members fled north to their Acholi communities in Gulu and Kitgum Districts, or further, into Sudan (see Map 11.1)

Map 11.1 **Ugandan districts affected by LRA activity, 1987–2005**

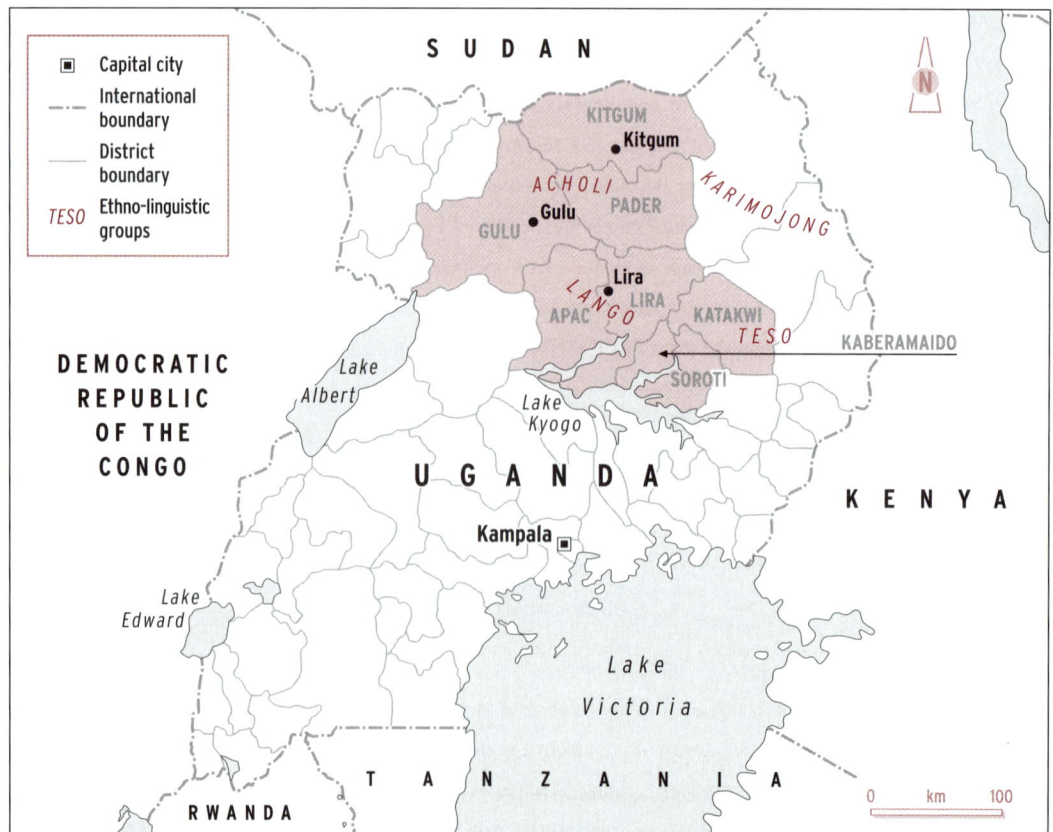

The Acholis' fears of reprisals were confirmed when the NRA mistreated the population.[3] Former UNLA soldiers formed the Ugandan People's Defence Army (UPDA) in opposition to the newly installed NRA, but the group degenerated into predation and violence against its own Acholi people (Behrend, 1999, p. 25).

The result was that a broad section of Acholi society acquired a number of grievances, which persist to this day. They felt strong opposition to the south; shame over 'their' military defeat; fear of reprisals; guilt over atrocities committed in the south; and high levels of insecurity. The situation was ripe for exploitation by anyone offering a way out of this collective predicament.

By late 1986 a popular uprising had started, centred around a woman named Alice, who claimed to be the medium for a spirit, the *Lakwena*. Alice Lakwena's Holy Spirit Movement (HSM) preached salvation through purification. Soldiers within the movement were told that, once purified, they would be immune from harm in battle. They were taught to walk straight at the enemy without taking cover, a tactic that was to prove extremely successful at first.[4]

Abduction became the predominant source of new recruits.

The HSM found many recruits in Acholi society, grew in numbers, and proceeded to inflict a series of defeats on the NRA. By January 1987, the Movement had advanced to within 80 km of Kampala. In October and November of that year, however, it was crushed in a series of defeats that government forces inflicted on it in and around Iganga District.

In the shadow of Alice's Holy Spirit Movement, Joseph Kony emerged as a spirit medium around January 1987. His movement, first known as the Lords' Army and later the Lord's Resistance Army,[5] began to attract a small number of followers from the UPDA and the local population. Like the HSM, it forcibly recruited some members (Behrend, 1999, pp. 179–80). However, over time, abduction became the predominant source of new recruits, because the LRA never achieved large-scale popular support.

Operations to dislodge the LRA, particularly Operation North of 1991, were intended to distance the local population from the group, but had the effect of precipitating attacks on the populace. Government troops forcibly displaced thousands into 'protected' settlements, or camps, and subsequently perpetrated numerous human rights abuses (HRW, 1997, p. 84). While these operations appear to have physically distanced the LRA from the local population, they also generated more intense opposition to the government and army among the Acholis. Moreover, the LRA increasingly operated as if the local population had colluded with the government.

From around 1991, the LRA began large-scale attacks on the ordinary civilian population, including raids on schools and clinics. Fighters massacred, abducted, and tortured people, cutting off limbs, ears, and lips and gouging out eyes (ICG, 2004, p. 6; HRW, 1997, p. 82).

The LRA also began to forge links with the Government of Sudan—Sudan's response to Uganda's support for the Sudan People's Liberation Army (SPLA). From 1997, the LRA received Sudanese weapons and ammunition, as well as the use of southern Sudan as a base for operations.[6] It abducted children and took them to Sudan for training. By 1997, the LRA was believed to field around 5,000 troops, of whom the majority were children (Nyeko and Lucima, 2002, pp. 18–19).

The year 2002 was a turning point for the LRA. Throughout the late 1990s and the first years of the next decade, the LRA had gained much from the poor efficiency of the recently renamed Ugandan People's Defence Force (UPDF).[7] However, the Nairobi Agreement of 2002 restored diplomatic relations between Sudan and Uganda, allowed the UPDF to conduct operations against LRA bases in Sudan, and apparently curtailed the bulk of Sudanese arms transfers to the LRA.[8] Since 2002, LRA fighters have suffered severe difficulties in moving arms and foodstuffs because of intense UPDF operations, such as Operation Iron Fist and its successors.

In Obalanga, northern Uganda, 46 bodies are exhumed from the bush for burial in June 2005. The incursions of LRA rebels into Teso sub-region in 2003 left thousands of people dead. © AFP/Getty Images

At first the LRA changed tactics and tried to expand its operations, but by 2003 this had resulted in high levels of disorganization (ICG, 2004, pp. 7–8). This change in circumstances is widely believed to be responsible for a series of peace talks between the LRA and government representatives since 2003. To date, the government has pursued a two-sided strategy of force and negotiation, albeit with more emphasis on the former.

The early months of 2005 saw continued erosion of the LRA's military capacity and the group again changed tactics. The estimated 500–1,000 remaining LRA fighters have dispersed into small groups, often far from contact with centralized command (ICG, 2005b, p. 3).[9] To date, this tactic has proved effective for attacking and abducting people from the local community, as well as for avoiding costly encounters with UPDF troops.[10]

In the latter months of 2005, a further change in LRA tactics was observable. For the first time in 19 years, the LRA launched a series of attacks against international aid agencies and workers, resulting in several deaths. Whether these attacks were a deliberate response to the International Criminal Court's (ICC) issuing arrest warrants for LRA leaders in October 2005 remains to be seen. It may well have signified to LRA commanders that aid agencies are a part of an international effort that is working explicitly against them and not simply providing aid to the affected population.

The LRA also changed its areas of operation in the latter months of 2005. While the LRA has long operated in southern Sudan, and has mounted attacks there in the past, it has recently done so more frequently. Most notably, it has been active on the west bank of the Nile, south of Juba; it raided Loka and Lainya on the Yei–Juba road, and launched attacks as far west as Yei. The latter represents the first time the group has attacked in Western Equatoria. These attacks have affected commercial access to Juba and NGO operations in the area. In particular, de-mining agencies suspended operations in the area. In September, a large group of fighters also entered Kivu province in the Democratic Republic of Congo (DRC). Attacks in both Sudan and the DRC appear to be on the increase—an ominous sign that the war may be spreading (see Map 11.2).[11] In one notable incident in the DRC, in January 2006, eight Guatemalan peacekeeping troops were killed in a clash with LRA fighters north of the town of Bunia (BBC, 2006; MONUC, 2006).

LRA MOTIVES AND ORGANIZATION

The LRA is the outward expression of the motives of Joseph Kony. Its fighters live hand to mouth, wear no trappings of modernity, and have no apparent economic motives. Lower-level fighters have no significant concerns other than to survive their time with the LRA. They take from the local population only enough for subsistence. Senior commanders also appear to have few material incentives. For many of them, 20 years or more of committing atrocities against their own people make the prospect of returning home bleak.

If there is a political project behind the LRA's campaign, it is Joseph Kony's alone. The LRA is an extension of Kony's apparent need to assert his power and to be a major threat to the people and Government of Uganda. Observers note that most major campaigns and attacks occur in response to some statement or action on the part of the government or local population that appears to detract from Kony's claim to potency.[12] The LRA therefore continues its campaign because it is Kony's wish. The group will desist only if Kony agrees to return from the bush, or is killed, or can no longer forcibly recruit and arm abductees.

Unlike the earlier HSM, the LRA has little popular support, so in order to survive it has to abduct recruits. The basis of Kony's control over the LRA is that between 80 and 90 per cent of the LRA's fighters are abducted children aged between 10 and 17 years. The group avoids abducting adults for two simple reasons: it is easier for them to escape and it is more difficult to indoctrinate them.[13]

Children, on the other hand, can be terrorized into becoming effective fighters. Indoctrination is complemented by the fact that small arms are small and light enough for children as young as seven years old to use to deadly effect. The LRA's ability to 'create' new fighters therefore hinges on two factors: the availability of potential abductees and the availability of arms with which to equip them.

The LRA has a simple formula for success in its attacks against government troops—fighters are indoctrinated into obeying absolutely the orders of their commanders, whose authority is backed up by the spirits.

Between 80 and 90 per cent of LRA fighters are abducted children.

The core tactic is one inherited from the HSM: 'The children just walk straight at the soldiers. In the long grass they are small and difficult to see and they just keep coming. The children are unstoppable and the UPDF usually run away.' 'The most important aspect of training is not learning how to shoot or ambush, but learning how to follow the spirits', explained one former LRA commander.[14]

Even for those children who may question the power of the spirits over them, the suffering inflicted upon those who attempt to escape, and the fact that the children themselves are forced to participate in exacting punishment, is a considerable disincentive to trying to desert.

Map 11.2 **Uganda and surrounding countries**

> ### Box 11.1 Abduction as recruitment
>
> There is no set pattern for LRA abductions. They take place whenever the LRA has access to people; some are planned, while others are opportunistic. However, when the LRA attacks to abduct children, it usually does so at night and targets isolated villages and poorly defended camps for internally displaced persons (IDPs) outside towns. It does so at night simply because children are concentrated in one place. Children are usually abducted singly or in pairs, but sometimes the LRA is able to round up upwards of 30 or even 100 children when settlements are poorly defended.[15]
>
> Indoctrination begins immediately upon abduction. The most comprehensive accounts are from those who have counselled returned children. According to them, the LRA commanders involve the children in atrocities at the place of abduction; as a result, the children feel that they cannot go home knowing what they have done. As one humanitarian worker described it, 'they make them kill family and local villagers who they know. They might, for instance, force them to chop off the arm of a mother or father.'
>
> Most of the children are beaten shortly after they are captured. Often this entails up to 200 hundred strokes of the cane and sometimes severe blows with pangas.[16] The commanders try to instil fear of escape from the start. They make the children kill those who try to escape by beating them with sticks and pangas, and even by biting one another to death. These atrocities are part of a deliberate policy designed to generate the utmost fear of the LRA in the children.
>
> Violence is accompanied by rituals, which are also designed to instil fear in the abductees. The children are anointed with shea nut oil crosses, which they are told contain the spirits. The oil, and the spirits it represents, is said to protect the children from bullets. The children are also told that the spirits will confuse them if they try to escape and that they will walk in circles and back to the LRA, who will then kill them. The fear of being in the LRA is perhaps exceeded only by the fear of being recaptured.
>
> **Sources:** Interviews conducted in northern Uganda in May 2005

In a force of mostly abducted fighters, there has never been any question that many wish to demobilize and return to their communities. The fact remains, however, that, despite numerous escapes and captures of fighters, many cannot leave due to the fear instilled in them by the LRA. The problem lies mainly with the leadership and with Joseph Kony in particular.

'The LRA don't control territory, they control people's minds.'

Most recruits have two or three days or one week of small arms training. During this period, all recruits are trained to operate and use an assault rifle, and to dismantle one. Behaviour under fire is simple. Fighters have to obey the commander absolutely under pain of death, firing only at close range when the chance of hitting the enemy is greatest.[17] This is a simple formula, but an effective one.

LRA USE OF SMALL ARMS

While LRA fighters are tightly controlled in their actions, they have great freedom of action in conducting campaigns of violence. As one senior humanitarian official put it, 'The LRA don't control territory, they control people's minds'.[18] LRA attacks are designed to do just that. Most assaults on the civilian population are therefore opportunistic, happening whenever fighters have the chance to attack the local population, and are usually accompanied by extreme brutality. Small arms enable or facilitate all of these attacks and are thus instrumental in the LRA's policies of sowing fear in the region.

The LRA undertakes a number of different activities. It abducts children as potential fighters; abducts local people to carry foodstuffs and munitions; targets vehicles on the roads; attacks poorly defended UPDF positions; assaults IDP camps; and engages UPDF mobile patrols. More often than not, the LRA does several of these things simultaneously. For example, an attack may take place in order to seize foodstuffs, but fighters may also abduct people at the same time to carry their seizure, or kill UPDF troops and take their munitions (see Table 11.1).[19]

Small arms are essential for the LRA, in contrast to larger weapons, because they enable a high degree of mobility. Most attacks involve only small groups of LRA fighters and usually a maximum of 10 to 15.[20] Attacks are fast and the LRA is quick to leave the area before the UPDF can bring in heavy weapons or aircraft to engage the LRA.

Security personnel of humanitarian agencies estimate from analyses of attack frequencies that LRA units may travel up to 50 km in one day. The LRA rely on this mobility to evade the UPDF and usually take the strongest and fittest fighters to attack larger targets because they can get clear quickly before the UPDF can launch a counter-attack. When confronted, the LRA's tactics are to split up into very small groups and disperse into the bush.[21]

Fighters usually refrain from firing at civilians. Humanitarian personnel testify that the LRA most frequently uses pangas and knives to commit atrocities. As one field security officer noted, 'In most cases, they kill people working in their gardens using their own tools'.[22]

Small arms are, nonetheless, pivotal in these acts. Atrocities are always committed at gunpoint, which prevents people from running away. Because of civilians' fears of being shot, by and large it is only when fighters engage UPDF troops that that they need to fire their weapons. Conserving ammunition, however, may not be a response to acute shortages: the LRA may choose to use farm tools or knives in its attacks simply for the traumatizing effect.

Former combatants report that, when engaging the UPDF, they rarely select automatic fire, but are taught to fire single shots at close range. Even then, former fighters relate how they are ordered to 'fire carefully' so that they might use only up to five rounds in a serious encounter.[23] Automatic fire is used only when attacking the UPDF in earnest, and this is rare. Fighters are taught to conserve stocks and, as one former fighter noted, it is a tactic that is rigidly enforced: 'If fighters lose ammunition, the punishment is death, no question.'[24]

In some situations there is a great danger of being heard and found by the UPDF.[25] Fighters are taught always to keep the safety catch on in tall grass, in case the weapon discharges and alerts the enemy. Although weapons are

Table 11.1 Types of LRA attack and use of small arms

Type of attack	Use of small arms
Road ambush	Shots fired from assault rifles to halt vehicles. Sometimes rocket-propelled grenade (RPG fire). Civilians may be attacked with other weapons.
Ambushes in the fields/bush	Assault rifles used to corral people for abduction or attack with tools or knives. Shots rarely fired.
Attacks on IDP camps	Assault rifles sometimes used directly to kill. More often used to corral people for abduction or attack with tools or knives.
Abduction	Assault rifles. Usually used to corral people. Shots rarely fired.
Attacks on UPDF bases	Assault rifles and sometimes light weapons used. Rarely sustained fire.
Attacks on UPDF foot patrols	Assault rifles used to kill troops. Rarely sustained fire.
Attacks on UPDF mobile patrols	Shots fired to halt vehicles and kill troops. Sometimes use light weapons.

Sources: Interviews conducted in northern Uganda in May 2005

Box 11.2 The human impact of the conflict

Today, around 1.3 million people in northern Uganda live in internally displaced persons (IDP) camps, some 90 per cent of the population in the region. The camps usually consist of hundreds of huts crammed together with perhaps only a metre between them. They are squalid and disease-ridden, with latrine access in many camps of only ten per cent. Cholera is a major problem.

People who once lived adjacent to their own fields, or 'gardens', now reside in camps many miles away. Some cannot tend their gardens because they are located too far from the camps; others have to walk for at least an hour to reach them. By military order, and for their own safety, people are not allowed out of the camps before 9.00 a.m. They have to return before 4.00 p.m. Tilling, sowing, husbandry, and harvesting therefore can take place, at most, only between 10.00 a.m. and 3.00 p.m. On some days, people are forbidden to leave the camps at all because of LRA activity. There is simply insufficient time to grow food effectively, and the majority of people are dependent on food aid.

Schooling is difficult. In some camps the teacher–pupil ratio approaches 1 to 700, forcing the children to stand outside the classroom where the teacher cannot be heard. Around 25 per cent of primary school-age children do not even attend school.

The population lives in constant fear of LRA attack. Fear is a difficult image to capture, and for the most part people going about their daily business display few signs of it. 'Fear is the biggest thing here: fear of the situation and fear of death', noted one local religious leader. Although the UPDF defends camps, people know that the LRA can bypass sentries, enter the camps, and abduct or kill them.

Families send their children into towns at night because they know that the LRA will otherwise abduct them. This 'night commuting' is perhaps the most evident expression of people's fears. In Gulu alone, the number of night commuters increased from 11,000 to 18,000 between mid-March and mid-May 2005. Night commuting across Uganda was estimated to total around 30,000 in March 2005 (USAID, 2005). The children are often unaccompanied and are at great risk, not only from the LRA but from all forms of predation by adults. Promiscuity among the children is rife, and sexually transmitted diseases, including HIV, are commonplace.

Weekly UN security reports detail numerous LRA attacks on various targets. In the two months between the end of May and the end of July 2005, 63 killings and 50 abductions came to the attention of UN staff in Gulu. Many incidents go unreported. Children under the age of 17 consti-tuted the greater part of the abductees. Some are taken as fighters; girls are more often taken to become 'wives' or sex slaves and are raped by LRA fighters. As one senior humanitarian official explained:

A child prepares to spend the night on the streets in Kitgum, northern Uganda, in November 2004. This youngster is one of Uganda's many thousands of night commuters who flee to the towns at night to avoid LRA abduction.
© Chris de Bode/Panos Pictures

I have been doing this job, in various places, for 25 years. Psychologically I cannot think of a worse effect on children in terms of terror in their minds.

While reporting is often incomplete, it is clear that every report adds to people's concerns, as local residents made clear in their statements in May 2005:

In 2004, in Pader, the LRA overran the UPDF. They cooked people in pots, ordering the LCs [local councillors] to eat the bodies of the cooked people.

Wednesday or Thursday last week the LRA is alleged to have abducted 129 people on the Lira-Pader border. The person who escaped said that Kony had ordered the LRA to capture 1,000 new recruits.

In Pabbo, they warned people they were coming, then came and killed.

One week ago in Koch Goma, the LRA killed 14 people, even though the UPDF is there. It is only 14 km from here.

Between mid-March and mid-May 2005, LRA activity was reported on every day but one. Although such activity included sightings and reports of looting, more often than not the LRA was involved in some form of violence or abduction. The LRA is thus a constant feature of everyday life in northern Uganda. Fighters appear daily, seemingly randomly, and virtually anywhere in the affected region. No one can be certain of his or her safety at any time.

The human cost is probably impossible to calculate. Thousands have died directly from the fighting. Many others have died of malnutrition and diseases. Some estimates put the conflict-related death toll at 1,000 per week (IRC, 2005). The majority of the northern population is affected, with around 1.3 million IDPs, over 200,000 refugees in Uganda from Sudan, and over 300,000 Ugandan refugees elsewhere. In total, some 25,000 to 30,000 children are estimated to have been abducted by the LRA during the conflict (UNICEF, 2005; USAID, 2005).

Sources: Interviews conducted in northern Uganda in May 2005; LRA attack and activity data derived from a compilation of UN daily, weekly, and monthly security summaries kindly provided by the UN Field Security Office, Gulu

---

Box 11.3   The regional trade in small arms

The borders between the Democratic Republic Congo (DRC), Ethiopia, Eritrea, Kenya, Somalia, Sudan, and Uganda are porous and allow unchecked small arms proliferation.

Sudan is perhaps the greatest route for small arms transfers in the north of Uganda. Southern Sudan hosts a heavily armed local population, but also routes to Ethiopia and from there to Somalia. Available weaponry may well be on the increase in southern Sudan because of the peace agreement signed between Sudanese government forces and the Ugandan-backed SPLA. Some reports suggest that the SPLA, or former SPLA fighters, trade weaponry in Kitgum, most notably at the Agora cattle market.

To the north-east, the border regions of Kenya, Sudan, and Uganda are the focus of an extensive market for weapons which links many actors in the region and those from further afield. Both the Turkana and Karimojong, which straddle the border between Kenya and northern Uganda, are heavily armed. Together with the Toposa of Sudan, the Karimojong and Turkana participate in a market for arms and cattle in the border triangle between Kenya, Sudan, and Uganda.

In the west, the DRC and Sudan border the West Nile region of Uganda. The region has, in the past, hosted four main rebel groups: the Former Uganda National Army (FUNA), the first Uganda National Rescue Front (UNRF), the UNRF II, and the West Nile Bank Front (WNBF). Arms procured in the DRC and Sudan, as well as those captured from government forces, have been a key element in facilitating the rise of these groups. While these forces are no longer active, small arms still proliferate in the region (RLP, 2004).

Sources: Interviews conducted in northern Uganda in May 2005

---

usually loaded, when in camp fighters always keep the safety catch of their weapons on, to prevent accidental discharge and potential disclosure of their presence.[26]

LRA ACQUISITION OF SMALL ARMS

Northern Uganda and its surrounding countries are awash with small arms (Box 11.3). The LRA is able to capture and trade weapons amid this plentiful supply. In many cases, it is able to do so because it is well armed. In short, as has been the case with armed groups elsewhere, arms beget arms (Small Arms Survey, 2005, pp. 186–7).

To judge from the accounts of former LRA commanders, and the number of fighters—around 5,000—fielded by the group in the late 1990s, Sudanese arms transfers to the LRA may have accumulated to tens of thousands.[27] These stocks are still accessible to the LRA, despite having been supplied nearly two decades ago. Some fighters report that their weapons were still wrapped and greased from the factory when they received them. Kony's battalion is reportedly equipped entirely with unused weapons from these stocks.[28]

Northern Uganda is awash with small arms. In addition to these supplies, however, the LRA continually acquires weapons by capturing them from UPDF troops and from local defence units. According to one former LRA fighter, on average each soldier in the UPDF 'detaches'[29] that his unit had overrun was equipped with ten full magazines.[30] This equates to around 300 rounds of ammunition per soldier. Even a conservative estimate suggests that the average eight-man mobile detach carries at least 1,400 rounds of ammunition and various weapons.[31] If an LRA unit of the same size overran a UPDF mobile patrol, for instance, and captured only half of its ammunition, it would gain twice its usual complement of ammunition.[32]

The LRA also uses its firepower to capture arms from various communities and to capture cattle, which it then exchanges for arms. One local religious leader reported that the LRA has traded cattle for weapons with the Karimojong in particular. LRA raids into southern Sudan have reportedly netted around 6,000 cattle solely for this purpose.

The LRA trades with the Toposa and Dinka of Sudan, but it also captures weapons in raids. As one former LRA commander noted, obtaining arms in Sudan was easy because 'everyone has a gun.'[33]

The LRA's supply of weapons is therefore ongoing and never static. It has a vested interest, not only in targeting the population, but also launching attacks to acquire weapons. This is a common tactic of armed groups that lack regular arms supplies.

LRA HOLDINGS

For the most part, the LRA now uses assault rifles and light machine guns, and uses larger weapons, such as RPGs, only when attacking light armoured vehicles. Even when the Sudanese government supplied heavier weapons, the LRA used mainly small arms because it needed to be highly mobile.[34] Current weapon holdings appear to reflect this, with the LRA stocking very few heavy weapons.

The LRA's primary weapons are Kalashnikov derivatives and, most commonly, the Chinese Type 56 assault rifle. However, the Ugandan military uses a number of weapons, and the same types are also used by the LRA. The most numerous after the Chinese Type 56 are Polish and Romanian AKM assault rifles and AK-74 assault rifles from the former Soviet Union. Some observers claim that NATO weapons, such as the Belgian FN FAL and the German G3, are also in service with some rebels, having formerly been in the arsenals of the Ugandan and Sudanese governments (IISS, 2005). However, several sources, including former LRA officers, note that, while the LRA sometimes used NATO weapons such as the G3 when operating in Sudan, it very rarely does now due to the shortage of NATO ammunition in the region.[35]

There have been reports of some large weapons, and perhaps the largest of these are Russian-designed B-10 82 mm or SPG-9 73 mm recoilless guns,[36] which senior Ugandan army officers and former LRA fighters report are in the LRA arsenal.[37] There have also been a few reports of the LRA using 60 mm and 81 mm mortars, although reports of the latter have been only for large groups of over 50 fighters.[38]

The LRA appears to have had access to anti-aircraft defences. While only one interviewee mentioned a 12.7 mm anti-aircraft gun, a number suggested that the LRA had been in possession of SA-7 man-portable air defence systems (MANPADS). To date, there have been no reports of the LRA using MANPADS, even though UPDF helicopters operate regularly against the LRA. If the LRA did or does have access to them, it is worth noting that the weapons require quite extensive training and must be stored and operated with some expertise (Small Arms Survey, 2004, pp. 77–97).

Table 11.2 **LRA weapon inventory: past and present**
Frequently used:
Kalashnikov derivatives (particularly Chinese Type 56/56-2 rifle)
Type 81/RPK light machine gun
PKM light machine gun
Less frequently used:
B-10
RPG
60 mm mortar
Past inventory/rarely used:
FN-FAL/SLR
G3
81 mm mortar
12.7 mm anti-aircraft gun
SPG-9
SA-7

Note: The LRA also stocks hand grenades and landmines.
Sources: Interviews with a wide range of persons in Gulu and Kampala, 21–27 May 2005

A former LRA commander made it quite clear that the types of weapons used by the LRA have changed over the years: 'in the old days we used RPGs, 60 mm mortars, and B-10s and SPG-9s.'[39] One person, who has been in a position to closely monitor the types of weapon that defecting fighters currently bring with them, claims that now 'Very rarely do we have returns of RPGs . . . . Most of their big stuff, such as B-10s, is not functioning.'[40] In August and September 2005, for instance, there were only two cases of the LRA using RPGs. In both, the targets were UPDF rapid response vehicles.[41]

This change to using smaller weapons may have several causes. Despite unsubstantiated claims of continued ammunition supply, most Sudanese arms transfers appear to have ceased following the 2002 Sudan–Uganda agreement.[42] The LRA may now be more dependent on what it can take from lightly armed UPDF detaches, which usually means assault rifles and light machine guns. Perhaps more important may be the change in LRA tactics, which most commentators in northern Uganda note has shifted from battalion-sized operations to small group actions. Such small groups often do not need, and indeed are unable to carry, heavy weapons.

LRA STOCKS

The LRA has two broad and connected systems in place for storing arms and ammunition. These reflect the two major sources of arms and ammunition that have been available to the LRA over the years: large transfers from Sudan and weapons captured from the UPDF.

Large caches

Fighters report that Sudanese-supplied weapons are buried, either in Sudan or in the far north of Uganda. Former fighters report that these stocks remain under Kony's control, and only he can give the orders to excavate arms and ammunition. The weapons are dismantled and their components stored separately, so that escapees or captured LRA members cannot reveal the location of entire stocks.

The LRA buries its weapons.

The stocks are reportedly still very large and far exceed other caches.[43] One former LRA fighter recalls how, in March 2004, his unit was ordered to stop some 5 km from the border with Sudan. Initially only the commanders were allowed to go forward, but later the rest of the unit came to a big hole filled with hundreds of arms and uniforms, and surrounded by the tyre marks of the machine that had excavated it.[44]

Some fighters have described how arms and ammunition have been supplied to them from these stocks while in Uganda. However, for the most part LRA combatants operating further south have their own caches.

Local caches

Local caches are small but numerous, and contain weapons that have been captured from the UPDF or local defence units. Because the LRA has to be highly mobile, it cannot carry arms and ammunition much in excess of those it carries already. Former LRA fighters report how, after successfully attacking UPDF troops, weapons and ammunition are distributed to each member of a unit to replenish stocks.

Responsibility for burying ammunition differs from one unit to another. In some units, once arms have been distributed, the balance is buried by the unit commander or one of the more experienced members of the group. As one junior LRA fighter stated, 'Only the hardened ones who have been there a long time are able to bury the ammunition'.[45]

In other units, each fighter buries captured ammunition on the orders of the unit commander. A former member of an LRA unit recalled, 'Whenever you have a lot of ammunition, you bury it on your own, and you remember where you buried it. When the order comes, everyone goes to fetch the ammunition from where they buried it, and it is distributed.' One described how, after excavating a cache, the fighters would fill three magazines and rebury the balance.[46]

Before being placed in holes in the ground, weapons are greased and wrapped in plastic sheeting. A typical burial contains two assault rifles and is approximately one metre long and 50 centimetres deep.[47]

These observations suggest that there are likely to be large caches of weapons in the far north of Uganda as well as in southern Sudan. While resupplying from these caches may be interrupted by UPDF

On the lookout for the LRA: A UPDF soldier from the 4th Infantry Division patrols a rural area of Patiko Sub-District in Lira, northern Uganda, in June 2005. © Chip Somodevilla/Getty Images

operations, fighters also maintain small, local caches throughout the north of Uganda. LRA units are likely to be able to resupply from these with greater ease because they need not transport arms and ammunition over long distances.

The quality of LRA small arms

Weapons are a durable commodity, but are subject to environmental conditions that can make them inoperable. LRA fighters, in contrast to many poorly organized armed groups, take good care of their weapons and take basic precautions to ensure they are well maintained and properly stored. Weapon longevity has serious implications for the numbers of weapons available to the LRA. If the group continues to acquire weapons but loses few due to poor maintenance, its chances of making a 'net profit' are increased.

While much of northern Uganda is dry for months, the rainy reason brings high levels of humidity that can irreparably damage poorly stored weapons. Weapons rust quickly when wet, and ammunition deteriorates.

LRA weapons, however, are generally well maintained and in good working order, despite being a mixture of ages. Some fighters are issued with brand new weapons, while others are given older weapons. All fighters that were interviewed reported that weapons were always in good working order. A senior Ugandan army official said that he would consider a ten-year-old weapon relatively new, which is a testament to the basic field maintenance skills of the LRA.[48]

As one senior intelligence official noted, 'There have been very few cases whereby LRA weapons have been handed in that are in a poor condition'. Individuals who closely monitor LRA returnees note that, with the exception of light rust on the barrels when it has been raining, the condition of returned weapons is always good. Moreover, the weapons appear to be in sufficiently good condition for the army to find it profitable to send the captured weapons to the small arms factory in Luwero for refurbishment.[49]

Although the LRA does not have access to armourers capable of maintaining weapons to the standards of a conventional army, some reports from the Ugandan army suggest that certain individuals in the LRA perform the function of armourers. One intelligence official noted that the LRA has the capacity to repair weapons found in the bush that are in very poor condition. The incentive to maintain weapons to a high standard is certainly strong because, as one army officer stated, 'The LRA kill their soldiers when they don't look after their guns'.[50]

In the field, fighters perform the normal cleaning and oiling tasks necessary to keep weapons serviceable. Captured UPDF weapons appear to be a good source of oil and cleaning kits, as most soldiers carry them on their person or in the butts of their rifles.[51] However, while some fighters report that they had been in possession of cleaning kits, others report that they often had to improvise. In these cases, LRA fighters clean their weapons with sponges made from mattresses, rags, or any other suitable material to hand. Any long implement that will fit down the barrel of a gun can be improvised to make a pull-through. When oil is not available, the LRA use engine oil, grease, or paraffin, and, if none of these are available, they drain and use the oil from a roasted chicken.[52]

These observations indicate that, while the LRA's care of weapons is often rudimentary, it appears to be effective. Weapons are a durable commodity, and more so when properly looked after. Not only are the weapons currently used by the LRA serviceable, but many of those cached are also likely to be so for many years to come.

The LRA received more arms and ammunition than it could use.

Alleged ammunition shortages

A number of sources claim that the LRA experiences ammunition shortages. As one senior Ugandan intelligence official noted, 'When the LRA was on good terms with Sudan, and ammunition was not a problem, they would attack with "full fire" [however] ammunition is now conserved'.[53] Others note that LRA fighters who have returned carry few weapons and little ammunition, drawing the conclusion that both must be in short supply.[54]

However, it is important to reiterate that the LRA does not consume much ammunition and chooses to conserve it wherever possible. There are also further reasons why the LRA is unlikely to experience damaging ammunition shortages.

Although many fighters return unarmed and without ammunition, this does not mean that the LRA is short of either. Many returnees cache their weapons in case of reprisals. In other cases, fighters abandon their weapons to the remaining members of their units, who then package and bury them for future use.[55] As one local leader commented:

> The [LRA] Brigadiers that have surrendered are open about the fact that weapons are left in the bush. In general, the returnees get very annoyed when we ask them where they have left their guns.[56]

LRA fighters also carry little ammunition anyway, because of their requirement for high mobility. Units will usually carry only the minimum ammunition required for the particular operation they intend to undertake. Junior fighters, who were interviewed separately, claimed they usually carried at most three full magazines, which is around 90 rounds of ammunition. Fighters try to keep at least one full magazine of ammunition on their person, although in times of scarcity fighters can be left with only 10 to 15 rounds. In this case, LRA fighters try to avoid contact with the UPDF until they have located a cache and replenished their ammunition.[57]

The supposed shortage of LRA arms and ammunition is further cast into doubt by claims made by former fighters. As one former LRA commander argued, even though Sudan had stopped supplying the LRA it didn't matter because so much was already stockpiled. Moreover, the LRA had captured much from SPLA and Ugandan government forces in southern Sudan. He summed up by saying that the LRA had received more arms and ammunition from Sudan alone than it could hope to use.[58] Junior fighters, furthermore, report that LRA stocks in Sudan far exceed those in Uganda. A recently returned LRA fighter commented, 'In Sudan they have more ammunition than they can use, but access has been difficult because the UPDF have been operating with gunships, so they move only a little at a time'. This statement suggests that the most serious difficulty the LRA faces with respect to ammunition is one of movement and resupply.[59]

A LONG ROAD TO STABILITY: SMALL ARMS PROBLEMS IRRESPECTIVE OF THE LRA

The UPDF has launched successive major operations against the LRA since 1991. Today some 40,000 Ugandan army troops are stationed in northern Uganda, most of whom are engaged in combating the LRA.[60] Not only have these troops provided a source of captured weapons for the LRA, but their presence and associated military policies have made the proliferation of small arms among the civilian population more severe. The pursuit of these strategies is a significant factor contributing to a small arms problem that affects not only the whole of northern Uganda but also surrounding regions.

The military option incorporates government and army policies of arming sections of the civilian population against the LRA. In some instances, issuing arms has been on a small, ad hoc, scale. In 2003, for instance, the military in Gulu town reportedly offered weapons to key local figures. In Teso and Lango Districts, however, arms have been distributed on a much larger scale. The recipients of these weapons are usually local government officials and their bodyguards, but other high-profile officials have been offered weapons. Some have used weapons, or their own armed escorts, for robbery.[61]

The formation of armed local defence units (LDUs) has added to the number of weapons in civilian possession. While these units have repelled some LRA attacks, they are also poorly trained and vulnerable to being overrun and losing arms to the LRA.[62] In addition, members of LDUs receive poor and irregular pay and, as a consequence, some members have simply left and sold their weapons to the local community. A number of UPDF troops have done likewise.[63]

Although the demand and supply of small arms in northern Uganda is closely related to the conflict with the LRA, the region also experiences high levels of insecurity whose origins lie beyond the conflict. These security problematiques encourage general arms acquisition, which further strengthens the LRA's ability to acquire arms through trade and capture. In addition, experience in

People on their twice-daily march from IDP camps to their fields—in constant danger of LRA attack. Kitgum, northern Uganda, November 2005.
© Thomas Morley/Reuters

Box 11.4   **Armed violence across northern Uganda**

Armed violence and insecurity, fuelled by the proliferation of small arms, is widespread across northern Uganda. The Karamoja region to the north-east on the Kenyan border is particularly affected. While the Karimojong in Uganda and other pastoralist groups in Kenya and Sudan have long practised cattle raiding with traditional weapons such as spears and bows and arrows, the use of small arms is now commonplace. As a result, low-intensity conflicts over livestock have escalated, causing deaths and displacement. In 2004, over 100 people were killed and thousands displaced as a result of armed cattle raids.

Criminality and lawlessness have also increased across the north. Petty crime, the looting of shops and homes, and armed roadside banditry and ambushes are becoming routine as local criminals take advantage of the security vacuum. In Acholi, armed local criminals, or *bookec*, pretend to be LRA rebels. They attack in the manner of the LRA to seize food and other produce, and usually kill if recognized to protect their anonymity.

In Kitgum local officials and the auxiliary police have hired out guns at night to people who use them to commit robberies. Elsewhere, officials have developed private armed militias ostensibly for the defence of dairy farms, but also for political ends.

In other cases, the burgeoning numbers of armed private security guards have themselves been the victims of attack. Very often attacks have been launched by local criminals explicitly to capture the guards' weapons.

Interpersonal armed violence is also commonplace. Women and girls have been singled out as victims of sexual violence and rape by UPDF forces and LDUs. This is often within the IDP camps which UPDF soldiers are responsible for protecting. In purely civilian disputes, there have also been reports of wives hiring armed assassins to kill in revenge for adultery, and weapons used for similar purposes.

In conjunction with the omnipresent threat of LRA attack, these activities have contributed to widespread insecurity among the local population.[64] Insecurity has led frequently to the acquisition of small arms. In Gulu, for instance, local residents report that most people in town now have access to arms, and the number of weapons circulating is believed to have made the trade in weapons in and around the town an unprofitable business.

Sources: Eavis (2002, pp. 252–53); CSOPNU (2004, pp. 9–11); Patrick (2005); USDOS (2005); interviews conducted in northern Uganda in May 2005

heavily armed post-conflict societies, such as El Salvador and Liberia, suggest that Uganda's small arms problem will also outlast the conflict with the LRA (Godnick, Muggah, and Waszink, 2002, pp. 10–18; Nichols, 2005).

Across the north of Uganda, small arms are used to steal, rape, intimidate, and settle scores between private individuals and between communities (Box 11.4).

The demand for small arms will certainly not end with the closure of the current LRA crisis. Political instability, in particular, is likely to remain a problem in the near future, of which the March 2006 elections will be a major test. Personal insecurity, the inability of the government to provide for adequate development and respect for human rights, and an absence of the rule of law in many parts of the north, all play their part. The end of the conflict would, however, eliminate a major source of insecurity and a major reason for the high demand for and supply of small arms in northern Uganda.

Uganda's small arms problem will outlast the LRA.

CONCLUSION

Taken together, the findings presented in this chapter suggest that the LRA is unlikely to be short of either arms or ammunition, even though recent UPDF operations have made it more difficult for the LRA to resupply. This observation suggests that a simple equation determines the future of the LRA: if it has weapons it can abduct, and, because it can abduct and arm fighters, it can continue fighting. Solving this equation requires either removing sources of small arms or stopping the LRA from abducting. To date, both approaches have failed.

Since 1991, the government has pursued a twin-track strategy of force and negotiations.[65] Of the two, the military option is predominant, but has not been particularly successful. Capturing Joseph Kony, killing him, or defeating the

LRA could bring the war to an abrupt end. Yet there is no certainty that the LRA can be defeated, or that Kony can be killed or captured.

The leadership can abduct at will and exerts near-absolute control over fighters through a mix of spiritualism, fear, and violence. It has no respect for the welfare of its troops. Fighters simply subsist—and no more—on the relatively fertile land of northern Uganda. They consequently require few resources other than weapons and ammunition.

Theoretically, the LRA's dependence on small arms is a powerful choke point at which to curtail its activities. In practice, however, supply-side measures to restrict the group's access to small arms are simply unfeasible while the LRA continues fighting.

Stocks of arms and ammunition, formerly supplied by Sudan, remain available to the LRA. The LRA also captures weapons from the numerous UPDF and local defence units. Trade and capture from the heavily armed regions in and around northern Uganda provide another fertile source of arms and ammunition, extending into Sudan and perhaps the DRC.

The LRA also stores its weapons and ammunition with care and observes basic field maintenance requirements that are sufficient to keep its weapons in good order. Taken together with the advantageous supply-side factors, these observations suggest that the LRA is able to maintain stocks in adequate quantities to equip its fighters.

The LRA's prospects are apparently, therefore, as healthy as perhaps they have ever been. It activities, which appear to have been on the increase throughout 2005, are commensurate with this appraisal. Where the LRA prospers, the population of northern Uganda suffers.

This is not simply due to LRA attacks. The people of northern Uganda suffer from being targeted by the LRA but also from being forcibly displaced into IDP camps. The LRA is not the only armed group that targets people in the region, but it is arguably the most destructive. Its activities, and those undertaken by the government in response, have effectively held an entire society hostage, landless and prone to depredation from a multitude of armed actors.

The LRA's prospects are as healthy as ever.

To compound this reality, the threat posed by small arms grows as the conflict progresses. People seek arms in response to the threat posed by the LRA, in addition to other armed groups. The government and the UPDF have met this demand by distributing arms to the local populace and by forming armed local defence units, thereby adding to a small arms epidemic in the north of the country. This has resulted in a cycle of small arms violence, leading to small arms-induced insecurity and further demand for small arms.

The most urgent response now appears to be to stop the LRA from targeting and abducting from the local population through a concerted effort to bring the war to a close. Yet removing the LRA from the equation will reduce only one crucial part of the security problematique in the region. It will not solve the wider small arms problem. To do so will require responsible small arms policies on the part of the states in the region that have been lacking to date.

The conflict should stand as a powerful reminder that small arms-related problems, when left to fester through poor government policies or an absence of international action, can destroy the fabric of entire societies. ◼

LIST OF ABBREVIATIONS

HSM	Holy Spirit Movement	SPLA	Sudan People's Liberation Army
ICC	International Criminal Court		(now SPLM)

ICRC	International Committee of the Red Cross	**SPLM**	Sudan People's Liberation Movement
IDP	internally displaced person(s)		(formerly SPLA)
LC	Local Council (Local Councillor)	**UNLA**	Ugandan National Liberation Army
LDU	Local Defence Unit	**UNRF**	Uganda National Rescue Front
LRA	Lord's Resistance Army	**UPDA**	Ugandan People's Defence Army
NRA	National Resistance Army	**UPDF**	Ugandan People's Defence Force
RPG	rocket-propelled grenade (launcher)	**WNBF**	West Nile Bank Front

ENDNOTES

1   This chapter presents findings from desk research and from interviews conducted in Uganda in May 2005. Among those interviewed were representatives of the Ugandan army and intelligence services, local government officials, and local community and religious leaders. Journalists, humanitarian assistance personnel, Ugandan and foreign non-governmental organizations, and other individuals closely involved with the conflict in the north shared their insights. Former fighters of the Lord's Resistance Army provided much of the detailed information presented herein, and interviews were conducted across a spectrum of the hierarchy, from the highest commanders, through mid-level leaders, to the children that make up the bulk of the organization's ordinary fighters. In addition to those interviewed on a formal basis, many people contributed greatly to this project by providing very valuable personal accounts and observations.

2   According to the vast majority of accounts, the political order is likely to be reaffirmed by elections scheduled for March 2006.

3   See ICG (2004) and HRW (1997) for details of atrocities committed in the north by the NRA.

4   For a rich analysis of the HSM and spiritualism in northern Uganda, see Behrend (1999).

5   The LRA was known as the Lord's Army from around late 1987, and the Ugandan People's Democratic Christian Army (UPDCA) from mid-1988. The movement was renamed the Lord's Resistance Army in the early 1990s.

6   Interview with a former major in the LRA, 26 May 2005, Gulu.

7   The name National Resistance Army was changed to Ugandan People's Defence Forces, following the 1996 presidential elections.

8   Interview with a former major in the LRA, 26 May 2005, Gulu.

9   Estimates of the total number of fighters vary considerably, from 500 to 1,100 or 2,000 fighters. The ICG (2004, p. 5) listed 5,000 members. An estimate of 500–1,000 is a conservative one. Sources include interviews conducted with Sudan People's Liberation Movement (SPLM) officials for the Small Arms Survey and interviews with UPDF officials, Uganda and Sudan, 2005.

10   Interview with a UPDF spokesman, 21 May 2005, Kampala; interview with a UPDF spokesman, 22 May 2005, Gulu; interview with a senior UN humanitarian official, 25 May 2005, Gulu.

11   E-mail correspondence with a foreign government security sector adviser based in southern Sudan, 14 October 2005.

12   Interview with a senior Ugandan intelligence official, 24 May 2005, Gulu; interview with a senior UN humanitarian official, 25 May 2005, Gulu; confidential interview with a knowledgeable source, 26 May 2005, Gulu.

13   Interview with a UPDF spokesman, 22 May 2005, Gulu; interview with a senior UN humanitarian official, 25 May 2005, Gulu; interview with a child support worker, 26 May 2005, Gulu.

14   Interview with a former major in the LRA, 26 May 2005, Gulu.

15   Data derived from a compilation of UN daily, weekly, and monthly security summaries kindly provided by the UN Field Security Office, Gulu.

16   A panga is a large heavy knife, used primarily to cut vegetation. It is synonymous with a machete. Interview with a child support worker, 26 May 2005, Gulu.

17   Interview with a former lieutenant in the LRA, 26 May 2005, Gulu; interview with a former major in the LRA, 26 May 2005, Gulu.

18   Interview with a senior UN humanitarian official, 25 May 2005, Gulu.

19   With the exception of a number of incidents in late 2005, the LRA does not explicitly target aid workers in the region. Only one incident, involving the hijack of three vehicles of the International Committee of the Red Cross (ICRC) in early 2004, is notable. In that case the attack was widely believed to have been launched in order to capture the long-range radios commonly installed on aid agency vehicles. It is, as yet, unclear whether recent attacks against aid agency personnel constitute a new long-term trend in LRA operations. Confidential interview with a knowledgeable source, 26 May 2005, Gulu.

20   Interview with a UPDF spokesman, 21 May 2005, Kampala.

21   Interview with a UPDF spokesman, 22 May 2005, Gulu; interview with a UN security official, 24 May 2005, Gulu.

22   Confidential interview with a knowledgeable source, 26 May 2005, Gulu.

23   Interview with a former lieutenant in the LRA, 26 May 2005, Gulu; interview with a child support worker, 26 May 2005, Gulu.

24   Interview with a former lieutenant in the LRA, 24 May 2005, Gulu.

25   Drinking and smoking are forbidden. As one interviewee noted, this is one of the reasons why the LRA consistently avoids UPDF ambushes, because fighters can the smell cigarette smoke of the soldiers. E-mail correspondence with a former LRA abductee, 22 September 2005, Geneva–Gulu.

26   Interview with a former major in the LRA, 26 May 2005, Gulu; interview with a former lieutenant in the LRA, 26 May 2005, Gulu; interview with a former LRA child fighter, 27 May 2005, Gulu.

27   This is based on the observation that the LRA fielded around 5,000 fighters in the late 1990s, and retains many weapons still in stock from those days.

28   Interview with a former LRA child fighter, 27 May 2005, Gulu.

29   The smallest detachment of UPDF troops, of around eight soldiers, is known locally as a 'detach'.

30   Interview with a former LRA child fighter, 27 May 2005, Gulu.

31   Most mobile detaches consist of around eight troops in a pickup truck. From observations made in Gulu town, it appears that the driver and commander usually sit inside the cab and are armed with assault rifles. The remaining six troops sit behind. Of these, five carry Kalashnikov derivatives and one a light machine gun. Most detaches were similar, although a few were equipped with a multi-shot grenade launcher. Each soldier is equipped with a chest pouch with three pockets, each holding two magazines. It was unclear whether more ammunition was carried in the vehicle. Each detach therefore carries a minimum of around 1,400 rounds of assault rifle ammunition.

32   This observation is based on reports that most LRA fighters carry at most three magazines on their person.

33   Since the 1980s, the local Acholi and Langi have had few cattle due to Karimojong raids. Although the government is currently trying to restock cattle, this shortage gives credence to reports that the LRA trades cattle for guns with the Karimojong. Moreover, former LRA fighters report extensive raiding against the Dinka in Sudan. Interview with a Muslim religious leader, 25 May 2005, Gulu; interview with a child support worker, 26 May 2005, Gulu; interview with a former captain in the LRA, 25 May 2005, Gulu.

34   Interview with a UN security official, 24 May 2005, Gulu; interview with a member of an international peace initiative, 23 May 2005, Gulu.

35   Interview with a senior UN humanitarian official, 25 May 2005, Gulu; interview with a former major in the LRA, 26 May 2005, Gulu.

36   The Chinese manufacturer Norinco also produces variants of the B-10, known as the Type 65 and Type 65-1, as well as later version, the Type 78. Both the Russian and Chinese variations are commonplace in Africa, and the former is known to be in service with the Egyptian army (Jones and Cutshaw, 2004, pp. 413, 439).

37   Interview with a UPDF spokesman, 21 May 2005, Kampala; interview with a former LRA child fighter, 27 May 2005, Gulu; interview with a former major in the LRA, 26 May 2005, Gulu.

38   Interview with a senior UN humanitarian official, 25 May 2005, Gulu; interview with a former major in the LRA, 26 May 2005, Gulu; interview with a former LRA child fighter, 27 May 2005, Gulu; e-mail correspondence with a knowledgeable source, 22 September, 2005, Geneva–Gulu.

39   Interview with a former major in the LRA, 26 May 2005, Gulu.

40   Interview with a former lieutenant in the LRA, 24 May 2005, Gulu.

41   Interview with a child support worker, 26 May 2005, Gulu; e-mail correspondence with a knowledgeable source, 22 September, 2005, Geneva–Gulu.

42   Interview with a former major in the LRA, 26 May 2005, Gulu.

43   Interview with a former major in the LRA, 26 May 2005, Gulu; interview with former LRA child fighters, 27 May 2005, Gulu.

44   Interview with a former LRA child fighter, 27 May 2005, Gulu.

45   Interview with a former LRA child fighter, 27 May 2005, Gulu.

46   Interview with a former lieutenant in the LRA, 26 May 2005, Gulu; interviews with former LRA child fighters, 27 May 2005, Gulu.

47   Interview with a former lieutenant in the LRA, 26 May 2005, Gulu; interview with a former LRA child fighter, 27 May 2005, Gulu.

48   Interviews with former LRA child fighters, 27 May 2005, Gulu; interview with a UPDF spokesman, 21 May 2005, Kampala.

49   Interview with a senior Ugandan intelligence official, 24 May 2005, Gulu; interview with a leading local figure and businessman, 27 May 2005, Gulu.

50   Interview with a UPDF spokesman, 22 May 2005, Gulu; interview with a senior Ugandan intelligence official, 24 May 2005, Gulu.

51   Interview with a UPDF spokesman, 21 May 2005, Kampala; interview with a former lieutenant in the LRA, 24 May 2005, Gulu.

52   Interview with a former major in the LRA, 26 May 2005, Gulu; interview with a former lieutenant in the LRA, 26 May 2005, Gulu; interview with a former lieutenant in the LRA, 24 May 2005, Gulu.

53   Interview with a senior Ugandan intelligence official, 24 May 2005, Gulu.

54   On average around three in ten of the returnees arrive armed. Interview with a leading local figure and businessman, 27 May 2005, Gulu; interview with a child support worker, 26 May 2005, Gulu.

55   Interview with a senior Ugandan intelligence official, 24 May 2005, Gulu; interview with a former lieutenant in the LRA, 26 May 2005, Gulu.

56   Interview with local religious representative, 24 May 2005, Gulu.

57   Interview with a former major in the LRA, 26 May 2005, Gulu; interview with former LRA child fighters, 27 May 2005, Gulu; interview with a former lieutenant in the LRA, 26 May 2005, Gulu.

58   Interview with a former lieutenant in the LRA, 26 May 2005, Gulu.

59   Interview with a former LRA child fighter, 27 May 2005, Gulu.

60   In September 2005, large numbers of UPDF troops were relocated to the DRC border, threatening to enter that country and attack the LRA there. It is unclear whether they have since been moved back to northern Uganda. Correspondence with a Ugandan NGO director, September 2005; interview with a senior UN humanitarian official, 25 May 2005, Gulu.

61   Interview with a senior Acholi leader, 23 May 2005, Gulu; interview with a leading local figure and businessman, 27 May 2005, Gulu; interview with an humanitarian programme officer, 23 May 2005, Gulu.

62   Reports of LRA–LDU encounters in UN daily, weekly, and monthly security summaries kindly provided by the UN Field Security Office, Gulu.

63   Interview with a senior Ugandan intelligence official, 24 May 2005, Gulu; interview with a Christian religious leader, 26 May 2005, Gulu; interview with local religious representative, 24 May 2005, Gulu; interview with a leading local figure and businessman, 27 May 2005, Gulu.

64   For an overview of the mistreatment of civilians (by both the LRA and government forces) and its impact, see HRW (2005).

65   For a concise update of the peace process and associated issues of justice, demobilization, and reintegration, see Rose, Sattarzadeh, and Baines (2005).

BIBLIOGRAPHY

AI (Amnesty International). 2004. 'Uganda: Government Cannot prevent the International Criminal Court from investigating crimes.' AI Index: AFR 59/008/2004. 16 November. New York: Amnesty International. Accessed 2 August 2005.
   <http://www.amnestyusa.org/countries/uganda/document.do?id=80256DD400782B8480256F4E005D6165>

BBC (British Broadcasting Corporation). 2006. 'Peacekeepers killed in DR Congo.' BBC News International Version. London: BBC.
   <http://news.bbc.co.uk/2/hi/africa/4639610.stm>

Behrend, Heike. 1999. *Alice Lakwena and the Holy Spirits: War in Northern Uganda 1985–1987*. Oxford: James Currey.

CSOPNU (Civil Society Organisations for Peace in Northern Uganda). 2004. *Nowhere to Hide: Humanitarian Protection Threats in Northern Uganda*. Kampala: CSOPNU. December. Accessed 19 December 2005.
   <http://www.oxfam.org.uk/what_we_do/where_we_work/uganda/downloads/nowheretohide.pdf>

De Temmerman, Els. 2001. *Aboke Girls: Children Abducted in Northern Uganda*. Kampala: Fountain Publishers.

Eavis, Paul. 2002. 'SALW in the Horn of Africa and the Great Lakes Region: Challenges and Ways Forward.' *Brown Journal of World Affairs*, Vol. 9, No. 1, pp. 251–60.

Godnick, William, with Robert Muggah and Camilla Waszink. 2002. *Stray Bullets: The Impact of Small Arms Misuse in Central America*. Occasional Paper No. 5. Geneva: Small Arms Survey. October.

HRW (Human Rights Watch). 1997. *The Scars of Death: Children Abducted by the Lord's Resistance Army in Uganda*. New York: Human Rights Watch.

——. 2005. *Uprooted and Forgotten: Impunity and Human Rights Abuses in Northern Uganda*. Vol. 17, No. 12(A). New York: Human Rights Watch. September.

ICG (International Crisis Group). 2004. *Northern Uganda: Understanding and Solving the Conflict*. ICG Africa Report No. 77. Nairobi/Brussels: International Crisis Group. 14 April. <http://www.crisisgroup.org/home/index.cfm?id=2588&CFID=11984841&CFTOKEN=21011998>

——. 2005a. *Shock Therapy for Northern Uganda's Peace Process*. ICG Africa Briefing No. 23. Kampala/Brussels: International Crisis Group. 11 April.

——. 2005b. *Building a Comprehensive Peace Strategy for Northern Uganda*. ICG Africa Briefing No. 27. Kampala/Brussels: International Crisis Group. 23 June. <http://www.crisisgroup.org/home/index.cfm?id=3523&l=1>

IISS (International Institute for Strategic Studies). 2004. *The Military Balance 2004–2005*. Oxford: Oxford University Press.

——. 2005. *Uganda: Military and Security Developments Overview*. London: IISS. Accessed 13 July 2005. <http://acd.iiss.org/armedconflict>

IRC (International Rescue Committee). 2005. 'IRC Calls for Urgent Action to Halt Northern Uganda Violence as New Data Reveals 1,000 Civilians Die Weekly in Conflict.' New York: IRC Communications. 9 September. Accessed 30 November 2005. <http://www.theirc.org/index.cfm/wwwID/2299>

Jones, Richard and Charles Cutshaw. 2004. *Jane's Infantry Weapons 2004–2005*. Coulsdon: Jane's Information Group.

JRS (Jesuit Relief Service). 2005. 'Sudan: insecurity in Nimule soars.' ReliefWeb, 15 May. Accessed 19 July 2005. <http://www.reliefweb.int/rw/rwb.nsf/0/5a26905ac618ce894925701b001c1f43?OpenDocument>

MONUC (Mission des Nations Unies en République Démocratique du Congo). 2006. 'Huit casques bleus de la MONUC tués dans le parc de la Garamba.' Communiqué de presse MONUC. Kinshasa: Monuc. <http://www.monuc.org/News.aspx?newsID=9719>

Nichols, Ryan. 2005. 'Disarming Liberia: Progress and Pitfalls.' In Nicolas Florquin and Eric Berman, eds. *Armed and Aimless: Armed Groups, Guns, and Human Security in the ECOWAS Region*. Geneva: Small Arms Survey, pp. 108–43.

Nyeko, Balam and Okello Lucima. 2002. 'Profiles of the Parties to the Conflict.' In Okella Lucima, ed. *Protracted Conflict, Elusive Peace*. Accord 11. London: Conciliation Resources, pp. 16–23. Accessed 19 December 2005. <http://www.c-r.org/accord/uganda/accord11/profiles.shtml>

Nzita, Richard and Mbaga Niwampa. 1998. *Peoples and Cultures of Uganda*. Kampala: Fountain Publishers.

Patrick, Erin. 2005. 'Little Protection in "Protected Villages": IDPs in Northern Uganda.' Migration Policy Institute. Washington, DC. May. <http://www.migrationpolicy.org/research/uganda_two.php>

RLP (Refugee Law Project). 2004. *Negotiating Peace: Resolution of Conflicts in Uganda's West Nile Region*. Refugee Law Project Working Paper No. 12. Kampala: Refugee Law Project.

Rose, Heidi, Irene Sattarzadeh, and Erin Baines. 2005. 'Northern Uganda—Human Security Update, Pursuing Peace and Justice: International and Local Initiatives.' Vancouver: Liu Institute for Global Issues, Conflict and Development Programme. May. Accessed 2 August 2005. <http://www.ligi.ubc.ca/admin/Information/480/Justice-Peace-Liu-May05.pdf>

Small Arms Survey. 2004. *Rights at Risk*. Oxford: Oxford University Press.

——. 2005. *Weapons at War*. Oxford: Oxford University Press.

Thusi, Thokozani. 2003. 'Assessing Small Arms Control Initiatives in East Africa: The Nairobi Declaration.' *African Security Review*. Vol. 12, No. 2. Pretoria: Institute for Security Studies. Accessed 9 January 2006. <http://www.iss.co.za/pubs/ASR/12No2/F2.html>

UNICEF (United Nations Children's Fund). 2005. 'Uganda: Background.' New York: UNICEF. Accessed 23 September 2005. <http://www.unicef.org/infobycountry/uganda_background.html>

UN OCHA (United Nations Office for the Coordination of Humanitarian Affairs). 2004. 'Humanitarian Update Uganda Volume VI: Issue III.' ReliefWeb, 31 March. New York: UN OCHA. Accessed 2 August 2005. <http://www.reliefweb.int/rw/RWB.NSF/db900SID/SKAR-647KF6?OpenDocument>

USAID (United States Agency for International Development). 2005. 'Uganda Complex Emergency Situation Report #2 (FY 2005).' ReliefWeb, 13 April. Washington, DC: USAID. Accessed 2 August 2005. <http://www.reliefweb.int/rw/RWB.NSF/db900SID/KHII-6BF35W?OpenDocument>

USDOS (United States Department of State). 2005. 'Uganda: Country Report on Human Rights Practices—2004.' Washington, DC: USDOS. 28 February. <http://www.state.gov/g/drl/rls/hrrpt/2004/41632.htm>

Veal, Angela and Aki Stavrou. 2003. *Violence, Reconciliation and Identity: The Reintegration of Lord's Resistance Army Child Abductees in Northern Uganda*. ISS Monograph No. 92. Pretoria: Institute for Security Studies. November.

ACKNOWLEDGEMENTS

Principal author

James Bevan

Contributors

David Capie, Anne-Kathrin Glatz, Anders Haugland, Denis Mwaka, Sharmala Naidoo, Stephen Okello, and Stella Sabiti

Few Options but the Gun

ANGRY YOUNG MEN

12

INTRODUCTION

Armed and angry young men are perhaps the most feared element of any society, but they also have the most to fear. Regardless of the countries in which they live, young men represent a disproportionately high share of the perpetrators and victims of gun-related, lethal violence.

The World Health Organization (WHO) estimates that boys are two to three times more likely than girls to get involved in fighting (WHO, 2002, p. 29). Young men—those aged 15 to 29—also account for half of global firearm homicide victims, or 70,000 to 100,000 deaths annually.[1]

While studies of youth violence have long considered the role played by young men, the issue has received little attention in the framework of small arms research. This chapter begins to fill the gap by examining the following questions:

- Why are young men the primary perpetrators of armed violence?
- What role do small arms play in this phenomenon?
- Have interventions designed to prevent or reduce armed violence adequately tackled the complex relationship that exists between young men and small arms?

The chapter first reviews the principal theories on why young men are more likely to turn to armed violence than other demographic groups. It finds that traditional biological and demographic arguments do not sufficiently take into account the multiple factors that encourage and prevent young men from resorting to violence. It argues that gender ideologies—particularly those that associate masculinity with power—offer crucial insight into why many marginalized young men see violence as an attractive means of achieving manhood and respect. The second section argues that small arms can be an important part of this complex social equation. It examines how the functional and symbolic attributes of small arms make them attractive for young men wishing to achieve power through association with or participation in violence. The final section reviews opportunities to address the problem by controlling young men's access to small arms and countering their espousal of a violent masculine ideology.

The following are among the chapter's most important conclusions:

- Young men frequently perceive violence—particularly small arms violence—as a means to reach positions of social or economic status that they feel entitled to.
- By offering empowerment in the face of exclusion from socially defined masculine roles, small arms can be strong symbols of power for marginalized young men.
- Curbing young men's access to firearms has proved an effective component of short-term strategies to reduce the number of deaths arising from youth violence.

• Countering socially constructed associations between guns, violence, power, and masculinity is a key component of any effective, long-term violence prevention strategy.

YOUNG MEN AND VIOLENCE

Most research identifies young men as the primary actors in contemporary violence—as well as other forms of anti-social activity. This gender and age distinction appears to apply across very different social strata. Moreover, age and gender can be more powerful determinants of levels of armed violence than geographical considerations (see Figure 12.1).

Young people in general and young men in particular comprise the largest group of perpetrators of most criminal activity. A review of more than 140 studies investigating a wide range of offences has found that people are most likely to commit a crime when aged between 12 and 30.[2] In Canada, for instance, 12–17-year-olds account for 8 per cent of the population, but as many as 21 per cent of all offenders; the overwhelming majority of these—almost 80 per cent of young offenders—are boys.[3] In Brussels, Belgium, 92 per cent of delinquent minors[4] in 1993 were male

Figure 12.1 **Age and gender as key determinants of vulnerability to small arms violence, 2000**

FIREARM MORTALITY RATE PER 100,000

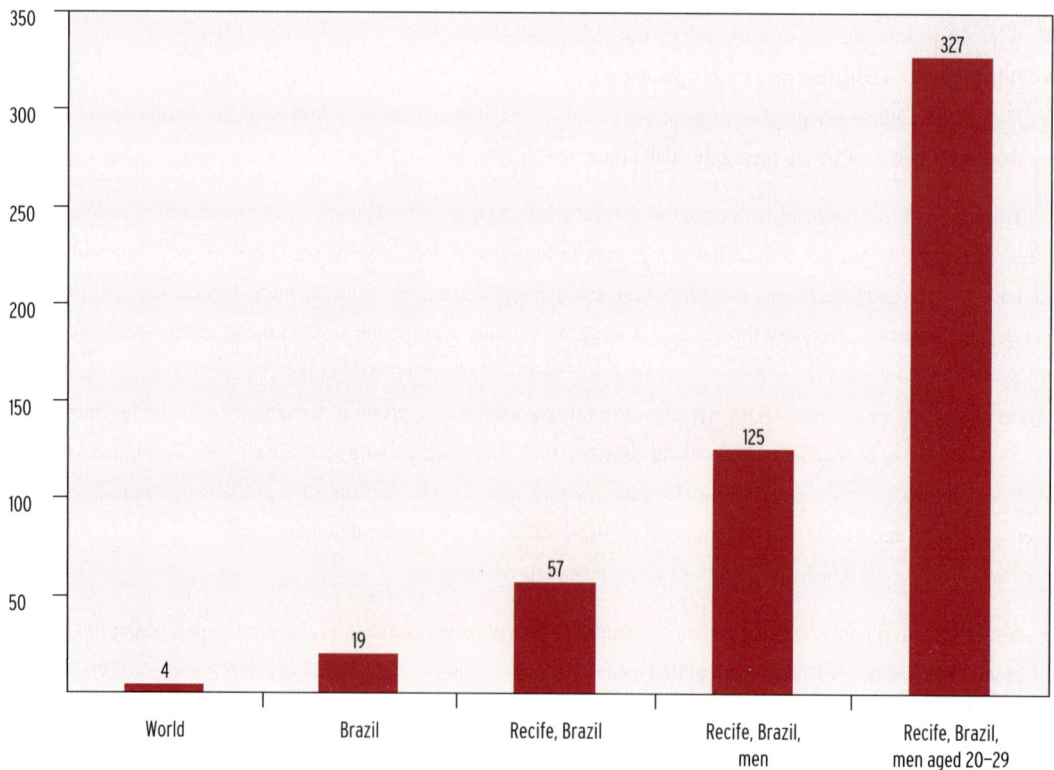

Sources: World: Small Arms Survey calculations based on Richmond, Cheney, and Schwab (2005, p. 348, using 229,000 annual non-conflict-related firearm deaths estimate) and UN Population Division (2005). Brazil and Recife: Peres (2004, pp. 129, 130, 132)

Figure 12.2 **Firearm homicide rates: victims per 100,000 among men aged 15–29 compared with the overall population (selected countries, latest year available)**

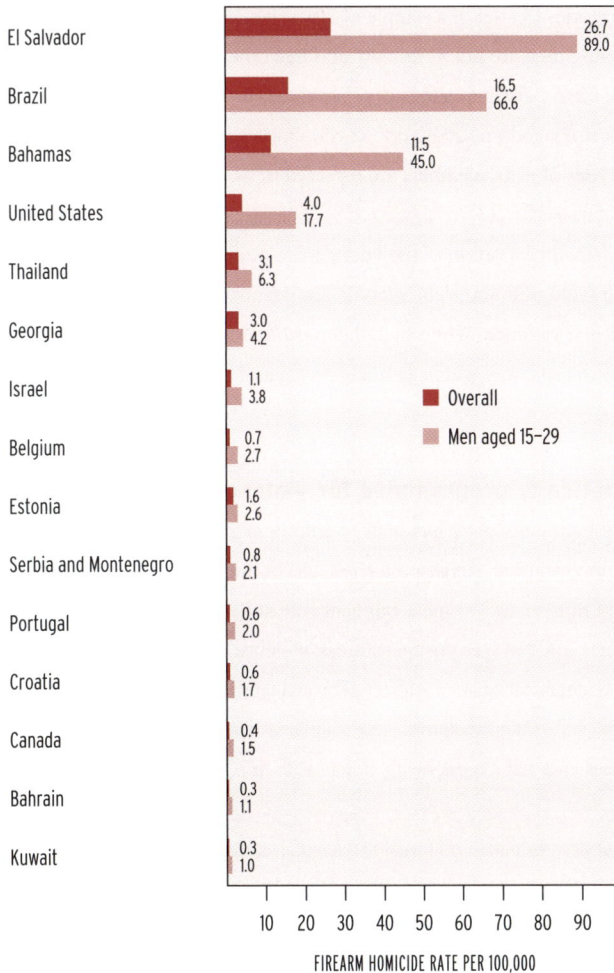

Country	Overall	Men aged 15–29
El Salvador	26.7	89.0
Brazil	16.5	66.6
Bahamas	11.5	45.0
United States	4.0	17.7
Thailand	3.1	6.3
Georgia	3.0	4.2
Israel	1.1	3.8
Belgium	0.7	2.7
Estonia	1.6	2.6
Serbia and Montenegro	0.8	2.1
Portugal	0.6	2.0
Croatia	0.6	1.7
Canada	0.4	1.5
Bahrain	0.3	1.1
Kuwait	0.3	1.0

FIREARM HOMICIDE RATE PER 100,000

Source: Small Arms Survey calculations based on WHO (2005) and UN Population Division (2002)

(Vercaigne, 2001, p. 285), while 83 per cent of minors arrested in the first six months of 2002 in Davao, Philippines, were boys (Templa, 2004, p. 18).

Young men are also more likely to use firearms when carrying out crime than any other demographic group. A 2001 study of US state and federal correctional facilities finds that 18 per cent of all inmates carried a gun for their last offence. The proportion rises to 29 per cent for inmates under the age of 25. Among inmates of all ages, men were almost three times as likely as women to have used small arms during their offence (Wolf Harlow, 2001). The same pattern applies in other contexts. In Montenegro in 2003, for instance, 99 per cent of the perpetrators of gun assaults were men, and almost half were aged 15–29 (Florquin and O'Neill Stoneman, 2004, p. 16).

Victims and perpetrators of armed violence typically know of each other and belong to the same demographic and social groups (Hemenway, 2004, p. 113; Kennedy, 1997, p. 457). Not surprisingly, young men are thus not only the main perpetrators of but also disproportionately vulnerable to violence involving small arms. In 2004, the Small Arms Survey reported that, while young men are the primary victims of violence in general, they account for an even greater proportion of gun violence victims (Small Arms Survey, 2004, p. 179). An analysis of WHO statistics confirms this assertion. Firearm homicide data available in 70 countries and territories shows that men aged 15–29 represent half of all firearm homicide victims, at a rate more than four times higher (22 per 100,000) than that of the general population (5 per 100,000) (WHO, 2005). This trend is consistent across regions and countries experiencing different rates of violence (see Figure 12.2). Extrapolating this ratio globally suggests that between 70,000 and 100,000 young men aged 15–29 die every year from gun homicides.[5]

It would be misleading to explain young men's disproportionately high involvement in armed violence with the theory that they are biologically more inclined to resort to violent behaviour. While some young men account for most of the world's armed violence, many more young men have no involvement in delinquency or violence. Recent surveys report that a mere 6–7 per cent of young men commit 50–70 per cent of all crime and 60–85 per cent of all serious and violent crime.[6] Similarly, young men are the principal victims of small arms violence in terms of overall numbers across the globe even though they are *not* the demographic group most at risk of dying of firearm homicide in a surprisingly high number of countries and territories (33 out of a sample of 70).[7] The experiences of these countries suggest that high rates of violence among young men are not a certainty and that other factors come into play.

Young men who do engage in armed violence often belong to gangs or other armed groups that tend to emerge in contexts of social and economic marginalization (Hagedorn, 2001, pp. 42–45). Such groups usually enjoy easy access to small arms. In the United States, 75 per cent of gang members are reported to own a gun, compared with only 25–50 per cent of non-gang youth.[8]

A mere 6–7 per cent of young men commit 60–85 per cent of all serious crime.

This section has demonstrated that a relatively small proportion of the young male population is responsible for most armed violence. The next sections evaluate biological and demographic arguments that seek to explain the disproportionate involvement of young men in armed violence. While such theories provide important insight, they fail to explain variations over time, among different cultures, and within a society—such as those along class, ethnicity, race, religion, and other lines.

The biological argument: men are genetically programmed for violence

Numerous researchers have sought to explain the high rates of violent activity among men through biological and genetic theories. The results, which are relatively inconclusive, suggest a limited and bi-directional relationship. For example, higher levels of testosterone (found in both men and women, but generally at much higher rates in men) have been linked to higher rates of aggression in men and boys. Exposure to stress, violence, and feelings of subjugation causes testosterone levels to rise. In other words, chemical balance is affected by changes in the social environment. This research generally suggests that, at most, higher levels of testosterone may trigger violent or aggressive behaviour in individuals who already exhibit violent tendencies and that experiencing violence, in turn, leads to higher levels of testosterone (Renfrew, 1997; Kimmel, 2003).

Brain research has also examined genetic differences in male and female styles of communication and reasoning, including traits that might be associated with aggression and violence. Yet the bulk of this research tends to indicate that there are greater differences within each sex than there are aggregate differences between the sexes (Kimmel, 2003). Furthermore, most researchers conclude that even if there is a biological or genetic basis for aggression and violence in men, this propensity is mitigated through the social environment and through higher cognitive functions. Indeed, some brain research confirms that neocortex functions and other higher brain structures are involved in reducing aggression (Renfrew, 1997). This provides some neurological basis for confirming what has already been confirmed in psychology, namely that humans can control their aggressive tendencies through more complex levels of cognition (Barker, 2005b). Any biological propensity or predisposition towards violence or aggression is therefore mediated by the social context and individual factors.

The demographic argument: too many young men

In recent years, various researchers have argued that young men are responsible for high rates of violence wherever they represent a disproportionately high segment of the population. This demographic theory implies that regions and countries with high proportions of young men (see Figure 12.3) are more likely to experience high rates of violence. A recent World Bank report states: 'Large-scale unemployment, combined with rapid demographic growth,

Figure 12.3 **Demographic bulges: Young men aged 15–29 as a proportion of total population, by income level and region, 2005**

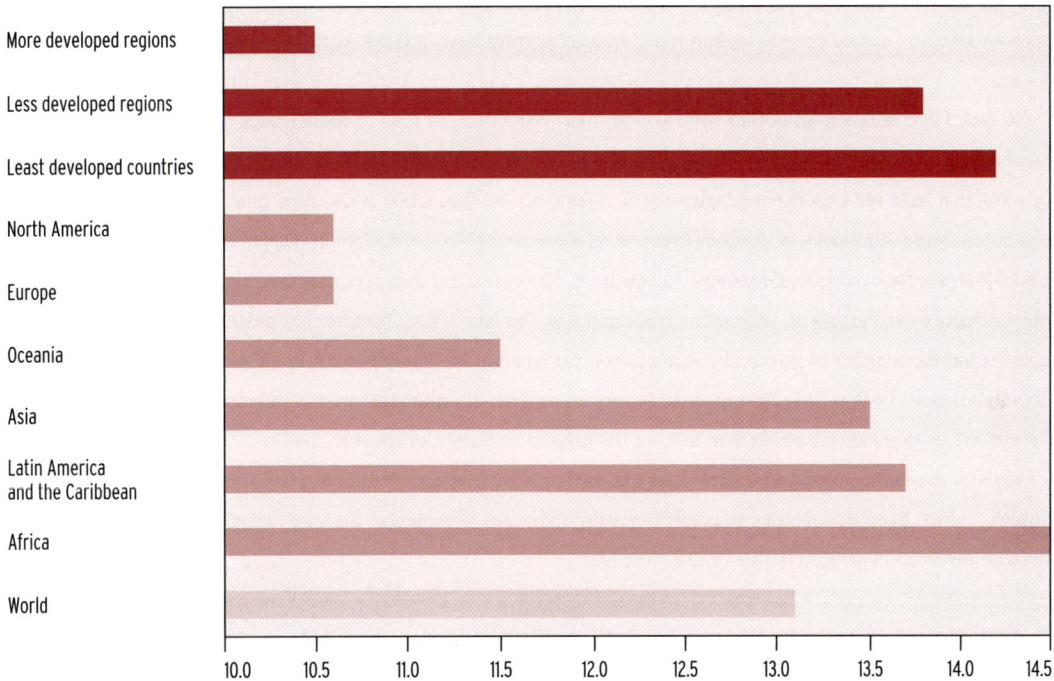

Source: UN Population Division (2005)

creates a large pool of idle young men with few prospects and little to lose' (Michailof, Kostner, and Devictor, 2002, p. 3). Mesquida and Wiener (1999) argue that one of the most reliable factors in explaining conflict (which they call 'coalitional aggression') is the relative number of young men (under 30 years of age) compared with men over 30. In analyzing data from more than 45 countries and 12 tribal societies, they find—even controlling for income distribution and per capita GNP—that the higher the ratio of 15–29-year-old men to older men, the higher the rates of conflict.

In a similar vein, Cincotta, Engelman, and Anastasion (2003, p. 44) ask:

Why are youth bulges so often volatile? The short answer is: too many young men with not enough to do. When a population as a whole is growing, ever larger numbers of young males come of age each year, ready for work, in search of respect from their male peers and elders. Typically, they are eager to achieve an identity, assert their independence and impress young females. While unemployment rates tend to be high in developing countries, unemployment among young adult males is usually from three to five times as high as adults' rates, with lengthy periods between the end of schooling and first placement in a job.

In reviewing demographic data and 207 conflict onsets in 1950–2000, Urdal also concludes that countries with large numbers of youths had a higher risk of conflict than countries with smaller numbers, and especially so under conditions of economic stagnation. He finds that an increase of 1 per cent in the proportion of 15–24-year-olds relative to the total adult population (i.e. aged 15 or more) increased the likelihood of conflict by 7 per cent (Urdal,

2004, p. 9). But he contends that there is no clear threshold as to the proportion of young men needed to render a country conflict-prone. Moreover, while the effect of youth bulges on conflict was clearly significant and positive during the cold war period, it is insignificant and even negative for the post-cold war era. In the latter period, factors such as levels of development, regime type, and geography have greater explanatory powers (Urdal, 2004, pp. 15–16).

In fact, large numbers of youths (and young men) only seem to lead to higher rates of armed violence when combined with economic stresses, and only in specific settings. There are, for example, countries with large numbers of youth that have not experienced high rates of violence or conflict, while at the same time, countries with relatively few youths suffer high rates of violence (the United States and United Kingdom in comparison with other industrialized countries, for example). Economist Steven Levitt has found, for instance, that the aging of the population had little influence on the fall of US violent crime rates in the late 1990s. Factors that did have an impact included increases in the number of police, the rising prison population, the waning crack epidemic, and the legalization of abortion[9] (Levitt, 1999, p. 581; 2004, p. 163). In sum, demographic arguments are not sufficient to explain the reasons for conflict, young men's participation in it, or the triggers or causes of specific conflicts.

Indeed, the demographic argument does not explain which young men in a given setting become involved in armed conflict. Even in countries in conflict or with high rates of violence, the vast majority of young men do not become involved in armed conflict or use weapons.

A situational approach: human ecology and gendered socialization

The biological and demographic arguments both neglect the fact that young men often differ greatly from one another in their behaviour. Some become involved in violence while others do not. It is therefore plausible that the involvement of young men in violence is 'situational'—conditioned by their interaction with the world around them (society, community, family).

Certain bundles of factors that young men are exposed to appear to have a particularly strong bearing on the likelihood that they will become involved in violence. Several major longitudinal studies in Western Europe and North America have sought to identify early childhood predictors of violent behaviour. These studies have consistently concluded that most violent behaviour is explained by multiple, interacting social factors arising during childhood and adolescence. Some of these factors figure in Table 12.1.

Table 12.1   Some of the factors leading to violent behaviour in young men

Being labelled as troublesome	Low school achievement
Coercive or violent parental controls	Limited social skills
Limited parental control	Holding more traditional or rigid views about gender
Having witnessed or experienced violence in the home or community	Having been shamed or experienced significant shame and humiliation as a child
Socializing with delinquent peers	Having been brutalized or violently subjugated
Perceiving hostile intentions in others	Having used violence and seen that violence produces respect

Sources: Elliott (1994); McAlister (2000); Sampson and Laub (1993); Barker (2005a); Rhodes (1999); Gilligan (1996)

Socialization into a violent masculinity? A child dressed as a combatant and armed with a replica assault rifle, Jerusalem, January 2006. © Central Imaging Agency/WPN

Taken as a whole, this research provides strong support for an 'ecological' model that emphasizes the interaction among multiple factors in explaining human development and human behaviour. These factors or levels include the family, local community, and the wider context of social, political, and cultural norms and realities (Bronfenbrenner, 1979).

In this ecological model, young men go through a process of socialization in which they are not merely passive receptors of social norms; rather, they participate actively in internalizing, reframing, and reproducing norms that they receive from their social settings, their families, and their peers. Further, a young man's behaviour—such as involvement as a combatant, use of small arms, or use of violence against a woman—is not attributed to one specific factor. External factors such as cultural norms are filtered through the community, family, peer group, and other close groups and influences before they are internalized and acted upon by the young man himself.

From this perspective, a young man's gender is not the sole determinant of his association with or willingness to use armed violence; in fact, his understanding and use of social and cultural ideologies of masculinity will largely determine whether he turns to armed violence. While the ideologies of masculinity and femininity are constructed in relation to each other, they are not equivalent. They reflect the dynamics of any given 'field of power'—in which men as a group have power over women as a group, and some men have power over other men (Kimmel, 2005, p. 6).

Young men who are marginalized, whether socially or economically, frequently lack power, despite being socially conditioned to seek it. Kimmel argues that this condition is a linchpin of young, male violence. Violence can allow a man who seeks to appear rich, powerful, and strong to counteract social marginalization and humiliation. For young men denied their place in society, violence—especially armed violence—evens the score (Kimmel, 2003).

SMALL ARMS: FUNCTIONAL AND SYMBOLIC

Where do small arms fit in the situational model? The following sections consider the functional and symbolic roles small arms play for some young men.

The appeal of small arms to young men can be expected to vary considerably across the world as a function of social and economic conditions. This chapter argues that the strength of their appeal is conditioned by the young man's social environment, both specific (his family, peers, and community) and broad (cultural norms).

From a functional perspective, societies, communities, and peers may legitimize young men's use of small arms to achieve certain (often shared) goals. This is particularly so when young men (and their communities) are excluded from non-violent avenues of social and economic advancement, or if they face discrimination or threats to their security. In these situations, violence can have a powerful functional appeal, with small arms making it even more effective.

Box 12.1   Small arm functions: the word from the street

Rites of passage and building reputation:

'I was 13 years old when they gave me a test for entering the group. They told me to give a gun to a guy leaving school... When I left the school some [rival] gang members began to follow me and I tried to lose them. They followed me and I had no alternative but to use the gun. I shot twice in the air and they took off...'[a]

13-year-old boy, Ecuador

Proof of masculine identity:

'One day I started to hang around with the men ... I started to carry a backpack, a bag full of bullets, and I continued hanging around with the men. Now I'm a *gerente de boca* [drug sales overseer] and I carry my own pistol.'[b]

16-year-old male gang member, Rio de Janeiro, Brazil

Status, wealth, and women:

'... they see you walking with a rifle all over the place, see you riding a motorbike, [wearing a] gold chain. These things influence [kids] a lot. So a youth will say "I want that, too." I want to have lots of women. I want a car. This influences minors to enter crime more and more each day: new clothes, new sneakers, new cap.'[c]

18-year-old male former gang member, Rio de Janeiro, Brazil

A means of defence/resistance:

'I had to use firearms. When your friend gets killed you have plenty of motives.'[d]

Young male gang member, Medellín, Colombia

'Gun is nice, gun is protection. When you have it no guy can disrespect you.'[e]

Young male former gang member, August Town, Jamaica

'... the government thought we were not serious and used our oil money to kill our people.... Now we have the guns they are beginning to respect and recognize us.'[f]

Young male armed group member, Niger Delta Region, Nigeria

Sources:

a Punctuation of original source edited for readability. COAV (2005).

b Dowdney (2003, p. 124).

c Dowdney (2003, p. 134).

d Dowdney (2005, p. 112).

e Dowdney (2005, p. 242).

f Dowdney (2005, p. 256).

From a symbolic perspective, small arms may be valued irrespective of their practical utility—for example, as symbols of resistance, defence, or self-sufficiency. Such characteristics are often connected to, or juxtaposed with, violent interpretations of manhood. For some young men, small arms also symbolize the transition from boyhood to manhood.

The functions of small arms[10]

Researchers have identified specific risk factors that appear to influence whether young men become involved in small arms violence. These are related to young men fulfilling—or attempting to fulfil—what they perceive to be socially expected of a man. In the most serious cases, young men may take violent action repeatedly to force peers and communities to recognize them as men or to generate the material and social wealth they perceive necessary for greater social status.

Risk-taking and rites of passage

A nearly universal aspect of male socialization is that manhood means having to publicly prove that one is a man. This may include rites of passage in some parts of the world, and informal tests of courage, bravado, and risk-taking in most regions. Outwardly risky or antisocial behaviour may be a way for a young man to prove himself, become part of a group, or simply be recognized—in short, to define himself as a 'real man' according to prevailing social standards.

Involvement with small arms can help young men define themselves as 'real men'.

Involvement with small arms is one such type of behaviour, as evidenced by a 2004 survey of 5,800 California adolescents. The study finds that respondents who engaged in high-risk behaviour such as smoking or binge drinking—often interpreted by young men as manly—were also more likely to have used a gun in self-defence or to have been threatened with one (Hemenway and Miller, 2004, p. 396).

Very young armed gang members from countries such as Brazil, El Salvador, and Jamaica report how their initiation into using firearms starts with, for instance, shooting into the air while 'hanging around' with the group, or borrowing a gun from a friend or other gang member. In other cases, they describe how carrying weapons and ammunition on behalf of older members brings important recognition (Dowdney, 2005, pp. 101, 222, 242).

The vast majority of young men, while just as intent on affirming their manhood, do not become involved in sustained delinquent behaviour or armed violence. What distinguishes them from those who turn to violence?

Building reputation

One of the chief hallmarks of manhood worldwide is achieving some level of financial independence, securing employment, and subsequently starting a family. Numerous studies have confirmed that when men are out of work, or otherwise unable to fulfil these requirements, their self-esteem suffers.

One hypothesis asserts that, when avenues to socially accepted manhood are closed or made more difficult, alternative sources of wealth generation and social status become more appealing. One such alternative is becoming associated with, or involved in, small arms violence.

Involvement with small arms can help young men define themselves as 'real men'. A 'bad boy' instils fear and wields power. He leaves behind anonymity and attracts immediate attention.

Emler and Reicher (1995) have suggested that violence and antisocial behaviour for these young men is a deliberate 'reputational project', an effort to affirm an identity as delinquent or violent in order to fit into an antisocial peer group. This strategy may be particularly attractive to young men who see mainstream goals and identities as beyond their reach, or who feel they have been rejected by mainstream social institutions. Small arms may be instrumental in their pursuit of wealth, respect, and security.

Proof of masculine identity

In many low-income areas, wielding a gun is often considered a sign of status, male affluence, and power. In many parts of South Africa, for instance, for much of the 1980s, 1990s, and with lingering effects today, both white and black young men have often been socialized into a militaristic version of manhood through the formation of a kind of brotherhood of combatants, usually involving small arms, whether for or against apartheid. As one commentator notes, 'The gun is a convenient peg on which to hang traditional notions of masculine power' (Cock, 2001, p. 49). In a similar vein, Wilkinson and Fagan (1996, pp. 81–82) conclude that for inner-city young men in the United States, guns provide a sense of power, even if they are not used. Likewise, in Rio de Janeiro, young men are reportedly drawn into the drug trade, not least because they are able to acquire small arms that they can openly display to the community. A nine-country[11] study of young men involved in organized armed violence found that they saw the carrying of guns as an effective way to gain respect and achieve status (Dowdney, 2003, p. 133).

If a young man carries a small arm, however, he may be called upon to use it. An analysis of homicides in Australia finds a common pattern of contests of honour and reputation, public challenges to men's reputation (and the associated perceived loss of self-respect), and unpredicted escalation to lethal violence in which the 'audience' provides important cues to appropriate behaviour (Polk, 1994). Similarly, in South Africa and the US city of Chicago, researchers note that when numerous young men carry small arms 'would-be fistfights in a less well-armed community become fire-fights' (Dowdney, 2005, p. 303; Cook et al., 2005, p. 7), leading to an arms race in which young men believe they need to be armed (DEMAND).

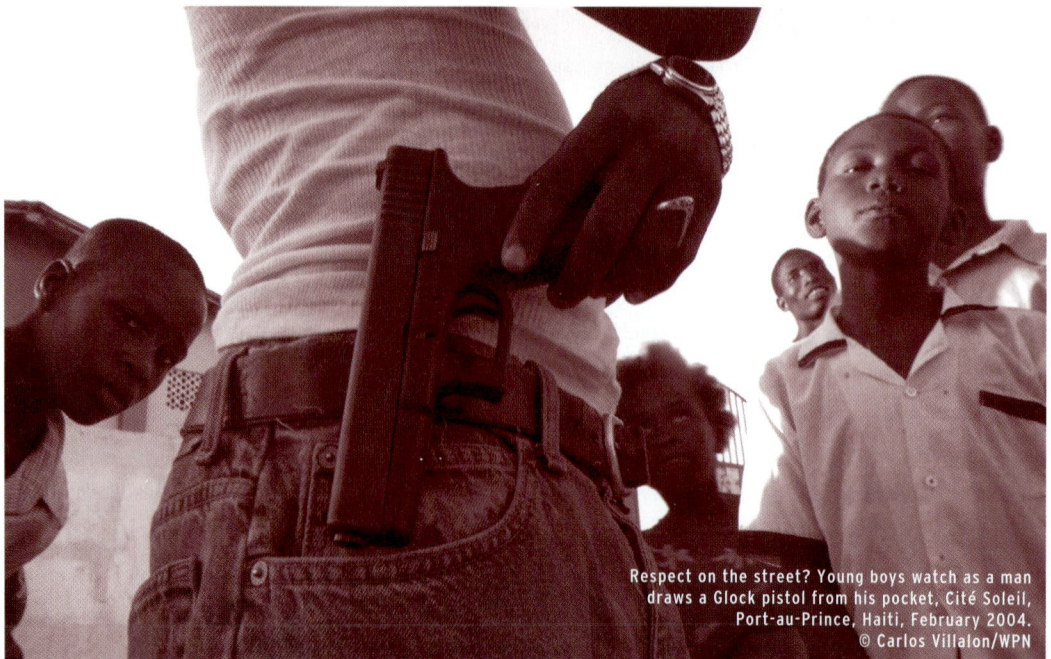

Respect on the street? Young boys watch as a man draws a Glock pistol from his pocket, Cité Soleil, Port-au-Prince, Haiti, February 2004.
© Carlos Villalon/WPN

Empowerment in the community

Small arms may enable young men, who would otherwise have little influence in their communities, to exert considerable control—even over traditional figures of authority. In some cases (see Box 12.2), young men simply follow in the footsteps of existing figures of community authority. In others, they seize authority through violence, reversing or destroying the existing social order.

Violence is more attractive where the peer group or broader community recognizes it as a viable, even acceptable, means of achieving material or social goals (see Box 12.2).

Boys involved in armed conflict in Liberia, Sierra Leone, and, more recently, the Democratic Republic of the Congo may be responding to socially accepted interpretations of manhood. By controlling a given setting and bringing violence to bear on those around them, they become, in essence, what are frequently known in Africa as 'big men' (Lindsay and Miescher, 2003). These young men may, however, have strayed too far from their communities and traditions. Observers of young male combatants in West and Central Africa note that the violence and fear tactics that they utilize take a socially accepted interpretation of manhood to its extreme (Ellis, 1997; Jourdan, 2004, pp. 162–63). As one study concludes, 'In many of these wars [in West Africa], both local and foreign observers have detected an element of youth out of control, adolescents and even children who, in societies with strong gerontocratic traditions, seize power by force' (Ellis, 1997, p. 110). These studies illustrate how, in certain settings, young men use small arms to bypass traditional structures and gain power. Further research is needed to improve our understanding of the social organization of violence and the role small arms may play in disrupting these controls.

Box 12.2   'Big men', armed violence, and social recognition in Belfast, Davao City, and Manenberg

Belfast in Northern Ireland, Davao City in the Philippines, and Manenberg, South Africa, differ in many respects. Yet they are all characterized by high unemployment, poverty far exceeding the national average, and relatively high levels of violence, both in the home and community-wide. These factors come together to produce environments in which young men, faced with few prospects, find violence an attractive, and sometimes necessary, course of action.

The three cases suggest that, for young men with few opportunities, being a member of an armed group is a rational activity. Young men often choose to adopt the violent means they recognize as inherent to their communities in the hope of achieving status and material gain. In the case of Belfast, however, few young men have access to small arms. Armed violence is consequently lower than in Manenberg and Davao City even though the underlying attractions of violence for young men do not differ significantly.

Where it is integrated into the fabric of the community, armed violence may bring social recognition. In Belfast, some people turn to paramilitaries to 'police' the community. The punishment attacks they inflict on perpetrators of 'unacceptable crimes', witnessed by young men growing up in the city, serve to normalize violence within the community (Fox, 2002; Smyth and Campbell, 2005, p. 5). Similarly, gangs in Manenberg have come to define everyday life through their violent control over territory and people. Some two-thirds of children interviewed in a school in Manenberg reported that they had seen someone be shot; this proportion rose to 79 per cent among 18-year-old men (Legget, 2005, p. 18). In Davao City, youths also witness high levels of violence, both on the street and at home. For example, about 90 per cent of Davao gang members report having experienced domestic abuse (Bonifacio et al., 2004).

In all three cities, violence receives additional community support as it is often oriented around defence of the family and ultimately the neighbourhood. Youths are thus linked to their communities, which can help them gain respect (BIP, 1998; Conway and Byrne, 2005, p. 14). In all three cases, young men receive social recognition for acting violently, whether from their peers or communities at large. Different levels of access to small arms yield different levels of armed violence in each community. Yet where weapons exist, they tend to confer status and respect on their (young) holders. Joining an armed group becomes part of the transition from subordinate child to dominant man.

Status, wealth, and women

Young men who use violence, and particularly armed violence, often seem to do so to gain status indirectly, namely by acquiring the kinds of material and social goods they are often denied because of poverty, exclusion, or a lack of respect from their communities.

Researchers have extensively documented the use of small arms by young men as a means of accruing wealth, such as in the illicit drugs trade (Cook et al., 2005; Dowdney, 2005; Wolf Harlow, 2001). Perceived benefits of being armed need not be linked to the use of firearms, but rather to their public display, which can produce a sense of respect, status, or the ability to inspire fear in others.

As one non-gang-affiliated young man in Chicago noted:

> When I bought [my .357], no, I didn't see if it was good [working]. Look man, I can get one of those guns that fires, but shit, sometimes you just need to show it, you know, and you get the respect you looking for. And, this thing was big man, I didn't give a shit if it fired or not, I could have killed somebody with it just hitting them over the head! (Cook et al., 2005, pp. 7–8)

One respondent to a survey in August Town, Jamaica, described how he had used a variety of weapons for reasons of status and, in particular, to attract women (COAV, 2005; Dowdney, 2005, p. 117). In a Rio de Janeiro *favela,* a young man observed, 'You see lots of hard workers without a woman, but you never see a gangster without a woman' (Barker, 2005a, p. 32). Also in Rio, a female youth made a similar comment: 'Sometimes guys will even borrow guns, just to walk around with them, to show off for the girls . . . . They use them because they know that pretty girls will go out with them.'[12]

Research in New York City neighbourhoods suggests similar connections between small arms and status. Possessing a gun is perceived as an important means of impressing peers, along with having a car or a girlfriend. The study concluded:

> For a generation of adolescents, gun violence has had instrumental value that was integrated into the social discourse of everyday life among urban youth. Guns were, and remain[,] salient symbols of power and status and strategic means of gaining status, domination, or material goods. (Wilkinson, 2003, p. 252)

'. . . notions of "manhood" are tied to gaining respect, women and guns.'

A study of armed groups in nine countries similarly found that, in many of the communities studied, 'notions of "manhood" are tied to gaining respect, women and guns; all of which are made available to adolescent boys when joining an armed group' (Dowdney, 2005, p. 117). Further research, including biographical research, would shed light on why young men become attracted to small arms and ultimately choose to use them. Key questions include how, exactly, small arms confer status. Does their principal value lie in their symbolic role (e.g. their association with a successful gangster or other authority figure) or in their direct use to acquire wealth and status through crime or confrontation? What effect does one type of use have on the other?

A means of defence and resistance

For some low-income men, violence is a way of maintaining status in their peer groups, but it is also a way of protecting themselves from violence.[13] Researchers writing about young men in the United States have concluded that violence, including gun-related violence, can be a form of self-protection—both physical and psychological. As one young Chicago gang member noted:

Who [is] going to fear me? Who [is] going to take me seriously? Nobody. I'm a pussy unless I got my gun. (Cook et al., 2005, p. 7)

Studies on gangs in Colombia, Ecuador, El Salvador, Jamaica, and the Philippines suggest protection, whether ultimately successful or not, is a prime reason why young men carry small arms (Dowdney, 2005, pp. 188, 204, 223, 242, 283).

Elsewhere in the world, small arms appear to serve similar functions. In the case of politically motivated armed groups, such as in the Delta region of Nigeria, small arms are often considered tools of resistance to oppression. Joining the group enables young men to acquire weapons with which to confront their enemies (Ukeje, 2001, p. 363). As often happens, however, once young men are armed and encouraged to use weapons, political motives may give way to potentially lethal hooliganism and harassment (ARMED GROUPS).

Several studies have pointed to the important defensive function that small arms have for some young men. There has been little consideration, however, of the relationship between this and other reasons for carrying weapons, including those outlined above. An especially important goal of future research would be to determine why the carrying of weapons may translate into actual use for crime or confrontation. In particular, this research needs to examine claims that the mere carrying of small arms can have an escalatory or 'arms-racing' effect (Cook and Ludwig, 2003; Cook et al., 2005, p. 7; Dowdney, 2005, p. 303).

The symbolic role of small arms

Young men who are denied regular paths for social and economic advancement appear especially susceptible to the symbolic—as opposed to the merely functional—appeal of guns.

One of the most noticeable aspects about small arms is that they pervade the types of media that are expressly marketed to young men. The link between these media and real-world violence is a contested one, but a growing body of evidence suggests it may well be real. As one survey of research into the influence of violence in popular culture on youth notes:

> Research on violent television and films, video games, and music reveals unequivocal evidence that media violence increases the likelihood of aggressive and violent behaviour in both immediate and long-term contexts. (Anderson et al., 2003, p. 81)

Across the board, young men are the prime consumers of violent movies (Fischoff, 1999), in which the gun is often the chosen instrument of violence. They are also the main consumers of music that features small arms violence and video games that involve the contestant in violent, armed scenarios.[14]

A number of psychological studies demonstrate that aggressive thought is activated or 'primed' in people when they are exposed to images (or concepts) of weapons, including small arms.[15] In turn, these people may be more prone to consider aggressive behaviour in certain situations.[16]

The degree to which aggressive responses are stimulated, however, differs among individuals. For instance, in one study of the 'weapons priming effect' in hunters (individuals with prior gun experience) and non-hunters (individuals with no direct gun experience), hunting guns were found to be more likely to trigger aggressive thoughts among non-hunters. By contrast, hunters were more likely to experience aggressive thoughts when exposed to pictures of assault rifles rather than hunting guns (Bartholow et al., 2005).

Small arms pervade the types of media that are expressly marketed to young men.

The results indicate that hunters associate hunting guns with non-aggressive activities (at least in relation to other people) and assault rifles with shooting at humans, while people unused to guns tend to associate all guns with aggression. The authors conclude that 'an object serves as a cue to aggression insofar as it is closely linked with aggression-related concepts in memory, regardless of how those links were established' (Bartholow et al., 2005, p. 57). This research therefore suggests that small arms can bring out aggressive thoughts depending on an individual's prior experience of guns.

A 15-year longitudinal study of children's exposure to television violence and violent behaviour in young adulthood concluded that children were receptive to violent images only to the extent that they could identify with the perpetrator, relate his or her circumstances to their own, and see rewards from violence (Huesmann et al., 2003, p. 218).

Firearms may also appeal to some young men for reasons other than social or material gain. One commentator has asserted that young men who grow up without fathers often seek violent, substitute role models.

Strong, powerful, high-status role models such as those offered in movies and on television fill the vacuum in their lives. We have taken away their fathers and replaced them with new role models whose successful response to every situation is violence. (Grossman, 1996, p. 322).

Anderson et al. (2003) admit that there are some inconsistencies in research to date. Nevertheless, they conclude that most studies point strongly to interactions between, on the one hand, the characteristics of the viewer, such as a tendency to identify with aggressive characters, and, on the other, content—in particular, whether violence is portrayed positively or negatively.

The debate over the influence of popular culture on the use of violence still rages. The idea that weapons, and small arms in particular, activate or stimulate aggression is also contested (Kopel, Gallant, and Eisen, 2002). Nevertheless, a growing body of evidence shows that guns are sometimes perceived as symbols of armed violence. As a result, a focus on violence in popular culture is now a prominent feature in proposals to prevent armed violence,

But being as this is a .44 Magnum, the most powerful handgun in the world, and would blow your head clean off, you've gotta ask yourself one question: Do I feel lucky? Well, do ya punk?
—Insp. Harry Callahan (Clint Eastwood) threatening a bank robber, *Dirty Harry*

My gun just went off, I dunno how.
—Vincent Vega (John Travolta) explaining a shooting death, *Pulp Fiction*

And I ask to see a shotgun. He brings me a Mossberg pump action shotgun. As soon as I held that baby in my hands, I knew what I was gonna do. It felt so good. It felt like it was a part of me. They had a mirror in the store. I looked at myself holding it, and [. . .] I immediately bought it.
—Mickey Knox (Woody Harrelson) telling his story, *Natural Born Killers*

such as the Armed Violence Prevention Programme of the UN Development Programme and WHO (UNDP and WHO, 2003, p. 14).

Visual culture frequently associates firearms with other popular symbols of male 'success', including fast cars, women, designer clothing, and jewellery. Young men involved in armed violence offer the same objects as proof of the better life that small arms make possible (Wilkinson, 2003, pp. 103, 252). High-risk groups of young men appear to relate closely to such images, as they justify, even glorify, their own use of small arms. The young men who are most prone to engage in armed violence, namely those who perceive success as unattainable through non-violent means, may be the very ones who tend to be responsive to popular representations—and particularly glorification—of gun violence.

Concomitant with widespread gun imagery is colloquial speech with well-developed firearm-related vocabulary. Wilkinson's glossary of slang terms used by New York minority youths lists ten different words for 'gun', while seven refer to the act of shooting. Only the words used for 'money' are more numerous (Wilkinson, 2003, pp. 283–86; see Box 12.3).

Over time, the gun imagery, style, and jargon that was formerly the purview of gang culture has permeated general popular culture. In Europe and the United States, for instance, the diffusion of gang culture into larger youth culture has led to the adoption of many of the symbols of gang life, including small arms (Klein et al., 2001, p. 3).

It appears that small arms are important to many young men, but the specifics of this relationship are not well understood. In examining the functional and symbolic aspects of small arms, the chapter has largely relied on research dealing with young male violence in general. Guns, though part of this broad canvas, are not usually the focus of analysis. There are exceptions, but these studies—where guns are brought squarely into the foreground—tend to concentrate on the United States (Wilkinson, 2003; Wilkinson and Fagan, 2001). In order to better understand why and under what circumstances young men around the world see weapons as functional and symbolic tools to achieve their goals, future research must be conducted across national, regional, and socio-economic boundaries.

Stills (left to right)

Die Hard (20th Century Fox, 1988)
Pulp Fiction (Miramax Films, 1994)
Rambo: First Blood Part 2 (Anabasis N.V., 1985)
Eraser (Warner Bros., 1996)
Natural Born Killers (Warner Bros. et al., 1994)

Scarface (Universal Pictures, 1983)
Dirty Harry (Warner Bros., 1971)
Reservoir Dogs (Dog Eat Dog Productions, 1992)
Taxi Driver (Columbia Pictures et al., 1976)
Die Another Day (United Artists et al., 2002)
All © Moviestore Collection

I want a .32 revolver.
And a palm gun. That .22 there.
—Travis Bickle (Robert De Niro) buying guns, Taxi Driver

My self-defence mechanism's right here.
—James Bond (Pierce Brosnan) referring to his
Walther PPK, Die Another Day

Box 12.3   Gun words in US hip-hop slang and street culture

22	Grip
38	Heat, Heater
380	Jammy
4 pound (.45 pistol)	Joint
40	Mossberg
9/9 mm	Oowop
AK/AK-47	Paddle
Biscuit	Pump
Blix	Shotty
Burner	Steel
Chrome	Strap
Cronze	Tec 9
Deuce, Duce Duce (.22 pistol)	Thompson
Gat, Gatt, Ghat	Toast/Toaster
Gauge	Toolie
Glock	Uzi

Note: The ten terms for guns listed by Wilkinson (2003, pp. 283–86) are marked in red. Wilkinson also lists blasted, barking, busting, bucking, spraying, wetting and letting off, as common terms used to describe shots and shooting. Other terms come from the *rapdict* online dictionary. The dictionary notes, 'Slang for gun and penis is almost always interchangeable.'

Sources: Rap Dictionary; Urban Dictionary; Wilkinson (2003, pp. 283–86)

YOUNG MEN AND SMALL ARMS: IMPLICATIONS FOR INTERVENTIONS

The following sections assess some of the programmes that have been launched worldwide to address the problem of armed violence perpetrated by young men. By and large, there are two types of programmes: those aiming at arms reduction and those seeking to change social, community, and ultimately individual attitudes towards armed violence. To date, the arms reduction approach has been dominant; however, results from both programme types indicate that small arms are central to the lives of some young men. Most violence reduction programmes have focused on guns as either a functional or a symbolic tool, with arms reduction initiatives more concerned with function and behavioural programmes paying more attention to symbolic aspects.

While the factors that lead to violence are numerous, complex, and difficult to influence in the short term, small arms may represent a potential choke point in violence prevention efforts, given their role in facilitating young men's recourse to violence. This has been the rationale behind a number of apparently successful initiatives that have tried to curb the availability of small arms to youth at risk. Perhaps most important for longer-term violence prevention, it is evident that some young men are resistant to involvement in criminal or violent behaviour, despite circumstances that would appear to push them in that direction.

Figure 12.4 **Trends in method of attack in homicides among youths aged 10–24, 1985–94**

PERCENTAGE OF HOMICIDES

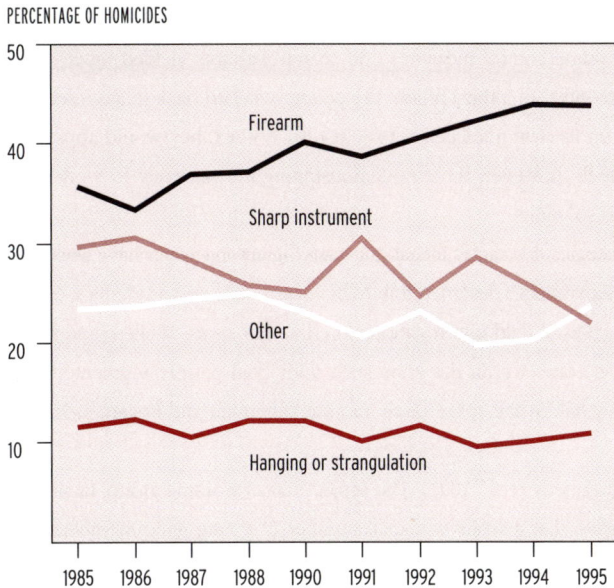

Source: WHO (2002, p. 27), based on WHO data for 46 countries

Restricting young men's access to firearms[17]

Given what appears to be the powerful functional and symbolic appeal of small arms, measures that seek to restrain young men's exposure and access to guns should be an important component of violence prevention initiatives. In most Western countries, however, little public policy, research, or action was devoted to the issue of gun use among children and youth, nor to its prevention, until the 1980s. With the exception of the United States, there was relatively little gun violence in criminal activity in these countries. To a large extent, the focus was on adults, and on the restriction of gun ownership.

This situation changed, especially in the United States, from 1985 to 1998, when an epidemic of gun violence committed by young men focused attention on an urgent need to develop prevention policies and programmes (Fagan, 2002, p. 134). At the peak of this epidemic, in 1994, nearly 6,000 young men under the age of 20 were killed by firearms in the United States, and many more were injured. Firearm homicide was largely responsible for the overall increase in youth homicide over the period, and there was often a high degree of overlap between victims and offenders (Braga, 2004, p. 7). While levels of gun use by young men have historically been far lower in other developed countries, from 2000 onwards there have similarly been reported increases in gun carrying and use by young men in a number of these countries (see Figure 12.4), including England and Wales (HMIC, 2004, pp. 32–33; Bullock and Tilley, 2002, pp. 33-34) and Canada—most notably the City of Toronto (Reuters, 2005).[18]

Given this situation, the great majority of youth gun violence prevention experience is US-based. It is also closely intertwined with strategies and programmes that aim to reduce or prevent general youth violence, gangs, and drug abuse. In the case of the United States, successful initiatives involved more than mere regulation; they attempted to identify and control the supply of guns used in crime by adopting a community-based, problem-solving approach.

One example is the 'Consent-to-Search' programme, implemented in St. Louis from 1994 to 1996. The programme's innovative approach emerged from consultations between police and residents of a crime-prone community (Decker and Rosenfeld, 2004, p. 1). It involved police officers knocking on doors in high-crime areas and asking parents for permission to search their homes for guns that their children might have hidden. Parents who opened their doors received a form, which clearly stipulated that any illegal guns found in their home would be confiscated without prosecution (Decker and Rosenfeld, 2004, pp. 9–10). Communities worked with the police to identify houses that should be offered a search. Parents and youths who requested help were referred to agencies or community-based groups that made appropriate services available.

The initiative, as originally implemented, was extremely successful, with 98 per cent of approached households consenting to the search. Within 18 months, 510 guns were recovered from half of these houses, at a rate of three guns per house (Decker and Rosenfeld, 2004, p. 12). By contrast, following the departure of the programme's founding police chief, a shift towards traditional coercive crime control measures and search warrants yielded much more limited results—netting only 31 firearms in nine months. In a third phase, the police switched back to a consent-to-search programme but this time relied primarily on its own intelligence to select the houses they would approach. Only 42 per cent of households agreed to a search, however, thereby demonstrating the necessity to work with communities to better identify problems and their solutions.

Carefully planned deterrence and punitive strategies that target illegal gun carrying among youth have also generated promising results. The Kansas City 'Gun Experiment', designed jointly by local police and academics, aimed specifically at increasing gun seizures through increased field interrogations in crime hot spots. While crime levels remained steady in areas around the intervention zone, within the zone such additional patrols augmented gun seizures by 65 per cent and reduced gun crime by half in the space of six months (Sherman and Rogan, 1995, pp. 677, 683–84, 691).

Another example is the Boston 'Gun Project' (Kennedy et al., 2001). The project's local working group, facilitated by a small team of Harvard researchers, established that a relatively small number of young gang members was responsible for and suffered most from violence in the city. The youths represented less than one per cent of Boston's population and could be easily identified. Through discussion forums and posters, the project issued strong warnings to gangs that violence would not be tolerated. The threat of prosecuting gang members under federal law—excluding any possibility of parole—if they did not respect initial calls for a 'ceasefire' made these warnings particularly credible. Yet members who wanted to renounce violence were given access to various support programmes made available by the community and government members of the working group.

Figure 12.5 **Monthly youth (under 24) homicide toll in Boston before and after the 'ceasefire' intervention***
NUMBER OF YOUTH VICTIMS

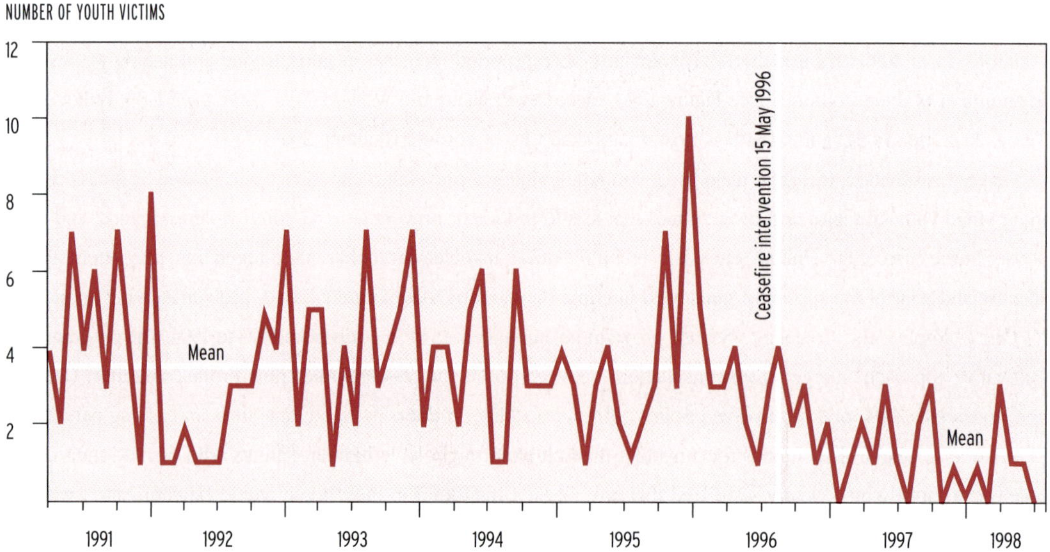

* Horizontal axis spans the period from January 1991 to May 1998.

Source: Kennedy et al. (2001, p. 58)

The Boston Gun Project resulted in significant reductions in youth homicide rates (see Figure 12.5). The initiative differed from traditional punitive strategies in that it targeted only gangs engaged in violence, establishing clear rules of behaviour for staying out of trouble (Kennedy, 1997, p. 463). Gun violence—and not gangs, people, or neighbourhoods—was the problem that needed to be resolved. The project's accomplishments suggest that successful youth gun crime reduction not only focuses on removing guns, but is also accompanied by credible carrots and sticks and the provision of non-violent alternatives. An attempt to replicate the Boston Gun Project in Los Angeles proved unsustainable, however. The lack of local ownership and accountability for the project, as well as insufficient funds for social services to balance law enforcement efforts, limited the project's performance and sustainability in this new setting (Tita et al., 2003).

The success of some of the strategies listed above highlights the importance of establishing a working partnership between local authorities, community workers, and academics to accurately define the problem to be treated and identify solutions that are appropriate for the local context. Primarily, however, strategies that target youth at risk involve criminal justice organizations and 'typically require a concerted concentration of police manpower and resources over an extended period of time' (Lab, forthcoming, p. 39). Implementation may thus be difficult where resources are scarce. Furthermore, measures are not designed to target the deep social and economic conditions that result in poverty and exclusion, nor can they necessarily help promote long-term prevention.

Strengthening protective factors

Long-term prevention of youth gun violence requires comprehensive initiatives that address both risk and protective factors (Lizotte and Sheppard, 2001, p. 1). Nevertheless, many youth violence prevention interventions focus heavily on high-risk youth or their families and pay less attention to the social and economic conditions that help generate youth violence, gangs, and gun use. Reducing violence requires a willingness to invest in young people and their communities (UN-HABITAT, 2004a, pp. 27–29; 2004b).

Reducing violence requires a willingness to invest in young people and their communities.

This perspective rests on the observation that some youths choose not to become involved in crime and violence. While young men are the primary actors in armed violence, many more young men who face the same risk factors are reluctant to participate in delinquency or violence.

Indeed, a number of studies illustrate how young men possess 'protective' factors that serve to reduce the risks of becoming involved in crime and violence. Research on young men who live in Brazilian and US communities where gang violence is prevalent has identified factors that reduce the probability of a young man's involvement in gangs. These factors include: (1) having a valued, stable relationship or multiple relationships with people (a parent, a grandparent, a female partner) whom they would disappoint by becoming involved with gangs; (2) having access to alternative identities or some other sense of self that was positively valued by the young man and by those in his social setting, particularly the male peer group (for example, being a good student, being a good athlete, having musical skills, having a good job); (3) being aware of the risks associated with the violent version of masculinity promoted by gang members; and (4) finding an alternative male peer group that provides positive reinforcement for non-gang-involved male identities (Barker, 2005a, pp. 146–57; Barker and Ricardo, 2005, pp. 53–54).

Preliminary analysis on what leads youths from Rio's *favelas* to join, or refrain from joining, armed gangs also points strongly to certain protective factors (Dowdney, 2005, p. 93). While all children interviewed in the research were exposed to the same risk factors—namely poverty, few economic options, social marginalization, exposure to violence, family problems, and a lack of non-violent recreational activities—it was their ability to respond to those risks, as conditioned by their personal environments, that kept them out of violence. Typical responses included staying in school, taking part-time work to continue studies, belonging to a sports team, receiving support from their grandparents, and learning to play an instrument. Reduced personal exposure to violence, which lessened young men's perceived need to seek revenge or protection, was another crucial protective factor.

The existence of such protective factors provides important opportunities for the design of community-based development projects that are extremely relevant to violence prevention (WHO, 2002, pp. 43–45; USDHHS, 2001, p. 57). Such projects will seek to provide youth with stable home environments, better economic options, and alternative sources of respect within their community. They will also act to negate perceptions among young men that guns and violence are associated with manliness and social status.

In recent years, campaigners for violence reduction have sought to uproot the identification of gun violence with popular conceptions of 'manliness'—thus targeting the symbolic appeal of small arms. A review of such campaigns reveals that there are two main strategies (see Box 12.4). The first involves working with young men to bolster their

Bad boy mystique? A young gang member embraces his girlfriend while holding a 9 mm pistol, Brooklyn, New York, December 2003.
© Boogie/WPN

Box 12.4 Relabelling guns: a review of campaigning strategies in Rio de Janeiro

Research on (mostly male) youth gun violence in Rio confirms that demand for guns is related to the desire for income, status, and women.[19] A certain 'bad boy mystique' or romanticism of gun-wielding males is assimilated and propagated by young women in the context of gun violence in Rio de Janeiro. This phenomenon is especially evident in *favela* communities, or urban shanty-towns, where opportunities are few and where heavily armed drug traffickers organized into factions or *commandos* vie for territorial control to conduct their illegal trade.

In these contexts, one of the few opportunities to 'be someone' is to become a drug trafficker–or to 'go out' with one. This is evident in the slang. For example, women who are attracted by gun-toting men are called *Maria AK47s* (an adaptation of *Maria Gasolina,* which is used for women attracted to men with nice cars). For some young people in these settings, the status and attractiveness of young men involved in the drug trade increases with their rank or position within the drug gang hierarchy. Likewise, the female partners of the higher-ranking drug bosses are called *Primeira Dama* or First Lady.

The Rio-based NGO ProMundo's 'Programme H' aims to transform violent constructions of masculinity that directly influence the behaviour and attitudes of young men with respect to gender equality, health, and violence. One component of Programme H involves discussion groups that use teaching materials to promote discussion and reflection on the 'costs' associated with traditional masculinity and the advantages of more equitable gender behaviour. Programme H also uses social marketing campaigns to promote changes in community norms related to notions of what it means to be a man. These show, for example, men actively involved in childcare and child rearing, and in supporting and caring for their partners. Their approach is innovative in that it places great emphasis on 'voices of resistance' (those who successfully resist traditional constructions of violent masculinity) and seeks to promote alternatives to violent or dangerous behaviour.

MV Bill, a renowned rapper from Rio's City of God *favela*–made infamous in the eponymous film–helped launch the Programme H campaign. MV Bill cuts a 'manly', 'cool' figure and enjoys considerable 'street cred' as a rapper. After his brother, a trafficker, was killed in a gun battle with rival drugs factions, however, he dedicated himself and his music to speaking out against violence. One of his most famous, and polemical, tracks was entitled *Soldado do Morro* (Favela Soldier), which he followed with *Soldado Morto* (Dead Soldier) on his next album. The lyrics of *Soldado do Morro* push young men and women to rethink traditional concepts of masculinities and violence:

Another baby left crying, another crazy guy goes down
That's it, war with no end
Too late for me to have regrets now
With no friends / With no family
Men don't cry–what a lie.

Drugs, guns in the sights of a young black man
With no respect, no money, no Nike
No life, no faith, no name
I was prepared to kill, but I wasn't prepared to die
It's been a long time since I've seen my mother cry.[20]

A 2001 campaign called 'Choose Gun Free! Its Your Weapon or Me', organized by the NGO Viva Rio attempted to target young women's attraction to men with guns. Focus groups conducted at the outset of the campaign told organizers that simply telling people that men with guns are more likely to die (something most of them already knew) would not achieve results. Instead the campaign decided to take a lighter, more humorous approach, using well-known and respected female television and music celebrities to transmit somewhat novel campaign messages. On nationally aired TV spots, a famous comedian said, 'Guys who use guns must have a little problem. . .', using a gesture to insinuate that they may be overcompensating for a small penis. Another campaign slogan was a play on words, 'A good man is one who does not expire before his time', mixing the ideas of premature ejaculation with dying young.

The campaign represented the first attempt to mobilize women around disarmament in Brazil. The funny, youth-specific messages were complemented by mobilizations by groups of mothers who had lost their sons to gun violence. The campaign was considered quite successful in galvanizing women's support for efforts to reduce gun violence, equipping them with supporting arguments and evidence. Nevertheless, campaigners saw a need to follow up these public awareness efforts with programmes designed to increase women's self-esteem, provide spaces for reflection, and generate opportunities for work and other meaningful forms of participation in society. This was particularly important in *favela* communities where opportunities for women outside of associations with drug traffickers are scarce. Research is currently being undertaken in Rio to address this gap.[21]

Source: Galeria (2005)

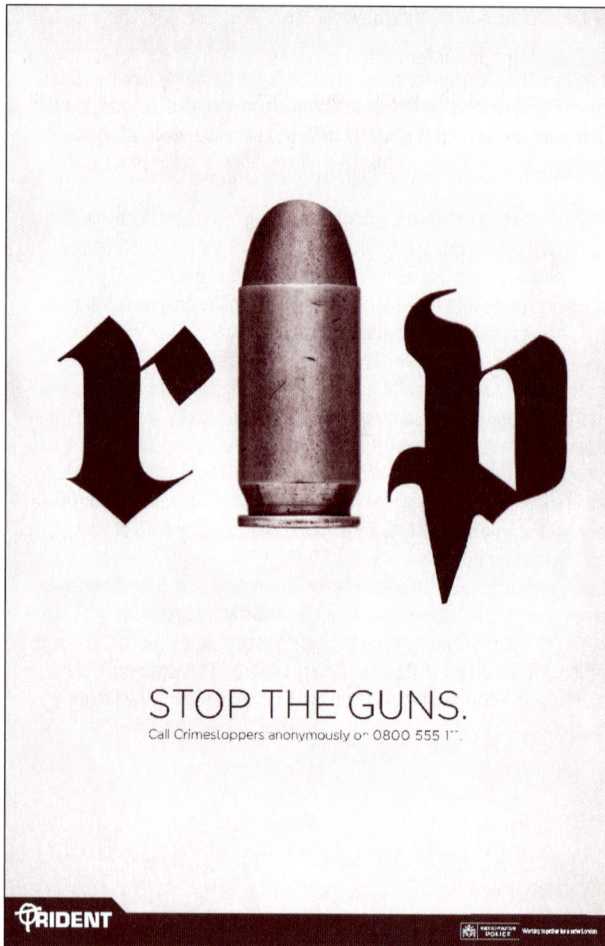

rip

STOP THE GUNS.
Call Crimestoppers anonymously on 0800 555 1''.

TRIDENT

A poster aimed at changing Londoners' perception of armed violence.
© Trident

resistance to becoming engaged in gun violence by providing space for groups to discuss and participate in alternative, non-violent kinds of masculine behaviour. The second strategy aims to convince young women that violent men (and men who carry or use guns) are not attractive, and to spread that message, for young women and men alike to assimilate.

In some cases, prominent personalities have been engaged to raise awareness and serve as role models—i.e. men who are seen as successful, admirable, and attractive to women, but who repudiate gun violence. These figures are chosen in a deliberate strategy to combat the 'big men' symbolism of firearms. In Chicago, for example, the Project for Violence Prevention's mentoring programmes have specifically targeted individuals at risk by using neighbourhood role models. Workers in such 'outreach' initiatives may seek to involve young men in sport, or help them with education as alternatives to gang life. Crucially, these initiatives employ workers who are 'from the street'—role models who already enjoy credibility and respect among the young (Diehl, 2005, pp. 9–11).

Nonetheless, it is important to stress that initiatives that are designed to encourage young men's resistance to becoming involved in armed violence are still very much in their infancy. Measuring their impact on youth violence levels will require patience and a reliance on an evidence-based model of violence prevention, at both the problem identification and programme evaluation stages. It can be said, however, that in recent years international agencies involved in traditional small arms control measures have increasingly supported such approaches.[22]

CONCLUSION

A large body of research links young male violence and small arms, though it has not focused specifically on the weapons themselves. By distilling the small arms-specific findings of this research, the chapter has drawn some initial conclusions about the role of small arms in young male violence.

The fact that the majority of young men do not become involved in armed violence suggests that the problem is probably social in nature, rather than biological or demographic. Violence and attitudes towards small arms violence are, in other words, learned. Young men take their lessons from the world around them. If unable to fulfil socially defined masculine roles, they may adopt violent alternatives as a means of asserting their place in society.

Functionally or symbolically, small arms resonate with young men who are tempted by violence. They are powerful tools with which young men can assert their masculinity, whether by acquiring the objects and status they are conditioned to seek, or by overturning the societies from which they are excluded. In many parts of the world, small arms hold out the power to change one's lot in life.

Measures to curtail armed violence need to recognize the serious threat some young men pose to society—and to each other. Measures that target at-risk youth—both victims and perpetrators—can successfully decrease violence levels in the short run through a careful mix of carrots and sticks. Targeting illicit gun ownership among young men is often an important component of such strategies as it tends to reduce young men's access to small arms and consequently prevents their misuse. Initiatives that tackle the many reasons that lead young men to become involved in armed violence need, in particular, to counter perceived associations between firearms and social status, ensuring that guns are no longer seen as a viable means of affirming one's masculinity.

In the long run, violence prevention efforts must focus on the various protective factors that seem to prevent the majority of young men from becoming involved in armed violence. Stable home environments, decent economic options, and alternative sources of respect within the community make young men—even those living in high-risk areas—more resistant to becoming involved in armed groups and armed violence. The 'angry young man' of popular lore is not an inevitable feature of modern life. ▰

LIST OF ABBREVIATIONS

GNP	gross national product	WHO	World Health Organization
UNDP	United Nations Development Programme		

ENDNOTES

1   See discussion below and endnote 3.

2   Ellis and Shaw, 2000 (pp. 107–10), as quoted in UNODC (2005, p. 6).

3   Statistics Canada (2000; 2001, as quoted in Shaw, 2001, p. 1).

4   Minors are classified as being under 18 years of age.

5   Based on the estimate of 140,000–200,000 annual firearm homicide victims worldwide (Small Arms Survey, 2004, p. 200). This figure does not include deaths in armed conflict situations. In this sample, men aged 30–44 are the second largest demographic group most affected by armed violence (12.6 gun homicides per 100,000). Women of all ages, in contrast, represent less than nine per cent of victims.

6   Tolan and Gorman-Smith (1998, cited in Shaw, 2001, p. 2).

7   Men aged 15–29 are more at risk of dying of firearm homicide than the average population, however. In the vast majority of these countries or territories (24 out of 33), the demographic group most at risk was men aged 30–44 (Belize, Czech Republic, Egypt, Estonia, Finland, French Guiana, Georgia, Germany, Greece, Hungary, Italy, Japan, Kyrgyzstan, Lithuania, Mexico, the Netherlands, Nicaragua, Peru, Poland, Republic of Moldova, Romania, Singapore, Spain, and Thailand). In nine other settings, groups most at risk included men aged 45–59 (Austria, Guadeloupe, Malta, Republic of Korea, Réunion, and Saint Lucia), men over 60 (Costa Rica and Iceland), and women aged 15–29 (Luxembourg). The 37 countries where men aged 15–29 were the group most at risk were Argentina, Australia, Bahamas, Bahrain, Barbados, Belgium, Brazil, Canada,

Chile, Colombia, Croatia, Cuba, Denmark, Dominican Republic, Ecuador, El Salvador, France, Ireland, Israel, Kuwait, Martinique, Mauritius, New Zealand, Norway, Panama, Paraguay, Portugal, Puerto Rico, Saint Vincent and Grenadines, Serbia and Montenegro, Slovakia, Slovenia, Sweden, United Kingdom, United States, Uruguay, and Venezuela. Small Arms Survey calculations based on WHO (2005) and UN Population Division (2002).

8   Huff (1998, quoted in Wilkinson, 2003, p. 19).

9   The abortion argument is not purely demographic as it includes a social dimension. Levitt argues that the legalization of abortion has led to a reduction in the number of unwanted births. Such unwanted children are at greater risk of engaging in criminal activity (Levitt, 2004, p. 182).

10   This section is based on Barker (2005b).

11   The countries included in the study were: Brazil, Colombia, Ecuador, El Salvador, Jamaica, Nigeria, the Philippines, South Africa, and the United States.

12   Interviews with young women aged 14–21, Complexo de Maré, Rio de Janeiro, January 2004. Conducted by Jessica Galeria, Viva Rio.

13   See Majors and Billson (1993), Anderson (1990), Archer (1994), and Schwartz (1987).

14   Buchman and Funk (1996, cited in Anderson and Bushman, 2001, p. 354).

15   Berkowitz and LePage (1967, cited in Anderson et al., 1998, p. 1).

16   Bartholow et al. (2005, p. 49) describe this process in the following terms: 'When a weapon concept is activated (e.g., through the identification of a gun in the environment), closely linked concepts (e.g., ideas related to aggression and hostility) also become activated via spreading activation . . . and are thus more accessible than they would be otherwise. Once these aggressive concepts become accessible, they can facilitate subsequent aggressive behavior in several ways. For example, highly accessible aggressive thoughts may color interpretations of ongoing social interactions, or they may make aggressive resolutions of a dispute seem more appropriate.'

17   This section is based on Shaw (2005). Most 'youth violence prevention initiatives' implicitly target young men as the group that is most at risk.

18   In both cases, this is largely restricted to one or two major cities, and levels of homicide are far lower overall than in the United States. In England and Wales less than half of one per cent of all reported crime is gun-related, and there were 68 such homicides in 2003–04. In Toronto, there were 64 gun-related homicides in 2004, compared with 450 in Chicago, a US city of a comparable size.

19   See, for example, Dowdney (2003); Barker (2005a); Lessing (2005).

20   Translation by Jessica Galeria. Lyrics and more about MV Bill on http://www.realhiphop.com.br/mvbill/

21   Research-action project by Viva Rio (Brazil) and the University of Coimbra (Portugal), *Women and girls in contexts of armed violence: a case study on Rio de Janeiro,* results forthcoming in 2006.

22   See, for example, UNDP and WHO's armed violence reduction programme (UNDP and WHO, 2003).

REFERENCES

Anderson, Craig and Brad Bushman. 2001. 'Effects of Violent Video Games on Aggressive Behavior, Aggressive Cognition, Aggressive Affect, Physiological Arousal, and Prosocial Behavior.' *Psychological Science,* Vol. 12, No. 5. September, pp. 353–59.

Anderson, Craig et al. 1998. 'Does the Gun Pull the Trigger? Automatic Priming Effects of Weapon Pictures and Weapon Names.' *Psychological Science,* Vol. 9, No. 4. July, pp. 308–14.

——. 2003. 'The Influence of Media Violence on Youth.' *Psychological Science in the Public Interest.* Vol. 4, No. 3. December, pp. 81–110.

Anderson, Elijah. 1990. *Streetwise: Race, Class and Change in an Urban Community.* Chicago: University of Chicago Press.

Archer, John. 1994. 'Violence between men.' In John Archer, ed. *Male violence.* London: Routledge, pp. 121–42.

Barker, Gary. 2005a. *Dying to be men: youth, masculinity and social exclusion.* London: Routledge.

——. 2005b. 'Young men and guns.' Unpublished background paper. Geneva: Small Arms Survey.

—— and Christine Ricardo. 2005. *Young Men and the Construction of Masculinity in Sub-Saharan Africa: Implications for HIV/AIDS, Conflict, and Violence.* Social Development Papers, Conflict Prevention and Reconstruction Paper No. 26. Washington, DC: World Bank. June.

Bartholow, Bruce et al. 2005. 'Interactive effects of life experience and situational cues on aggression: the weapons priming effect in hunters and nonhunters.' *Journal of Experimental Social Psychology.* Vol. 41, pp. 48–60.

Berkowitz, Leonard and Anthony LePage 1967. 'Weapons as Aggression-Eliciting Stimuli.' *Journal of Personality and Social Psychology.* Vol. 7, pp. 202–07.

BIP (Belfast Interface Project). 1998. *Young People on the Interface.* Belfast: BIP.

Bonifacio, Julie Anne et al. 2004. 'Uncovering the Hidden World of Youth Gangs.' *Minda News.* 12 March.
<http://www.mindanews.com/2004/03/12ftr-gangs.html>

Braga, Anthony. 2004. 'Gun Violence among Serious Young Offenders.' *Problem-Oriented Guides for Police, Problem-Specific Guides Series No. 23.* Washington, DC: US Department of Justice. <http://www.cops.usdoj.gov/mime/open.pdf?Item=1078>

Bronfenbrenner, Urie. 1979. *The ecology of human development.* Cambridge, MA: Harvard University Press.

Buchman, Debra and Jeanne Funk. 1996. 'Video and computer games in the '90s: Children's time commitment and game preference.' *Children Today*. Vol. 24, pp. 12-16.

Bullock, Karen and Nick Tilley. 2002. *Shootings, Gangs and Violent Incidents in Manchester: Developing a Crime Reduction Strategy*. Crime Reduction Series Paper 13. London: Home Office, Research, Development and Statistics Directorate.

Cincotta, Richard, Robert Engelman, and Daniele Anastasion. 2003. *The Security Demographic: Population and Civil Conflict after the Cold War*. Washington, DC: Population Action International.

COAV (Children and Youth in Organized Armed Violence). 2005. Research: Youth Testimonies. Rio de Janeiro: Viva Rio, COAV. <http://www.coav.org.br>

Cock, Jackie. 2001. 'Gun violence and masculinity in contemporary South Africa.' In Robert Morell, eds. *Changing men in Southern Africa*. Durban: University of Natal Press/Zed Books.

Conway, Mary and Jonny Byrne. 2005. 'Interface Issues: An Annotated Bibliography.' Belfast: Institute for Conflict Research. <http://www.conflictresearch.org.uk/documents/Interface%20Bibliography.pdf>

Cook, Phillip and Jens Ludwig. 2003. *Does Gun Prevalence Affect Teen Gun Carrying After All?* Working Paper Series. Durham, North Carolina: Terry Sanford Institute of Public Policy, Duke University. 15 August. <http://www.pubpol.duke.edu/people/faculty/cook/san03-04.pdf>

Cook, Phillip et al. 2005. *Underground Gun Markets*. Working Paper 11737. NBER Working Paper Series. Cambridge MA: National Bureau of Economic Research. November. <http://www.nber.org/papers/w11737>

Decker, Scott and Richard Rosenfeld. 2004. *Reducing Gun Violence: The St. Louis Consent-to-Search Program*. Washington, DC: NIJ.

Diehl, Digby. 2005. 'The Chicago Project for Violence Prevention.' In Stephen Isaacs and James Knickman, eds. *To Improve Health and Health Care: The Robert Wood Johnson Foundation Anthology, Volume VIII*. 2nd edn. Hoboken: Jossey-Bass. December. <http://www.rwjf.org/files/publications/books/2005/chapter_06.pdf>

Dowdney, Luke. 2003. *Children of the Drug Trade: A Case Study of Children in Organized Armed Violence in Rio de Janeiro*. Rio de Janeiro: 7 Letras.

——. 2005. *Neither war nor peace: international comparisons of children and youth in organized armed violence*. Rio de Janeiro: COAV.

Elliott, David. 1994. 'Serious violent offenders: onset, developmental course and termination—The American Society of Criminology 1993 Presidential Address.' *Criminology, Vol.* 32, No. 1, pp. 1–21.

Ellis, Stephen. 1997. *Young soldiers and the significance of initiation: some notes from Liberia*. Leiden: Afrika-Studiecentrum. <http:/www.asc.leidenuniv.nl/pdf/>

Ellis, Lee and Anthony Walsh. 2000. *Criminology: A Global Perspective*. Boston: Allyn and Bacon.

Emler, Nicolas and Stephen Reicher. 1995. *Adolescence and delinquency: the collective management of reputation*. Oxford: Blackwell Publishers.

Fagan, Jeffrey. 2002. 'Policing guns and youth violence.' *The Future of Children*. Vol. 12, No. 2, pp. 133–51.

Fischoff, Stuart. 1999. 'Psychology's Quixotic Quest For the Media–Violence Connection.' Address at the Annual Convention of the American Psychological Association, Boston, 21 August. <http://www.calstatela.edu/faculty/sfischo/violence.html>

Fox, Aine. 2002. 'Youth Culture and Anti-social Behaviour.' *Other View*, No. 8, Spring. <http://www.theotherview.net/>

Florquin, Nicolas and Shelly O'Neill Stoneman. 2004. *A house isn't a home without a gun: SALW Survey, Republic of Montenegro*. Belgrade: South Eastern Europe Clearinghouse for the Control of Small Arms and Light Weapons (SEESAC) and Small Arms Survey. <http://www.smallarmssurvey.org/copublications/survey-montenegro-eng.pdf>

Galeria, Jessica. 2005. 'Gun violence prevention initiatives in Rio de Janeiro.' Unpublished background paper. Geneva: Small Arms Survey. November.

Gilligan, James. 1996. *Violence: Reflections on a National Epidemic*. New York: Vintage Books.

Grossman, Dave. 1996. *On Killing: The Psychological Cost of Learning to Kill in War and Society*. New York: Back Bay Books and Little, Brown and Company.

Hagedorn, John. 2001. 'Globalization, Gangs, and Collaborative Research.' In Klein et al., pp. 41–58.

Hemenway, David. 2004. *Private Guns, Public Health*. Ann Arbor: University of Michigan Press.

—— and Matthew Miller. 2004. 'Gun Threats against and Self-defense Gun Use by California Adolescents.' *The Archives of Pediatric Adolescent Medicine*, Vol. 158. April, pp. 395–400.

HMIC (Her Majesty's Inspectorate of Constabulary). 2004. *Guns, Community and Police. Thematic Report into the Criminal Use of Firearms*. May. London: HMIC.

Huesmann, Rowell et al. 2003. 'Longitudinal Relations Between Children's Exposure to TV Violence and Their Aggressive and Violent Behavior in Young Adulthood: 1977–1992.' *Developmental Psychology*, Vol. 39, No. 2, pp. 201–21.

Huff, C. Ronald. 1998. *Comparing the criminal behavior of youth gangs and at-risk youths*. Washington, DC: National Institute of Justice.

Jourdan, Luca. 2004. 'Being at War, Being Young: Violence and Youth in North Kivu.' In Koen Vlassenroot and Timothy Raeymaekers, eds. *Conflict and Social Transformation in Eastern DR Congo*. Ghent: Academia Press, pp. 157–76.

Kennedy, David. 1997. 'Pulling Levers: Chronic Offenders, High-Crime Settings, and a Theory of Prevention.' *Valparaiso University Law Review*. Vol. 31, No. 2. Spring, pp. 449–84.

—— et al. 2001. *Reducing Gun Violence: The Boston Gun Project's Operation Ceasefire*. Washington: National Institute of Justice. October. <http://www.ojp.usdoj.gov/nij/pubs-sum/188741.htm>

Kimmel, Michael. 2003. *The gendered society*. 2nd edn. Oxford: Oxford University Press.

——. 2005. *The history of men: essays on the history of American and British masculinities*. New York: SUNY Press.

Klein, Malcom et al. (eds.). 2001. *The Eurogang Paradox: Street Gangs and Youth Groups in the U.S. and Europe*. Dordrecht, the Netherlands: Kluwer Academic Publishers.

Kopel, David, Paul Gallant, and Joanne Eisen. 2002. 'No Choice: "Weapons-effect" paralysis.' National Review Online. 17 April.
    <http://www.nationalreview.com/kopel/kopel041702.asp>

Lab, Steven. Forthcoming. *Unresolved Issues for Crime Prevention Research*. Washington, DC: Office of Research and Evaluation, National Institute of Justice.

Legget, Ted. 2005. 'Terugskiet (returning fire): growing up on the street corners of Manenburg, South Africa.' Rio de Janeiro: Viva Rio, COAV.
    <http://www.coav.org.br/publique/media/Report%20%C1frica%20do%20Sul.pdf>

Lessing, Benjamin. 2005. 'The Demand for Firearms in Rio de Janeiro.' In Rubem César Fernandes, ed. *Brazil: The Guns and the Victims*. Rio de Janeiro: Viva Rio, pp. 202–20.

Levitt, Steven D. 1999. 'The limited role of changing age structure in explaining aggregate crime rates.' *Criminology,* Vol. 37, No. 3, pp. 581–98.

——. 2004. 'Understanding why crime fell in the 1990s: four factors that explain the decline and six that do not.' *Journal of Economic Perspectives,* Vol. 18, No. 1. Winter, pp. 163–90.

Lindsay, Lisa and Stephan Miescher, eds. 2003. *Men and Masculinities in Modern Africa*. Portsmouth: Heinemann.

Lizotte, Alan and David Sheppard. 2001. 'Gun Use by Male Juveniles: Research and Prevention.' *Juvenile Justice Bulletin*. Washington, DC: US Department of Justice, Office of the Juvenile Justice and Delinquency Prevention. July.

Majors, Richard and Janet Mancini Billson. 1993. *Cool pose: the dilemmas of black manhood in America*. New York: Touchstone.

McAlister, Alfred. 2000. *La violencia juvenil en las Américas: estudios innovadores de investigación, diagnóstico y prevención* (Youth violence in the Americas: Innovative studies, diagnosis, and prevention). Washington, DC: Pan American Health Organization.
    <http://www.paho.org/Spanish/HPP/HPF/ADOL/violence.pdf>

Mesquida, Christian and Neil Wiener. 1999. 'Male age composition and severity of conflicts.' *Politics and the Life Sciences,* Vol. 18, No. 2, pp. 181–89.

Michailof, Serge, Markus Kostner, and Xavier Devictor. 2002. 'Post-conflict recovery in Africa: an agenda for the Africa region.' *Africa Region Working Paper Series No. 30*. Washington, DC: World Bank.

Peres, Maria Fernanda Tourinho. 2004. *Firearm-related violence in Brazil: Country Report*. Sao Paulo and Geneva: Center for the Study of Violence, University of Sao Paulo, Small Arms Survey, and WHO.

Polk, Kenneth. 1994. 'Masculinity, honour and confrontational homicide.' In Tim Newburn and Betsy Stanko, eds. *Just boys doing business? Men, masculinities and crime*. London: Routledge.

Rap Dictionary. <http://www.rapdict.org>

Renfrew, John. 1997. *Aggression and its causes: a biopsychosocial approach*. Oxford: Oxford University Press.

Reuters. 2005. 'Toronto shooting spree harms city's "good" image.' 22 August.

Rhodes, Richard. 1999. *Why They Kill: The Discoveries of a Maverick Criminologist*. New York: Vantage Books.

Richmond, Therese, Rose Cheney, and C. William Schwab. 2005. 'The Global Burden of Non-conflict Related Firearm Deaths.' *Injury Prevention,* Vol. 11, pp. 348–52.

Sampson, Robert and John Laub. 1993. *Crime in the Making: Pathways and Turning Points through Life*. Cambridge, Mass.: Harvard University Press.

Schwartz, Gary. 1987. *Beyond Conformity or Rebellion: Youth and Authority in America*. Chicago, Ill.: University of Chicago Press.

Shaw, Margaret. 2001. *Investing in Youth 12–18: International Approaches to Preventing Crime and Victimization*. Montreal: International Centre for the Prevention of Crime.

——. 2005. 'Youth Violence Prevention: Programme Evaluations.' Unpublished background paper. Geneva: Small Arms Survey.

Sherman, Lawrence and Dennis Rogan. 1995. 'Effects of Gun Seizures on Gun Violence: "Hot Spots" Patrol in Kansas City.' *Justice Quarterly,* Vol. 12, No. 4, pp. 673–93.

Small Arms Survey. 2004. *Small Arms Survey 2004: Rights at Risk*. Oxford: Oxford University Press.

Smyth, Marie and Patricia Campbell. 2005. 'Young People and Armed Violence in Northern Ireland.' Rio de Janeiro: Viva Rio, CAOV.
    <http://www.conflictresearch.org.uk/>

Statistics Canada. 2000. *Youth Court Statistics 1998–99: Highlights Juristat*. 20(2). Ottawa: Statistics Canada, National Criminal Justice Reference Service.

——. 2001. *Crime Statistics in Canada, 2000: Juristat*. 21 (8). Ottawa: Statistics Canada, National Criminal Justice Reference Service.

Templa, Mae Fe Ancheta. 2004. *Understanding children in conflict with the law: contradictions on victimisation, survivor behaviour and the Philippine justice system*. Quezon City: Save the Children UK Philippines Programme.

Tita, George et al. 2003. 'Unruly Turf: The Role of Interagency Collaborations in Reducing Gun Violence.' *RAND Review*. Santa Monica: RAND. Fall.

Tolan, Patrick H. and Deborah Gorman-Smith. 1998. 'Development of serious and violent offending careers'. In Rolf Loeber and David P. Farrington, eds. *Serious and violent juvenile offenders*. Thousand Oaks: Sage Publications.

Ukeje, Charles. 2001. 'Youths, violence and the collapse of public order in the Niger Delta of Nigeria.' *Africa Development,* Vol. 26, Nos. 1 and 2, pp. 338–66.

UN Population Division. 2002. *World Population Prospects: The 2002 Revision.*
    <http://www.un.org/esa/population/publications/wpp2002/wpp2002annextables.PDF>

——. 2005. *World Population Prospects: The 2004 Revision.* Population Database. Accessed 22 November 2005.<http://esa.un.org/unpp/>

UN-HABITAT. 2004a. *Youth, Children and Urban Governance.* Policy Dialogue Series No. 2. Nairobi: UN-HABITAT Global Campaign on Urban Governance. <http://www.eya.ca/wuf/docs/Downloads/WUF/UN%20HABITAT/Final%20Policy%20Di%C9%20Governance.doc>

——. 2004b. *Strategy Paper on Urban Youth in Africa: A Focus on the Most Vulnerable Groups.* Policy Dialogue Series No. 2. Nairobi: UN-HABITAT Safer Cities Programme. <http://www.gpean.org/aaps/strategypaperenglish.pdf>

UNDP (United Nations Development Programme) and WHO (World Health Organization). 2003. 'Armed Violence Prevention Programme (AVPP): Support to Community Based Violence Prevention Programmes.' Project Document. Project Number INT/03/MXX. <http://www.who.int/violence_injury_prevention/violence/activities/en/avpp_overview.pdf>

UNODC (United Nations Office on Drugs and Crime). 2005. *Crime and development in Africa.* Vienna: UNODC. June. <http://www.unodc.org/pdf/African_report.pdf>

Urban Dictionary. <http://www.urbandictionary.com>

Urdal, Henrik. 2004. 'The devil in the demographics: the effect of youth bulges on domestic armed conflict, 1950–2000'. *Social Development Papers,* No. 14. Washington, DC: World Bank. <http://www-wds.worldbank.org/servlet/WDSContentServer/WDSP/IB/2004/07/28/000012009_20040728162225/Rendered/PDF/29740.pdf>

USDHHS (United States Department of Health and Human Services). 2001. *Youth Violence: A Report of the Surgeon General.* Washington, DC: USDHHS. <http://www.surgeongeneral.gov/library/youthviolence/report.html>

Vercaigne, Conny. 2001. 'The Group Aspect of Youth Crime and Youth Gangs in Brussels: What We Do Know and Especially What We Don't.' In Malcolm Klein et al., eds., pp. 283–98.

WHO (World Health Organization). 2002. *World Report on Violence and Health.* Geneva: WHO. <http://www.who.int/violence_injury_prevention/violence/world_report/en/>

——. 2005. Mortality Database. Accessed 1 September 2005. <http://www3.who.int/whosis/mort>

Wilkinson, Deanna. 2003. *Guns, Violence, and Identity among African American and Latino Youth.* New York: LFB Scholarly Publishing.

—— and Jeffrey Fagan. 1996. 'Understanding the role of firearms in violence "scripts": The dynamics of gun events among adolescent males'. *Law and Contemporary Problems,* Vol. 59, No. 1, pp. 55–90.

—— and Jeffrey Fagan. 2001. 'What We Know About Gun Use Among Adolescents.' *Clinical Child and Family Psychology Review,* Vol. 4, No. 2, pp. 109–32.

Wolf Harlow, Caroline. 2001. *Firearm Use by Offenders.* Bureau of Justice Statistics Special Report. NCJ 189369. Washington, DC: United States Department of Justice, Office of Justice Programs. November. <http://www.ojp.usdoj.gov/bjs/abstract/fuo.htm>

ACKNOWLEDGEMENTS

Principal authors

James Bevan and Nicolas Florquin

Contributors

Gary Barker, Jessica Galeria, Glenn McDonald, and Margaret Shaw

INDEX